The Fishing and Oil Industries in Coexistence

A. Espinoza-Tenorio • D. E. Ramos-Muñoz
E. Núñez-Lara
Editors

The Fishing and Oil Industries in Coexistence

The Socio-ecological System of the Southwestern Gulf of Mexico

Springer

Editors
A. Espinoza-Tenorio
Department of Sustainability Sciences
El Colegio de la Frontera Sur
Campeche, Campeche, México

D. E. Ramos-Muñoz
Department of Society and Culture
El Colegio de la Frontera Sur
Villahermosa, Tabasco, México

E. Núñez-Lara
Universidad Autónoma del Carmen
Ciudad del Carmen, Campeche, México

ISBN 978-3-032-02620-0 ISBN 978-3-032-02621-7 (eBook)
https://doi.org/10.1007/978-3-032-02621-7

This Springer imprint is published by the registered company Springer Nature Switzerland AG
The registered company address is: Gewerbestrasse 11, 6330 Cham, Switzerland

For those who've interweaved their fate with the Gulf of Mexico—your knowledge, commitment, resilience, and generosity inspire every page of this book.

Foreword

As we approach the end of the first quarter of the twenty-first century, the latest global estimates reported by the Food and Agriculture Organization (FAO) suggest that a quarter of the global population cannot afford a healthy diet. Aquatic foods are increasingly recognized as part of the solution given their nutrition and poverty alleviation potential. At the same time, according to the International Energy Agency, global demand for energy production like oil and natural gas will continue to increase for the next few decades at least. For instance, in the last twenty years, the North American region including Mexico, Canada, and the USA has experienced a remarkable production increase of up to 75%. Some forms of energy production, like oil and natural gas, can threaten the health of aquatic ecosystems and their food production potential. As these global and regional trends illustrate, food and energy production are deeply intertwined, and finding sustainable ways to produce both is an urgent topic that requires careful consideration. It is in this context that *Fishing and Oil in Coexistence: The Social-Ecological System of Southwest Gulf of Mexico* offers a timely exemplar. What are the main interactions between the oil and aquatic food industries? and what are their outcomes for local inhabitants and the environment on which they depend on to sustain their livelihoods? This book explores these questions through the case of the Southwestern Gulf of Mexico, a region that brings together diverse symbols and visions of nationalism, pride, self-determination, and sustainable development. Here, energy production and fisheries co-exist in the same space and in contradiction with one another. This region is an epicenter of modernity in terms of oil production and extractivism ideologies, but it is also cradle to the Olmec culture, one of the most influential civilizations in the early Americas. It is also home to the deltas of the Grijalva and Usumacinta rivers, and main arteries of the majestic Centla wetland Biosphere reserve, recognized for its high endemism and biodiversity. All of these overlapping cultural, historical, biological, and economic realities constitute the rich canvas on which the interrelationships between fishing and oil described in the book play out. In its pages you will find ample references to how fishing resources are increasingly scarce, yet also about how fishing is a livelihood characterized by the sense of freedom, joy, and purpose it brings to fishers and their communities. It offers a contrast to the better-known and dominant narrative of oil production as a benign motor of economic progress. Indeed, both activities, extractive in principle, offer very different sets of benefits and costs to local inhabitants. Yet, both intertwine as a source of environmental change. Fishers talk about the pride they feel as providers of food, and oil workers talk about the important role they have played in Mexican modern history. At the outset it seems to be a story of David against Goliath. The book however quickly reminds us that it is not one or the other, but coexistence. In this region, coexistence is not a choice but a reality for decades to come. So, what is the future for fishing, biodiversity, and environmental change? The book premises that fishing and oil interactions cannot be understood through fragmented disciplinary approaches, taking an interdisciplinary approach to dissect how coexistence plays out. Through this process it provides the building blocks to imagine future paths forward for the region.

The Editors' understanding of the complexity and multidimensionality of the topic can be quickly appreciated by their choice of chapters. Chapters which are authored by locally-based topical experts, often working with early career Mexican scholars, acknowledging the need to look toward the future. Two final observations, or two additional ways in which this book can be read: First, the book constitutes a rich compendium of information about small-scale fisheries. This is easier said than done. Generating data about this livelihood-based activity is challenging, and when available it is often not put in explicit contrast with other competing uses of the ocean space (other than tourism or large-scale fishing). Second, this book can also be read as a product of an ambitious national project of knowledge production, from the identification of research priorities by the National Council for the Humanities, Science and Technology, which provided the funding, to the laborious work of assembling a multi-disciplinary team, executing the science, and writing the book. Providing to all of us with a tangible successful example of regionally-relevant science at the service of society.

Stanford University
Santa Cruz, CA, USA
Aug 10, 2025

Xavier Basurto

Preface

This book synthesizes interdisciplinary research on the coexistence of fisheries and the oil industry in the southwestern Gulf of Mexico (SWGoM) over the past five decades, examining how this interaction has been managed amid competing environmental and socioeconomic interests.

As editors, our intention is for readers to gain a comprehensive understanding of a crucial region that has historically supported fishing and, since the 1970s, oil and gas production. A significant portion of Mexico's Gross Domestic Product derives from the oil-rich *Sonda de Campeche* and Tabasco's coastal areas, which explains the substantial governmental interest in the SWGoM. The energy policy reform of 2013 permitted private companies, both domestic and international, to participate in oil extraction, while current government policies have emphasized maintaining energy sovereignty, making both milestones in Mexico's approach to marine resource management. However, these ever-changing energy interests also risk unintended consequences for vulnerable coastal fishing communities and the conservation of endangered species and ecosystems in the region.

Given that the fishing and oil industries in the SWGoM face complex and interconnected socioeconomic, political, and ecological challenges, we employ a socio-ecological systems (SES) framework to examine and address the trade-offs emerging between these two pillars supporting Mexico's food security and energy needs. Unlike previous discipline-specific works focusing on isolated sectors and disciplines, our collaborative work unravels how both extractive industries interact, conflict, and adapt within a coupled SES. Our authors comprise interdisciplinary scholars, alongside representatives from state governments, who employ innovative, participative, and knowledge co-production methodologies to provide critical insights for improving governance and sustainability processes in the SWGoM.

The resulting compilation primarily targets researchers, graduate and undergraduate students, policymakers, and professionals interested in the sustainable coexistence of the fishing and oil industries in contested marine spaces. The book is also relevant for international scholars and policymakers, especially those from nations (Mexico, the USA, and Cuba) facing the challenges threatening the Gulf of Mexico—notably climate change, biodiversity loss, overexploitation, and invasive species. Moreover, its empirical insights from case studies will resonate particularly with emerging economies across the Americas, Africa, and Asia facing similar circumstances.

Campeche, Mexico A. Espinoza-Tenorio
Villahermosa, Tabasco, Mexico D. E. Ramos-Muñoz
Campeche, Mexico E. Núñez-Lara

Acknowledgments

Most contributors to this volume are affiliated with the Secretariat of Science, Humanities, Technology, and Innovation (SECIHITI) of Mexico, which supported the research project "The Challenging Coexistence of Coupled Social-Ecological Systems: The Fishing and Oil Industries in Campeche Sound (CF-2019-1564454, CONACyT)." We extend our deepest gratitude to the fishers, government authorities, academics, students, oil industry representatives, and non-governmental organization participants who generously contributed to the project's workshops and interviews, forming vital learning communities through their engagement.

Special thanks to Deysi G. Cupido-Santamaria for her exceptional editorial support, along with English language editors such as Andrea MacTavish and John Lidsay. For quality assurance, we employed large language models (LLMs) for initial grammar and spelling checks, with all content subsequently reviewed and refined by human editors.

Overview

- This work presents a comprehensive analysis of the southwestern Gulf of Mexico—a region central to Mexico's fishing and hydrocarbon production and home to one of the world's largest shallow-water oil reserves—and examines its socio-ecological dynamics through an interdisciplinary lens.
- The book synthesizes empirical and theoretical discussions on the unintended consequences and the trade-offs emerging from the spatial overlap between industrial fisheries and offshore oil extraction, addressing ecological, economic, and governance challenges.
- Designed for academics, policymakers, and stakeholders, the book provides accessible yet rigorous insights into managing competing resource-use activities in contested marine spaces.

Contents

About the Editors and Contributors

Editors

A. Espinoza-Tenorio Ph.D. is a Senior Researcher at *El Colegio de la Frontera Sur* (ECOSUR) and member of Mexico's National System of Researchers (Level 2). As the co-founder of the Transdisciplinary Laboratory for Sustainability, he develops integrative approaches for coastal socio-ecological systems, combining fisheries co-management, ecological modeling, geographic information systems (GIS), remote sensing, serious games, and local knowledge to advance collaborative, ecosystem-based strategies. With over 80 peer-reviewed publications, his work focuses on tropical nearshore ecosystems and the design of holistic methodologies for integrating scientific innovation with community and stakeholder needs. His publications center on transdisciplinary research, technological solutions to socio-environmental challenges, and participatory governance, and he actively collaborates with communities, governments, and non-governmental organizations to bridge science, policy, and traditional practices. For more details on his work, visit: https://www.ecosur.mx/academico/aespinoza

D. E. Ramos-Muñoz is an industrial engineer by training (*Instituto Tecnológico y de Estudios Superiores de Monterrey*, 1989) and obtained her Ph.D. in Ecology and Sustainable Development from *El Colegio de la Frontera Sur* (ECOSUR) in 2002. Her publications address women's work, heritage, the hydrocarbon industry, artisanal fishing, and information technologies. She has conducted research in Mexico and Guatemala and has taught in Mexico, Guatemala, Germany, and Finland. She is a Senior Researcher in the Department of Society and Culture at ECOSUR and a member of Mexico's National System of Researchers. For more details on her work, visit: https://www.ecosur.mx/academico/dramos

E. Núñez-Lara Ph.D. is a Researcher and teacher at the Faculty of Natural Sciences of *Universidad Autónoma del Carmen* and a member of Mexico's National System of Researchers since 2008. He has led six research projects on environmental topics, including sustainability indicators, coral reef ecology, climate change effects, and species distribution modeling, founded by the Secretariat of Science, Humanities, Technology, and Innovation. His research and teaching are focused on methodological designs applicable to tropical ecosystems, sustainability assessments, and reef fish community ecology. He is currently a teacher in the Master's program of Restoration Ecology and the undergraduate program of Marine Biology.

Contributors

Anabel Alejandra Ramírez Pacheco PhD holds a Bachelor's degree in Ethnology with honors from the National School of Anthropology and History, as well as a Master's degree in Natural Resources and Rural Development from El Colegio de la Frontera Sur. Her research is guided by an interest in rural youth, with a focus on identity construction, technology use, and territorial perceptions. She specializes in qualitative and participatory methodologies, and her fieldwork has been primarily conducted in southeastern Mexico, particularly in the state of Tabasco. She also has extensive experience as an external consultant for the private hydrocarbon industry, where she fosters community engagement and collaboration. Her most recent publication explores the experiences of young people in the Pantanos de Centla Biosphere Reserve, questioning whether formal education has adequately prepared them to contribute to the conservation of this Protected Natural Area.

Rodrigo Armada-Ramírez is an ethnologist and social anthropologist who graduated from the National School of Anthropology and History. He has conducted research in the upper Lerma Basin, on the Yucatan Peninsula, and in the high and low mountains of the state of Guerrero. He is a Professor of Intercultural Development and Management, Faculty of Philosophy and Letters and Communication Sciences of the Open University and Distance Education System of the Faculty of Political and Social Sciences of *Universidad Nacional Autónoma de México*. He is currently a Ph.D. student within the research line of Anthropology of Violence, coordinated by Dr. Maria de la Paloma Escalante Gonzalbo and Dr. Florence Rosemberg. His research interests center on social-environmental conflicts, with a focus on environmental recovery processes, agroecological alternatives, and forestry and water management in indigenous and rural communities.

Oscar de J. Ruiz-Blanco is a technician (level C) in the Information and Communications Technology Unit at *El Colegio de la Frontera Sur*. He is a Mechatronics Engineer by training, with a Master's degree in Software Management and Engineering. His experience includes data analysis, IoT design and security, software development, and network management. He has participated in technological innovation projects, citizen science, and data analysis to identify trends from a gender perspective in the hydrocarbon industry. For more details on his work, visit: https://www.ecosur.mx/academico/oscarruiz1806 or https://orcid.org/0009-0001-9631-1545

Sebastián Caña-Hernández is a marine biologist who graduated from *Universidad Autónoma del Carmen* with a Master's degree in Natural Resources and Rural Development. He is currently a Professional Service Provider at Mexican Institute for Sustainable Fisheries and Aquaculture Research (IMIPAS), through the Gulf of Mexico Large Marine Ecosystem Project (Component 2), where he collaborates in assessments of the project's target fishery resources. He has conducted studies on fishery biology, as well as the age and growth of elasmobranchs, in the southeastern Gulf of Mexico. He has also served as the Lead Researcher on fishery research projects at IMIPAS. For more details on his work, visit: https://www.researchgate.net/profile/Sebastian-Cana-Hernandez?ev=hdr_xprf.

Ricardo Cavieses-Núñez is a Fisheries Engineer with a Master's and Ph.D. in Marine and Coastal Sciences from *Universidad Autónoma de Baja California Sur* (UABCS). He is currently conducting a postdoctoral stay at the Institute of Oceanological Research, where he has served as a subject Professor in the academic department of Fisheries Engineering of UABCS since 2013. He is a Member of the National System of Researchers (candidate). He is a coordinator at *Clubes de Ciencia México A.C.* He has served as an independent consultant for fishing cooperatives and civil associations, working with data analysis and visualization and implementing artificial intelligence technology.

Eva Coronado Ph.D. is an Adjunct Professor (teaching staff) at the National School of Higher Studies (Mérida) of *Universidad Nacional Autónoma de México*. She is the co-founder and Researcher at ECOMORPHOSES Independent consultancy. She holds a Bachelor's degree in Biology, a Master's degree in Marine Biology, and a Ph.D. in Marine Sciences. Her line of research focuses on transdisciplinary assessments in coastal and rural areas, socio-ecological analysis, governance indicators, mapping of the agrifood value chain, social responsibility assessments, and business-social connections. She collaborates with social and private producers, government agencies, non-governmental organizations, and academic groups. Her work has contributed to management and public policies for fisheries systems, as well as the development of marine protected areas, fish refuges, ecolabeling, and management plans. For more details on his work, visit: https://www.researchgate.net/profile/Eva-Coronado

Leopoldo Palomo is a Professor and Researcher at Universidad Marista de Mérida. He received a degree in Biology from the *Universidad Autónoma de Yucatán*, a Master's degree in Aquaculture Business Management, and a Ph.D. in Fisheries and Aquaculture Bioeconomics, both from *Universidad Marista de Mérida*. He has participated in research projects related to commercially important fishery resources in the state of Yucatán, as well as in studies on recreational fishing, fish refuge areas, and the economic valuation of ecosystem goods and services. His practical and research- oriented experience focuses on the social, economic, and ecological factors affecting coastal communities. His current research line centers on social-ecological systems in fisheries and aquaculture. He is currently a Professor-Researcher at the School of Natural Resources of *Universidad Marista de Mérida* and contributes to the institutional Ph.D. program in Fisheries and Aquaculture Bioeconomics. For more details on his work, visit: https://www.researchgate.net/profile/Leopoldo-Palomo

Moisés Frutos-Cortés has a Master's degree in Social Sciences from the Latin American Faculty of Social Sciences (FLACSO). He is a Ph.D. candidate in Political and Social Sciences (specialization in sociology) at National Autonomous University of Mexico and Ph.D. Student in Sustainability. He is a Full-time Professor and Member of the Center for Research in Environmental Sciences at the Autonomous University of Carmen (CICA-UNACAR). His lines of research are social, economic, and environmental processes in southeastern Mexico; vulnerable groups and social violence; management and use of natural resources. He has been coordinator of the University Observatory of Social Violence since 2014. She is currently working on the research project "Socio-environmental vulnerability in the region of Laguna de Terms, in the state of Campeche." ID: https://orcid.org/0000-0002-1381-9504. Mail Contact: mfrutos@pampano.unacar.mx

Johnny Bryan Cruz-Pech is a Marine Biologist who graduated from *Universidad Autónoma del Carmen* (UNACAR) in 2023. Currently, he is studying for a Master's degree in Natural Resources and Rural Development from El Colegio de la Frontera Sur (2024) in the Sustainability Sciences academic department. His research focuses on detecting and classifying oil on the sea surface using satellite imagery. He has participated in numerous conferences as a co-author of papers and posters. His experience in geographic information systems, applied to the detection of hydrocarbons on the sea surface, has led him to work as a technical assistant and provide scientific support. In addition to his experience as an instructor in hydrocarbon detection and monitoring courses, he contributes to scientific publications and research projects.

Eduardo Cuevas Ph.D. is a biologist from *Universidad Autónoma de Yucatán*. He has led the development of national programs for the recovery of priority species in Mexico, helped justify of the creation of National Parks and Protected Areas (NPAs), and worked to develop and update their management programs. He is currently developing a line of research in Operational Ecology at the Institute of Oceanological Research of the *Universidad Autónoma de Baja California*. Operational Ecology integrates and links multiple sectors to address environmental problems and monitor regions of interest. He has trained undergraduate and graduate students and is fully active in publishing JCR articles, books, and chapters. He is a member of the National System of Researchers (Level 2).

Deysi Guadalupe Cupido-Santamaria is an Environmental Engineer and graduated from the Technological Institute of Villahermosa (ITVH) in Tabasco (2016). She obtained her Master of Science in Natural Resources and Rural Development *from El Colegio de la Frontera Sur* (2021). She is the project coordinator at *El*

Colegio de la Frontera Sur for the Society and Culture academic department. She has worked since 2017 as a field technician or project coordinator in border municipalities and southeastern Mexico in projects focused on human rights, gender studies, race, and ethnic discrimination, as well as gender studies in fishing communities alongside civil associations and social studies with oil consultancies. She also coordinates and integrates information for a fisheries and oil projected and conducts spatiotemporal analysis to study effects and changes in the marine space used by the fishers of Tabasco and Campeche.

John W. Day Jr. is a Distinguished Professor Emeritus at Louisiana State University, specializing in coastal and wetland ecosystems, particularly the Mississippi Delta. With over 400 peer-reviewed publications, he has conducted extensive research on ecological management. He has co-edited multiple works on estuarine ecology and coastal sustainability. Earning his Ph.D. in marine and environmental sciences from the University of North Carolina in 1971, he studied under ecologist H.T. Odum. His research spans over four decades, including collaborations with Mexican institutions and studies on Mediterranean wetlands affected by climate change. He has been involved in nitrogen removal strategies for the Mississippi River, served on committees for delta restoration, and contributed to discussions on sustainability concerning twenty-first-century megatrends. He has held visiting professorships at universities across Mexico, Europe, and the UK, with a focus on ecological conservation and urban sustainability.

Juan de Dios Valdez-Leal is a Researcher and Professor in the Academic Division of Biological Sciences at *Universidad Juárez Autónoma de Tabasco* (UJAT). He is a biologist with a Master's degree in Wildlife Management and Conservation and a Ph.D. in Ecology. His research work focuses on biodiversity conservation, addressing topics such as landscape ecology, with a focus on connectivity, environmental assessments, and the design of protected natural areas. He also teaches and disseminates scientific knowledge, sharing his passion for wildlife conservation in the tropics. He was a Visiting Professor at the International Institute for Wildlife Conservation and Management (ICOMVIS) at *Universidad Nacional de Costa Rica*. He is also a co-founder of the Human Infrastructure and Biodiversity Impact Group (GIT) of the Mesoamerican Society for Biology and Conservation, which belongs to the Connectivity research network in Mexico. His comprehensive approach combines applied research, teaching, and environmental management to promote the sustainability and well-being of Neotropical ecosystems. For more details on his work, visit: https://www.researchgate.net/profile/Juan-De-Dios-Valdez (Orcid: 0000-0002-0315-2400).

Angelina del Carmen Peña-Puch Ph.D. is a postdoctoral Researcher at the EPOMEX Institute of the Universidad Autónoma de Campeche (UACAM), working on the project "Advances in Vertical and Horizontal Governance in Coastal Socio-Ecological Systems with Artisanal Fisheries and Community-Based Tourism in the Protected Natural Areas of Campeche, Mexico." Her research focuses on the socio-environmental governance of coastal resources and sustainable development in coastal socio-ecological systems. She has published her research as peer-reviewed articles, book chapters, and international reports. Additionally, she has taught undergraduate and graduate courses at UACAM and *El Colegio de la Frontera Sur* and has supervised and evaluated theses at both academic levels.

Ana Carolina Esteves-Dias holds a Ph.D. in Social and Ecological Sustainability from the University of Waterloo, Canada. Her work is centered on environmental governance in coastal communities, marine protected areas, and small-scale fisheries in the Global South. She is currently a fellow of the Intergovernmental Platform for Biodiversity and Ecosystem Services (IPBES), collaborating as an author of the Nexus Assessment. As a postdoctoral fellow of the V2V Global Partnership, she collaborated with research teams in Africa and Asia, leading a situational analysis of small-scale fisheries in the region alongside over 70 researchers, students, and practitioners. She has conducted extensive work in fisheries and marine protected areas in Brazil and collaborates with research groups in Brazil, Argentina, and Uruguay.

Miguel Ángel Díaz-Perera holds a Bachelor's degree in History from *Universidad Juárez Autónoma de Tabasco*, and a Master's and Ph.D. in History from *Centro de Estudios Históricos de El Colegio de Michoacán*. He completed a postdoctoral fellowship in the Section of Theory and Methodology of Science at *Centro de Investigación y de Estudios Avanzados*. Since 2009, he has worked at *El Colegio de la Frontera Sur* as a Senior Researcher, a member of the National System of Researchers (Level II), and part of the academic group on Migration Studies and Cross-Border Processes. He is also the founder of the Centro Documental de Estudios Sobre el Agua (CDEA). He led a project under the National Strategic Programs, focusing on diagnostics and scenarios in the territories affected by the Mayan Train. He has published 16 scientific articles, eight peer-reviewed book chapters, and six peer-reviewed books (three as sole author), as well as 12 popular science publications, including extended book chapters.

Claudia María Fumero-Andreu is a biologist with a Ph.D. in Fisheries and Aquaculture Bioeconomics from *Centro Interdisciplinario de Ciencias Marinas-Instituto Politécnico Nacional*. Her work focuses on governance networks, sustainability, and fisheries resilience. She has participated in international research initiatives led by the Food and Agriculture Organization of the United Nations, Duke University, and Pronatura Noroeste and is the co-author of peer-reviewed articles in journals such as *Ocean & Coastal Management* and *Ecological Economics*. She combines academic research with applied work in marine conservation, environmental consulting, and the development of innovative artificial reef systems. She is one of the founders of Satoumi Reefs, a company dedicated to supporting artisanal fisheries through reef-based technologies.

Lilia María Gama-Campillo Ph.D. is a Senior Researcher and Professor in the Laboratory of Landscape Ecology and Global Change at the *Universidad Juárez Autónoma de Tabasco* and a National System of Researchers member. She is the co-founder of the Academic Network on Disasters in Tabasco. Her experience in territorial diagnosis, ecological planning, and community-based sustainable solutions has led her to develop environmental planning instruments and natural areas programs, participating in the state government as undersecretary of Environmental Policy. Her main lines of research are territorial diagnoses, coastal dynamics, analysis of the impacts associated with global warming and biodiversity conservation, and the use of traditional and local ecological knowledge in natural resources management. For more details on his work, visit: https://www.researchgate.net/profile/Lilia-Gama?ev=hdr_xprf (Orcid: 0000-0002-5417-9697)

Manuel Mendoza-Carranza is a senior researcher in the Department of Sustainability Sciences at ECOSUR and a member of the National System of Researchers. Mendoza-Carranza is an expert in tropical small-scale marine and freshwater research, fish and elasmobranch biology, ecology population dynamics and management. Currently, he is developing research on women's participation in small-scale fisheries and in marine and freshwater fishery refuge zones, and is developing collaborations based on statistical data analyses. More information can be found at https://www.ecosur.mx/academico/mcarranza.

Delázkar Noel Rizo-Gutiérrez obtained a Ph.D. in Social Anthropology from *Centro de Investigaciones y Estudios Superiores en Antropología Social* (Southeast) in 2019. He is currently attached to *Universidad Autónoma Chapingo*, Chiapas (2022–2026) and is also a National System of Researchers candidate. He was a postdoctoral fellow at *Universidad Nacional Autónoma de México* (2020–2022) and postdoctoral fellow of the Secretariat of Science, Humanities, Technology, and Innovation (SECIHITI) (2022–2024). His research projects include the future of common lands on the Tzeltal ejido of Selva Lacandona, Chiapas; imagined futures of amber: labor exploitation, community vindication, and global positioning, within the Amber Worlds Project (UiO); and youths, futures, hopes, and negligence of the past. His topics of interest include ethnography, futures, utopias, environmental narratives, temporalities, ethics, and humor. He is a member of working groups and seminars such as the Network of studies on Communities, Utopias and Futures (RIOCOMUN), the Working Group of the Latin American Association of Anthropology, the Seminar on Social Studies of Outer Space, and the Working Group on humor, laughter, and hierarchies. (ORCID: 0000-0002-6246-0905)

Mitzy Tizel Zarate-Herrera is a biologist from the *Universidad Nacional Autónoma de México* (UNAM) and holds a Master's in Coastal and Marine Zone Management from *Universidad Autónoma de Campeche*, where she graduated with honors and received the University Merit Medal. She is currently pursuing a Ph.D. in Ecology and Sustainable Development at *El Colegio de la Frontera Sur* (ECOSUR), Campeche Unit. Her research focuses on the biology and fisheries assessment of coastal sharks, particularly the Atlantic sharpnose (*Rhizoprionodon terraenovae*), bonehead (*Sphyrna tiburo*), and blacknose (*Carcharhinus acronotus*) sharks in the southeastern Gulf of Mexico. She has collaborated with the Mexican Institute for Sustainable Fisheries and Aquaculture Research (IMIPAS) and has presented her work at national and international conferences. For more details on his work, visit: https://www.researchgate.net/profile/Mitzy-Zarate-Herrera?ev=hdr_xprf

Rosa Isela Hirales-Cota is a Fisheries Engineer with a Master's degree in Marine and Coastal Sciences with a Sustainable Management Orientation from *Universidad Autónoma de Baja California Sur* (UABCS). She has worked as a geographic information systems (GIS) analyst of marine and coastal ecosystems in government agencies and civil society organizations. She joined UABCS in 2020 as a Full-time Professor, training undergraduate students in GIS and sustainable development. She is a Ph.D. student in Marine and Coastal Sciences (CIMACO)-UABCS.

Rachael G. Hunter Ph.D. has extensive experience in inland and coastal wetland ecology, particularly in the use of natural and constructed wetlands to remove nutrients and sediments from point and non-point sources. She has experience quantifying vegetation dynamics (species composition, percent cover, and biomass) and biogeochemical processes (denitrification, nitrogen uptake, decomposition, and phosphorus sorption) in Louisiana's freshwater forested wetlands, brackish marshes, and saltwater marshes. For several projects, she has assessed relationships between ecosystem biogeochemical functions and specific site characteristics such as soil nutrient and carbon concentrations, vegetation productivity, hydrology, and precipitation. In recent years, her work has concentrated primarily on assessing the impacts of anthropogenic activities on the function and structure of bald-cypress-tupelo swamps and fresh-to-saltwater marshes. Much of her work has focused on the impacts of oil and gas exploration, production, and transportation on Louisiana's coastal wetlands.

Raúl E. Lara-Mendoza is the main coordinator of Component 2 of the Project: Implementation of the Strategic Action Program for the Gulf of Mexico Large Marine Ecosystem (IMIPAS-GEF-UNOPS) and a member of the National System of Researchers. He was also the Head of Department of Fisheries Resource Assessment in the Atlantic at the Mexican Institute for Fisheries and Aquaculture Sustainable Research. He has been responsible and collaborated on national and international projects, conducting research on the biology, ecology, fisheries, population dynamics, and management of marine resources in the Gulf of Mexico, Mexican Pacific, and areas beyond national jurisdiction. He has served as a lead, coordinator, or liaison with various national and international entities for the development of numerous Initiatives, Strategies, Agreements, Conventions, and Projects related to fisheries research and management. He has particularly promoted the development and publication of Fisheries Management Plans, the National Fisheries Charter, National Action Plans, Technical Opinions and Assessments, and international agreements. He has also advanced various initiatives related to stock assessments, fishery refuge zones, fishery improvement projects, climate change, food sovereignty, foreign fisheries and exports, and various CITES-related issues, among others. For more details on his work, visit: https://www.researchgate.net/profile/Raul-E-Lara-Mendoza

Hilda María Díaz-López is an Academic Technician at the Landscape Ecology and Global Change Laboratory in the Academic Division of Biological Sciences at *Universidad Juárez Autónoma de Tabasco*. She is a collaborator in the Consolidated Academic and Research Group: Resilience to Global Change. She has participated in research projects funded by national and international institutions such as the Secretariat of Environment and Natural Resources (SEMARNAT), the National Commission for the Knowledge and Use of Biodiversity (CONABIO), the National Institute of Fisheries and Aquaculture (INAPESCA), and the Inter-American Development Bank (IDB), among others. She is co-author of various articles, book chapters, and papers. Her research interests are focused on landscape ecology, spatial analysis, geographic information systems (GIS) applications, environmental diagnosis, and the ecology and conservation of the ecology and conservation of Mexican primates. For more details on his work, visit: https://orcid.org/0000-0002-7754-9072.

Luis Malpica-Cruz Ph.D. is a Researcher at the Oceanographic Research Institute (IIO) of *Universidad Autónoma de Baja California* (UABC). His research focuses on marine ecology, natural resource management, and the resilience of coastal social-ecological systems. His work examines trophic interactions in marine food webs, including predator ecology in the northeastern Pacific, invasive mesopredators in the Caribbean and Gulf of Mexico, and phase shifts of invasive macroalgae in Baja California, using isotopic techniques. He is also studying how global change stressors affect fishing communities and analyzing adaptive strategies like tourism and aquaculture to enhance social-ecological resilience. As part of the MexCal research consortium (Ecosystem Management for the Californias), researchers from UABC and *Centro de Investigación Científica y de Educación Superior de Ensenada* (CICESE) collaborate with fishers, coastal communities, and policymakers to develop monitoring tools for marine resource management. Their interdisciplinary approach bridges ecological and social sciences to address sustainability challenges in marine systems. For more details on his work, visit: https://www.malpicacruzlab.com/

Abigail Uribe-Martínez is a computer systems engineer with a Master's in Marine Sciences and Limnology and a Ph.D. in Geography and is a member of the National System of Researchers (Level 1). She has extensive experience in geomatic sciences (remote sensing and geographic information systems) to evaluate the distributions of ecosystems, habitats, and marine and coastal species. She worked as a specialist in remote sensing in geomatics at the National Commission for the Knowledge and Use of Biodiversity (CONABIO). She was a research assistant for 3 years in the remote sensing and geographic information systems (GIS) laboratory of the Center for Research and Advanced Studies (CINVESTAV) in Mérida and spent 3 years as a Full-time Professor at the Faculty of Natural Sciences, *Universidad Autónoma del Carmen*. She has completed two postdoctoral stays and is currently a Senior Researcher at the Institute of Oceanological Research of the *Universidad Autónoma de Baja California*.

Jovana Martínez-Amezcua holds a degree in Intercultural Development and Management, with a specialization in Social Mediation, from *Universidad Nacional Autónoma de México* (UNAM). Her thesis was funded by the project "The Challenging Coexistence of Coupled Socioecological Systems: The Fishing and Oil Industries in the Sonda de Campeche" at *El Colegio de la Frontera Sur*. Her research interests are technologies, social mediation, and social and environmental issues.

Cynthia Nayeli Martínez-Fernández has a Ph.D. in Educational Research from the Institute for Research in Education at Universidad Veracruzana. She is a biologist and holds a Master's degree in regional development from El Colegio de Veracruz. Her academic and professional trajectory has been centered on environmental education, and her research and publications address university policy and the role of higher education institutions in sustainability while exploring strategic approaches and opportunities. She contributed to the project "Communication Strategy for the New Port of Veracruz" during stages II, III, and IV. Currently, she is a Professor at the Faculty of Natural Sciences at *Universidad Autónoma del Carmen*, where she teaches in the Marine Biology program. She also coordinates the Environmental Agenda communication project. Since 2018, she has been recognized as a member of the National System of Researchers.

Ena Edith Mata-Zayas is an Associate Researcher and Professor in the Biological Science Academic Division at *Universidad Juárez Autónoma de Tabasco* (UJAT). She is a member of the Landscape Ecology and Global Change Laboratory and the National System of Researchers. Her primary research focuses on biological conservation in human-modified landscapes, biodiversity, and ecosystem services in agroecosystems, as well as the impacts of climate change, among other topics. More information about her work can be found here: https://www.researchgate.net/profile/Ena-Mata-Zayas?ev=hdr_xprf (ORCID: 0000-0001-7673-3081).

Maria Azahara Mesa-Jurado is a Senior Researcher at the Department of Sustainability Sciences at El Colegio de la Frontera Sur (ECOSUR) and co-founder of the Transdisciplinary Laboratory for Sustainability (established in 2014). Her research explores the complex interplay between humans and nature, with a focus on how values, emotions, and perceptions shape environmental decision-making across diverse scales and cultural contexts. She works closely with local communities, fostering the co-construction of knowledge and co-designing context-specific solutions that aim to build desirable and sustainable futures. Her approach integrates participatory methods, critical reflection, art-based methods, and transdisciplinary collaboration to support inclusive and meaningful sustainability transitions. https://www.ecosur.mx/academico/mmesa

Eduardo Javier Moguel-Ordóñez is a Professor in the Academic Division of Biological Sciences at *Universidad Juárez Autónoma de Tabasco*. He has participated in research groups on traditional agriculture, agrometeorology, climate change, environmental management, and conservation and has worked on topics such as restoration and reforestation in tropical areas, carbon sequestration in agrosystems, soil evaluation, environmental assessment and impact, community planning, and the development of university-run companies that provide services. He participates in research on biodiversity and ecosystem conservation, the use of plant resources, cropping systems, and the impacts of ecosystems. He is also interested in applying university research results to various aspects of social, productive, and environmental issues. For more details on his work, visit: https://www.researchgate.net/profile/Eduardo-Moguel-Ordonez (ORCID: 0000-0002-1641-6794)

Claudia Monzón-Alvarado is a biologist from Universidad del Valle de Guatemala and holds a Master's degree in Natural Resources and Rural Development from *El Colegio de la Frontera Sur* and a Ph.D. in Geography from the University of Florida. Since 2015, she has worked for the Secretariat of Science, Humanities, Technology, and Innovation (SECIHTI) at *El Colegio de la Frontera Sur*, Unidad Campeche. Her research focuses on the governance of social-ecological systems, with emphasis on knowledge flows and knowledge co-production, which is reflected in her interest in designing platforms that integrate people, data, information, and knowledge.

Jorge Sergio Narro-Ríos Ph.D. is a person preoccupied by global warming and the challenges experienced in the world. For this reason, he studied to become a specialist in solar energy, with a particular focus on thin film solar cells. He has been working in the lab, producing solar cells using CVD (Chemical Vapor Deposition) and PVD (Physical Vapor Deposition) processes, including spray pyrolysis, chemical bath deposition, sputtering, evaporation, and co-evaporation. He then designs, dimensions, and installs photovoltaic systems. Currently, he is in charge of the Subsecretariat for Energy Development in the Secretariat of Environment, Biodiversity, Climate Change, and Energy of the state of Campeche.

Adrian Núñez M.C. is an academic technician responsible for the Fisheries Ecology Laboratory at *El Colegio de la Frontera Sur* (ECOSUR), Campeche. She holds a degree in Fisheries Biology from the Universidad Autónoma de Sinaloa and a Master's in Marine Biology from the Center for Research and Advanced Studies in Mérida. Her research focuses on socio-ecological fishery systems and the application of interdisciplinary approaches to examine vulnerability, risk, and well-being in coastal communities engaged in small-scale fisheries. She has collaborated on multiple projects related to the fisheries in the Yucatán peninsula contributed to assessments of fisheries resource management. For more details on her work, visit: https://www.ecosur.mx/academico/adriannunez

Avik Núñez-Laffón is a student of Computer Systems Engineering at the Faculty of Information Sciences of *Universidad Autónoma del Carmen*. He collaborates in research projects on the application of information technology and digitalization to environmental problems. His practical experience focuses on implementing systems technology in educational and everyday activities. He is an enthusiastic student and is moving toward digitalization and the practical implementation of artificial intelligence.

Anja K. Nygren Ph.D., focuses on global extractivism and altered socio-environments, with special interests in resource access, state-making, the politics of scale, and the dynamics of socionature. She conducts research on hydrocarbon extractivism and oil conflicts, the political ecology of resource frontiers, environmental justice, disasters and displacements, urban politics and planning, and certified coffee and cacao. She also has long-term field research experience in Mexico, Honduras, Nicaragua, and Costa Rica. Her research has been published in various scientific journals, including *Development and Change*, *World Development*, *Journal of Latin American Studies*, *Journal of Peasant Studies*, *Annals of the American Association of Geographers*, *Geoforum*, *Journal of Urban and Regional Research*, *Journal of Agrarian Change*, *Society and Natural Resources*, *Journal of Hydrology*, *Environmental Values*, *Journal of Environmental Policy and Planning*, *Agriculture and Human Values*, *Applied Geography*, *Environmental Sociology*, *Latin American Perspectives*, and *Urban Planning*, among others.

Coral Jazvel Pacheco-Figueroa is a Researcher and Professor in the Academic Division of Biological Sciences at the Universidad Juárez Autónoma de Tabasco (UJAT). She has been trained in Veterinary Medicine and Animal Husbandry and holds a Master's degree in Wildlife Management and Conservation and a Ph.D. in Ecology. Her work focuses on biodiversity conservation, addressing topics such as landscape ecology, road ecology, environmental assessments, and the design of protected natural areas. She also teaches and disseminates scientific knowledge, sharing her passion for wildlife conservation. She has collaborated as a Visiting Professor at the International Institute for Wildlife Conservation and Management (ICOMVIS) at the Universidad Nacional de Costa Rica. She is also the Founder of the Human Infrastructure and Biodiversity Impact Group (GIT) of the Mesoamerican Society for Biology and Conservation. She is a Road Ecology Specialist at the International Union for Conservation of Nature and co-founder of the Institutional Mentoring Program for Women in STEM Careers at UJAT. Her comprehensive approach combines applied research, teaching, and environmental management to promote the sustainability and well-being of Neotropical ecosystems. For more details on her work, visit: https://www.researchgate.net/profile/Coral-Pacheco/research (ORCID: 0000-0001-5281-9251)

Esther Solano-Palacios is a Master's in Regional Studies from the Dr. José María Luis Mora Research Institute and a Graduate in Sociology from the Juarez Autonomous University of Tabasco (UJAT). Since 2002, she has been a Full-time Professor at the Autonomous University of Carmen (UNACAR), where she develops the line of research on social, economic, and environmental processes in the southeast of Mexico, particularly in the state of Campeche, on topics such as fishing, oil, and tourism. Mail Contact: esolano@pampano.unacar.mx

Juan Manuel Palacios-Hernández is an Environmental Engineer with over 5 years of experience in managing environmental permits and licenses for large-scale EPC (Engineering, Procurement, and Construction) projects in the oil, gas, and energy sectors. He worked as a Supervisor of Permits and Licenses at Samsung Engineering for the Olmeca Refinery project in Tabasco, where he coordinated environmental compliance and permitting for five key process plants and over 20 subcontracted companies. He is currently the Environmental and Quality Coordinator at the Combined Cycle Power Plant project with Quantum Energía in San Luis Potosí. Originally from a fishing community in Centla, Tabasco, and raised in a family dedicated to fishing, his background and upbringing inspired him to focus his work on sustainable development. As part of his undergraduate thesis, he conducted a study on the impact of sustainable fishing on local communities. He is committed to promoting sustainable development through regulatory compliance and environmental responsibility in industrial projects.

Martha Alicia Perera-García is a Doctor of Science in Ecology and Management of Tropical Systems, Juárez Autonomous University of Tabasco/Academic Division of Biological Sciences. Research area: Ecology and Biology of Aquatic Organisms. Studies aimed at estimating the basic biological parameters necessary to develop population dynamics models, such as migration, growth, and mortality rates, size at first maturity, and fecundity. Evaluation of physical and biological processes that influence the dynamics of species, populations, and communities at different spatial and temporal scales. The fieldwork carried out extends from the coastal zone to deep waters. Evaluation and understanding of the coupling between physical, chemical, and biological parameters and different ecosystem behaviors, from low trophic levels (early life stages, zooplankton, macrobenthos) to the highest levels of predators, with special emphasis on key species for fisheries. Research with ecological approaches to assess aquatic resources by providing biological benchmarks based on trophic flows and bioeconomic models that complement standard population assessment methods.

Angela Viviana Rabelo-Avalos is a historian specializing in environmental history and social studies related to the oil industry in southeastern Mexico. She graduated from the Universidad Juárez Autónoma de Tabasco and earned her Master's degree from *El Colegio de la Frontera Sur* with a thesis on the Ixtoc-I oil spill and its social and environmental impacts. She has published research on the history of oil in Tabasco and has participated in participatory diagnosis projects focused on sustainable livelihoods. Her work stands out for combining documentary, oral, and cultural sources to analyze the relationship between communities and extractivism.

Lenin Rangel-López is currently a Professor and Researcher in the Aquaculture Engineering Educational Program of the Academic Division of Agricultural Sciences at the Juárez Autonomous University of Tabasco. Academically, he holds a Ph.D. in Agricultural Sciences (UAEH), a Master's degree in Aquaculture Sciences (ITBOCA), and a degree in Fisheries Engineering with a specialization in aquaculture. In general, his line of research focuses on the use of alternative products, such as plants, extracts, and essential oils, to determine their antibacterial activity against bacteria of importance in aquaculture, in addition to carrying out activities related to engineering and societal impact. He currently has six published articles addressing topics related to his research interests. He also has a YouTube channel where his primary focus is to have conversations with people who work in aquaculture, discussing their experiences in the field.

Evelia Rivera-Arriaga Ph.D., has been a Research Professor at *Universidad Autónoma de Campeche* for 31 years, with additional teaching roles at institutions in Mexico and the United States. She served as Secretary of Environment and Sustainable Use for Campeche (2009–2015) and has consulted on coastal management projects for international organizations. She has advised the Inter-American Development Bank on coastal management initiatives in Latin America and the Caribbean and collaborated with the United Nations Development Programme and the National Institute of Ecology and Climate Change as National Coordinator for the Sixth Communication on Climate Change. Her expertise encompasses marine policy, legislation, governance, and coastal management for sustainable development, as well as international ocean treaties and environmental and climate change policies. She has led 40 research projects, delivered over 100 conferences, taught more than 150 university courses, and mentored 20 students. She played a pivotal role in developing Campeche's environmental policy and is currently an active member of international scientific committees, contributing to global research on ocean and climate change.

Jaime R. Ruiz-Blanco is a technician (level C) of the Information and Communications Technology Unit at El Colegio de la Frontera Sur. He is a computer systems engineer with a Master's degree in Computer Science. His focus is on research and technological development based on artificial intelligence and data science. He has participated in innovation and national security projects, developing biometric models, simulation, image processing, and natural language processing. For more details on his work, visit: https://www.ecosur.mx/academico/jaimeruiz (ORCID: 0000-0001-9985-5102)

Camilo Ruiz-Méndez has a degree in Physics from UNAM (2000), a Ph.D. in Physics from the University of Salamanca (2005), and has been a postdoctoral Researcher at the Max Planck Institute for Complex Systems (2006–2007) and at Imperial College London (2008). He was a Ramón y Cajal Ramón fellow (2009–2014) at the Centro de Láseres Pulsados (CLPU). He achieved the recognition of I3 Researcher (2014). Since 2016, he has been a Professor at the Department of Didactics of Mathematics and Experimental Sciences at the University of Salamanca. He is the Principal Investigator of the GIR Education, Mathematics, Sciences, and Climate Change (EMC3), whose lines of research are the interaction of Climate Change and Education. He has participated in 20 national and European projects. Currently, he is PI of the national projects "Education for Climate Change and Sustainability, a longitudinal study of intergenerational learning. EduC3." PID2020-114358RB-I00, "Climate impacts on education and nature-based adaptation measures in education. EduHeat." He has published 105 papers in JCR and Scopus journals (https://orcid.org/0000-0001-9538-5780. H=23 citations 1700. He has supervised eight doctoral theses and two more in progress.

Daniela Matías-Sánchez Ph.D., is a Social Anthropologist and received her degree from *Universidad Veracruzana* in 2008 and her doctorate in the same specialty from *Centro de Investigaciones y Estudios Superiores en Antropología Social* (Southeast) in 2019. Her research and publications focus on the hydrocarbon industry from sociocultural and gender perspectives. She examines the impact of energy restructuring in Mexico on labor and family conditions, with a special focus on female oil workers. She is currently a postdoctoral Researcher at *El Colegio de la Frontera Sur* and a member of the National System of Researchers. For more details on his work, visit: https://www.ecosur.mx/academico/danielamatias (ORCID: 0000-0002-3723-4690).

Luis Manuel Uribe-Sandoval has a Bachelor's degree in Environmental Sciences from the Universidad Nacional Autónoma de Mexico, with his thesis entitled *Use of the Fishing Space of an Artisanal Cooperative and Areas of Recurrent Presence of Oil in the Campeche Sound* (2024). He is currently part of the remote sensing team of the Fisheries and Petroleum Project of Secretariat of Science, Humanities, Technology, and Innovation (SECIHITI) and a technician for research projects.

Saúl Jaime Serrano-Guzmán is an aquatic biologist and Master of Science in Marine Ecology. He is Professor-Researcher at the Universidad del Mar (UMAR, Puerto Ángel *campus*, 1998–2025). His main areas of knowledge and research include reproductive biology, early life, and the integral cultivation of marine bivalve mollusks in the tropical Pacific of America, monitoring of physicochemical variables in the coastal zone, distribution of microseston and particulate organic matter, succession, and eutrophication of coastal lagoons, dispersal of holoplankton and chlorophylls in the neritic zone, coastal ecological planning studies and environmental impact statements, and fisheries and aquaculture management plans (Mariculture) for the Oaxaca Coast Region. The cultivation of bivalve mollusks depends on the environmental and socioeconomic conditions of the inhabitants in the tropical regions of Mexico and Latin America.

Pauleth Soria-Gómez has a Bachelor's degree in Environmental Sciences and is a Master's student in Ecosystem Management of Arid Zones of *Universidad Autónoma de Baja California* (UABC), Faculty of Sciences. Her academic background and interests focus on environmental governance, socio-ecological systems management, and coastal conservation. She has worked on the evaluation of governance mechanisms in protected natural areas and the analysis of red snapper fisheries as part of her undergraduate thesis. Her approach combines social and ecological tools to contribute to the design of more integrative and participatory management strategies.

Sandra Manuela Suárez-García is a Professor and Researcher at the *Instituto Tecnológico Superior de Centla*. She holds a Master's degree in Materials and Energy Engineering (MIME) and is currently pursuing a Ph.D. in Engineering Sciences at the Universidad Juárez Autónoma de Tabasco. Her academic work focuses on developing internally and externally funded research projects, primarily related to environmental monitoring. She has worked on air quality analysis, specifically detecting pollutants such as PM10 particles, as well as on water quality studies, evaluating physicochemical parameters and the presence of heavy metals, including chromium. She is a member of the Academic Body "Conservation and Preservation of Tropical Ecosystems," where she promotes research focused on the application of environmental engineering for the diagnosis and recovery of ecosystems degraded by human activities. Her publications seek to provide technical and scientific solutions to the region's environmental challenges by addressing issues of environmental diagnosis, conservation, and sustainability. His approach integrates academic knowledge with the ecological and social needs of the tropical environment.

Mario Vásquez-Ortíz holds a Bachelor's degree in Geography from *Universidad Nacional Autónoma de Mexico* (UNAM). He has served as a research technician at the Mexican Institute for Sustainable Fisheries and Aquaculture Research (IMIPAS). He has participated in multiple research cruises aboard the R/V Dr. Jorge Carranza Fraser, covering areas such as the Gulf of California, Baja California peninsula, and Gulf of Tehuantepec. His contributions include spatial mapping of biomass distributions to support of sustainable fisheries management in Mexico. He actively collaborates with IMIPAS's strategic efforts in modernizing fisheries monitoring through acoustic technologies and geospatial analysis. For more details on his work, visit: https://mx.linkedin.com/in/mario-vásquez-ortiz-383046153

Claudia Villanueva-García is an Associated Researcher and Professor in the Laboratory of Ecology and Tropical Conservation at the Universidad Juárez Autónoma de Tabasco and a member of the National System of Researchers. She is a biologist with a Master's degree in Health Sciences, Production, and Animal Welfare and a Ph.D. in Biodiversity and Environmental Management. Her main lines of research include conservation genetics, wildlife conservation, ethology, and landscape ecology. For more details on her work, visit: https://www.researchgate.net/profile/Claudia-Villanueva-Garcia-2 (ORCID: 0000-0001-5026-2947).

Nicolás Vite-García has been a Professor-Researcher in the Academic Division of Agricultural Sciences at the Juárez Autonomous University of Tabasco since 2013. As head of the Specialized Aquaculture Research Laboratory, he develops lines of research in ecophysiology and the aquaculture of mollusks, crustaceans, and marine fish. He developed the culture of the seahorse *Hippocampus erectus* during his doctoral studies. He has collaborated on various research projects and has been responsible for the project "Technological development for the cultivation of Mexican seahorse species." He has experience as a researcher at the Mexican Institute for Research in Fisheries and Sustainable Aquaculture, within the shrimp fishery program in the Mexican Pacific. He has published several articles and book chapters. He is currently developing a line of research that relates climate change to the education of cooperative societies in fisheries and aquaculture production, pursuing a second Ph.D. at the University of Salamanca, Spain. Professionally, he has participated in commercial aquaculture projects involving bivalve mollusks, scallops, pearl oysters, and marine fish, farming fish in floating cages offshore. He is part of an aquaculture production cooperative farming scallops in the Mexican Pacific.

José Alberto Zepeda-Domínguez is a Professor in the Marine Sciences Faculty of *Universidad Autónoma de Baja California* (UABC) in Ensenada, Mexico. He is a marine biologist with a specialty, Master's, and Ph.D. in Marine Sciences. Since 2003, he has worked on the Mexican coasts as an extensionist, consultant, non-governmental organization member, and academic. As a Professor, he facilitates the development of an integrated perspective on the management of social-ecological systems, particularly to cope with resource scarcity, variable environments, and an uncertain future. As a researcher, his approach is inter- and transdisciplinary, advocacy-based, and action-oriented, focusing on understanding and promoting the factors that condition and trigger the effectiveness of managing social-ecological systems, particularly their resilience. He seeks to create knowledge from different methodological approaches (qualitative, quantitative, and mixed techniques) and areas of social-ecological systems management (resilience-based management, adaptive co-management, polycentric governance, nature-based solutions, and blue transformation). For more details on his work, visit: https://www.researchgate.net/profile/Jose-Zepeda-Dominguez?ev=hdr_xprf.

Abbreviations

AGEB	*Área Geoestadística Básica*—Basic Geo-Statistical Area
CANAINPESCA	*Cámara Nacional de las Industrias Pesquera y Acuícola*—National Chamber of the Fishing and Aquaculture Industry
CC	Climate change—*Cambio Climático*
CCE	*Consejo Coordinador Empresarial*—Business Coordinating Council
CIGOM	*Consorcio de Investigación del Golfo de México*—Gulf of Mexico Research Consortium
CINESTAV	*Centro de Investigaciones y de Estudios Avanzados del Instituto Politécnico Nacional*—Center for Research and Advanced Studies of the National Polytechnic Institute
CEMIE-Océano	*Centro Mexicano de Innovación en Energía Océano*—Mexican Center for Innovation in Ocean Energy
CFE	*Comisión Federal de Electricidad*—Federal Electricity Commission
CNH	*Comisión Nacional de Hidrocarburos*—National Hydrocarbons Commission in Mexico
COBATAB	*Colegio de Bachilleres de Tabasco*—Tabasco High School level institution
COBI	*Comunidad y Biodiversidad A.C.*—Community and Biodiversity NGO
CONAHCYT	*Consejo Nacional de Humanidades, Ciencia y Tecnología*—National Council for the Humanities, Science and Technology
CONANP	*Comisión Nacional de Áreas Naturales Protegidas*—National Commission of Natural Protected Areas
CONAPESCA	*Comisión Nacional de Acuacultura y Pesca*—National Commission of Aquaculture and Fishing
CONEVAL	*Comisión Nacional de Evaluación de la Política de Desarrollo Social*—National Council for the Evaluation of Social Development Policy

CRE	*Comisión Reguladora de Energía*—Energy Regulatory Commission
DACA	*División Académica de Ciencias Agropecuarias*—Agricultural Sciences Faculty, UJAT
DENUE	*Directorio Estadístico Nacional de Unidades Económicas*—National Statistical Directory of Economic Units
DSDG	UN Division for Sustainable Development Goals—*División de las Naciones Unidas para los Objetivos de Desarrollo Sostenible.*
ECOSUR	*El Colegio de la Frontera Sur*—The College of the Southern Border
EEZ	Economic Exclusive Zone—*Zona Económica Exclusiva*
EIA	Environmental Impact Assessment
EPOMEX	*Instituto de Ecología Pesquerías y Oceanografía del Golfo de México*—Gulf of Mexico Institute of Ecology, Fisheries, and Oceanography
FIFOPESCA	*Fideicomiso del Fondo de Reconversión Pesquera del Golfo de México*—Trust Fund for Fishing Reconversion of the Gulf of Mexico
FIP	Fishery Improvement Project—*Proyecto de Mejora Pesquera*
FNR	Fishing National Register—*Registro Nacional Pesquero*
GEF	Global Environmental Facility of the UN Environmental Program—*Fondo para el Medio Ambiente Mundial del Programa de las Naciones Unidas para el Medio Ambiente*
GIZ	Deutsche Gesellschaft für Internationale Zusammenarbeit—German Agency for International Cooperation
GHG	Greenhouse gases—*Gases de Efecto Invernadero*
GPS	Global Positioning System—*Sistema de Posicionamiento Global*
HDI	Human Development Index—*Índice de Desarrollo Humano*
ICT	Information and Communication Technology—*Tecnología de la Información y Comunicación*
ICRI	International Coral Reef Initiative
IMIPAS	*Instituto Mexicano de Investigación en Pesca y Acuacultura Sustentables* (before INAPESCA)—Mexican Institute of Fisheries and Sustainable Aquaculture
IMO	International Maritime Organization—*Organización Marítima Internacional*
IMSS	*Instituto Mexicano del Seguro Social*—Mexican Social Security Institute
INEGI	*Instituto Nacional de Estadística y Geografía*—National Institute of Statistics and Geography
INSABI	*Instituto de Salud para el Bienestar*—Health Institute for Well-being
INPESCA	*Instituto de Pesca y Acuacultura del Estado de Campeche*—Fisheries and Aquaculture Institute of Campeche

IPN	*Instituto Politécnico Nacional*—National Polytechnic Institute
ISIF	Industrial and semi-industrial fleets—*Flotas Industrial y Semi-Industrial*
LAN	*Ley de aguas nacionales*—National Water Law
LGEEPA	*Ley General de Equilibrio Ecológico y la Protección al Ambiente*—General Law of Ecological Equilibrium and Environmental Protection
LGPAS	*Ley General de Pesca y Acuacultura Sustentables*—General Law of Sustainable Fisheries and Aquaculture
LGCC	*Ley General de Cambio Climático*—General Law on Climate Change
LGVS	*Ley General de Vida Silvestre*—Law General of Wildlife
MSP	Marine spatial planning—*Ordenamiento Ecológico Marino*
MSY	Maximum sustainable yield—*Rendimiento Máximo Sostenible*
NGO	Non-Governmental Organization—*Organización No Gubernamental*
NOMs	*Normas Oficiales Mexicanas*—Official Mexican Standards
NPA	Natural protected area—*Área Natural Protegida*
NREL	National Renewable Energy Laboratory
OBM	Obras de Beneficio Mutuo—Mutual benefit works
OECM	Other effective area-based conservation measures
PACMA	*Programa de Apoyo a la Comunidad y Medio Ambiente*—Community and Environmental Support Program
PAH	Polycyclic aromatic hydrocarbons—*Hidrocarburos Aromáticos Policíclicos*
PEMEX	*Petróleos Mexicanos*—Mexican petroleum parastatal
PERMARGO	*Perforaciones Marinas del Golfo*—Gulf Offshore Drilling
PAN	*Partido de Acción Nacional*—National Action Party
PARM	*Partido Auténtico de la Revolución Mexicana*—Authentic Party of the Mexican Revolution
PPS	*Partido Popular Socialista*—Popular Socialist Party
PROA	Programas, Obras y Acciones—Programs, works, and actions
RII	Relative Importance Index—*Índice de Importancia Relativa*
SAGARPA	*Secretaría de Agricultura, Ganadería, Desarrollo Rural, Pesca y Alimentación* (now SADER)—Secretariat of Agriculture, Livestock, Rural Development, Fisheries and Food
SAV	Submerged Aquatic Vegetation—*Vegetación Acuática Sumergida*
SCT	*Secretaría de Comunicaciones y Transportes*—Secretariat of Communications and Transportation
SDG	Sustainable Development Goals—*Objetivos de Desarrollo Sostenible*
SEMABICCE	*Secretaría de Medio Ambiente, Biodiversidad, Cambio Climático y Energía*—Secretariat of Environment, Biodiversity, Climate Change and Energy

SEMAR	*Secretaría de Marina*—Secretariat of the Navy
SEMARNAT	*Secretaría de Medio Ambiente y Recursos Naturales*—Secretariat of Environment and Natural Resources
SENER	*Secretaría de Energía*—Secretariat of Energy
SES	Social-ecological system—*Sistemas Socioecológicos*
SSF	Small-scale fisheries—*Pesquerías de Pequeña Escala*
SWGoM	Southwestern Gulf of Mexico—*Suroeste del Golfo de México*
UJAT	*Universidad Juárez Autónoma de Tabasco*—Juárez Autonomous University of Tabasco
UNACAR	*Universidad Autónoma del Carmen*—Autonomous University of Carmen
UNAM	Universidad Nacional Autónoma de México—National Autonomous University of Mexico
UPGM	Universidad Politécnica del Golfo de México—Polytechnic University of the Gulf of Mexico
VMS	Vessel monitoring system—*Sistema de Monitoreo de Buques*
VPI	Vulnerable Population Index—*Índice de Población Vulnerable*
WWII	World War II

Chapter 1
Interwoven Destinies: A Coupled Socio-Ecological System in the Southwestern Gulf of Mexico

A. Espinoza-Tenorio, E. Núñez-Lara, D. E. Ramos-Muñoz, D. Matías-Sánchez, and D. N. Rizo-Gutiérrez

1.1 Introduction

The socio-ecological systems (SES) concept has evolved over the last decades to improve our understanding of the dynamic and complex human-nature interaction by linking different disciplines and knowledge and providing a comprehensive approach to overcome current challenges associated with achieving sustainability. This hierarchic, nested framework consists of four first-level core subsystems: (1) the resource system (e.g., a fishery, lake, or park), (2) resource units (e.g., trees, fish, or water), (3) users, and (4) governance systems (Ostrom, 2009). Thus, due to their holistic vision committed to problem-solving (Biggs et al., 2021; Holling, 2001),

A. Espinoza-Tenorio (✉)
El Colegio de la Frontera Sur, Unidad Campeche, Campeche, Mexico
e-mail: aespinoza@ecosur.mx

E. Núñez-Lara
Universidad Autónoma del Carmen, Ciudad del Carmen, Mexico
e-mail: elara@pampano.unacar.mx

D. E. Ramos-Muñoz
El Colegio de la Frontera Sur, Villahermosa, Villahermosa, Tabasco, Mexico
e-mail: dramos@ecosur.mx

D. Matías-Sánchez
Centro de Investigaciones Interdisciplinarias en Ciencias Sociales, Universidad Veracruzana, Minatitlán, Veracruz, México
e-mail: daniela.matias@ecosur.mx

D. N. Rizo-Gutiérrez
Universidad Autónoma Chapingo-Sede Chiapas, San Cristobal de Las Casas, Chiapas, México
e-mail: dgips.invest16@chapingo.mx

A. Espinoza-Tenorio et al. (eds.), *The Fishing and Oil Industries in Coexistence*,
https://doi.org/10.1007/978-3-032-02621-7_1

SES approaches have the potential to address complex global human threats, including climate change and biodiversity loss, as well as their associated wicked problems (Rittel & Webber, 1973).

Due to the growing demands placed on marine resources by humans, persistent, multi-causal issues are becoming increasingly more common and complex. For instance, faced with depletion, geopolitical conflicts, and environmental degradation in terrestrial mines, the mining sector is now pursuing seabed mineral extraction to meet the growing demand for decarbonization-critical materials (Levin et al., 2020). In those cases in which exploited natural resources differ, the coexistence of two or more systems in the same seascape has allowed researchers to study coupled SES.

Marine habitats are particularly prone to forming coupled SES, as multiple exploitation and management processes often target resources within shared spaces and timeframes (Hilborn, 2007). These systems form complex structures with adaptive attributes and processes that emerge from the interactions among their original systems (Miller et al., 2010). Overlapping economic activities in three-dimensional seascapes are indicative of an intricate SES (García-de-Vinuesa et al., 2025), which includes the attributes of each activity, the society that depends on it, and the new capacities and characteristics that emerge when activities coexist. Thus, the nature of coupled SESs facilitates the generation of new knowledge and highlights the relationships between regulations, coastal communities, and marine ecosystem health. The coexistence interactions in a coupled SES are reflected in a governance process that seeks to manage common spaces, which contributes to establishing a shared objective among multiple social actors (Kooiman, 1993).

1.2 Attached Agendas: Food and Energy

Food and energy are two critical global priorities that create competing demands for seascapes and open the door for the formation of coupled SES. A particular coupled SES arises from fishing and offshore oil industries (Fig. 1.1). While these extractive activities (users) target distinct marine resources, they operate within the same seascape (resource system). Despite each sector having its own sector-specific institutions, processes, and formal or informal rules, their spatiotemporal coexistence in this shared space gives rise to an interdependent governance system.

In marine resources management, SES applications have primarily focused on small-scale fisheries (SSF), which represent a vital common-pool resource with significant global social, economic, and ecological implications. According to the Food and Agriculture Organization of the United Nations (FAO, 2024), the world annual fisheries production for 2022 was 92.3 million tons, of which 36.9 million tons (40%) corresponded to SSF, which supported 90% of the capture fisheries workforce and the livelihoods of some 500 million people around the world. SSF provide employment for 60.2 million people, 20.9% of whom are women, thereby contributing to cultural and socioeconomic cohesion in communities. In low- and middle-income countries, the total net trade (exports minus imports) of aquatic animal products reached USD 45 billion, surpassing the total of all other agricultural

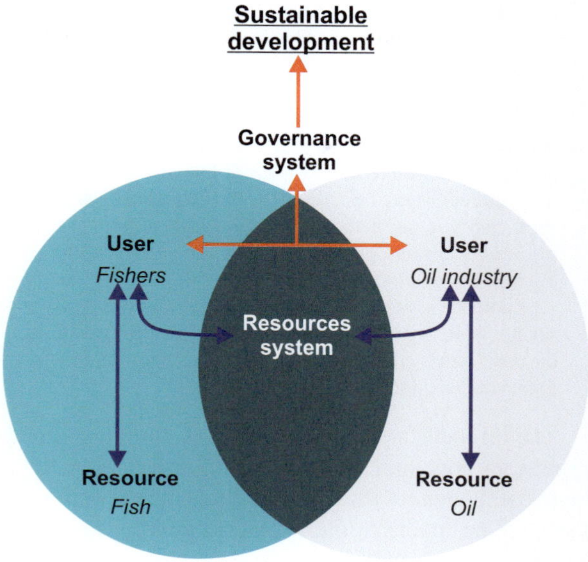

Fig. 1.1 Conceptual model of a coupled socio-ecological system formed by fisheries and the oil and gas industries competing for the same seascape

products combined. Fourteen of these countries ranked among the top 20 producers (with annual production exceeding 1,000,000 metric tons) (Espinoza-Tenorio et al., 2011).

SSF constitute a dynamic sub-sector of fisheries characterized by labor-intensive harvesting, processing, and distribution methods for marine and inland water resources. SSF operations, which may be conducted either full- or part-time, typically supply fish and fishery products to local and domestic markets and are often classified as subsistence fisheries. Nonetheless, the SSF sub-sector encompasses a diverse range of organizational levels, from self-employed individuals to informal micro-enterprises and formal businesses. The contributions of small-scale fishers to global catches and human welfare, including nutrition, food security, poverty alleviation, and development, remain systematically underrecognized (Basurto et al., 2025).

Despite challenges, such as overfishing, climate change, and pollution, SSF is an activity with strong cultural roots in coastal communities (Escamilla-Pérez et al., 2021). Here, the pride of fishing livelihoods and a profound connection to the sea are intergenerational traditions (Saldívar-Moreno et al., 2022). Indeed, it is a passion, a way of life, as a 19-year-old fisher from Sabancuy, Campeche,[1] conveys:

[1] The fishers' narratives presented here have been carefully edited from original interview transcripts to craft their anecdotes into literary form while preserving their authentic voices. This editorial process focused on stylistic enhancement—removing interview interruptions and refining narrative flow—without introducing external content or exaggerations. We maintained strict fidelity to each storyteller's distinctive language patterns, speech rhythms, and cultural expressions throughout. The interview extracts were provided by Adrian Núñez and Armando Hernández de la Cruz, both researchers from *El Colegio de La Frontera Sur*. We are grateful for their collaboration.

> I feel joyful, happy to be a fisher. I feel good; I bring something home, something for my family, but, also, because when you're fishing, you're outside, it's cool, and the waves and everything you see is beautiful, and when a fish comes along that you've never seen before, it gives you joy and all that. There are fish that you don't know, and sometimes they get caught in the nets or on the hooks, and you take photos, record videos, and say, this one is like that, this one is called that. Like August, right now, I like it because there's octopus, because when you're there with your line and you're pulling and pulling, pulling and pulling, you feel happy because they give you a loan, and you know you're going to pay them back, and you put more effort into it, and it gives you joy because when you see large, medium, and small octopus coming, and you're pulling and grabbing, you should see the joy it brings to all the people who go with you.
>
> Look, the joy to be a fisher, because a fisher brings joy as well as… let's say, how do you say it…well, as a kind of bitterness. He brings joy and he also brings sadness because when the north winds come, they take you down to the coast, where there is land, so that you don't drown or something like that.

The oil industry, on the contrary, is dominated by particular individuals, private groups, or governments, and its exploitative activities are often a decisive and strategic component of the global political and economic order. This industry is among the most profitable in the world, accounting for approximately 31% of the global energy supply in 2023 (IEA, 2025), a share that is likely to continue, as the production of fuel powers millions of automobiles, aircraft, and ships, as well as other products integral to modern life.

Over the course of a century, oil companies have developed significant economic and political power, enabling them to facilitate the exploitation of hydrocarbon resources. The profits generated from oil extraction and processing sustain the economies of oil-producing States, underpinning them through the use of large oil revenues as subsidy sources and allowing them to obtain primary, low-priced inputs (e.g., electricity, fuels, and materials) to develop other industries. Moreover, in states whose income depends largely on hydrocarbons, democracy tends to be "hidden" and the economy to be stifled (Ross, 2001), which reinforces the presence of autocratic states and non-state powers, where secrecy is entrenched as a means of power, but without the checks and balances of institutional norms (Bobbio, 2013; Cvitanovic et al., 2013; Espinoza-Tenorio, 2020).

The oil industry's operational preference for isolation proves particularly efficient on offshore platforms, which account for roughly 30% of global oil production (Paolo et al., 2024). Crude oil is pumped from wells on marine platforms or coastal lands and transported by pipelines or tanker ships to refineries where it is transformed into fuels and other industrial products. Offshore oil extraction differs significantly from onshore operations due to its geographic remoteness, technological complexity, and reliance on specialized labor. As Appel et al. (2015) point out, these platforms function as highly regulated spaces where state, corporate, and technical systems converge to sustain production far from inhabited areas. These operations also create distinct working environments, shaped by isolation, rotating shifts, and strict safety protocols. These authors describe them as "hydrocarbon lifeworlds" governed by industrial rhythms that rarely intersect with coastal livelihoods.

Thus, the sea becomes a space that is organized for oil and gas extraction and structured by legal, environmental, and economic considerations that differ from those on land. While offshore oil contributes to the economic development of the coastal zone, it also raises challenges related to environmental impact and the limited integration of local voices in decision-making processes. Furthermore, increasing evidence suggests that, in addition to the severe ecological deterioration it causes (e.g., oil spills), offshore oil production creates a complex network of circumstances unique to the territory and its associated societies. For instance, slow economic growth, high costs of living, the deterioration of quality of life, inefficient benefit programs, bureaucracy, corruption, insecurity, accidents, and the exclusion from other activities in productive areas, which are particularly relevant to fishing (Andrews & Siakwah, 2020; Obi, 2010; Owusu et al., 2023).

1.3 The Shared Southwestern Gulf of Mexico

The Gulf of Mexico, a semi-enclosed marine basin shared by the United States, Mexico, and Cuba, maintains high productivity through dynamic interactions between its terrestrial and marine zones. The *Sonda de Campeche* and the coast of Tabasco in the southwestern Gulf of Mexico (SWGoM) are particularly influenced by rivers, which transport abundant freshwater and nutrients into estuarine ecosystems, supporting a high level of natural diversity (Fig. 1.2). The region supports extensive biodiversity, including populations of birds, marine mammals, marine turtles, and pelagic fish species. Important natural protected areas (e.g., *Pantanos de Centla* Biosphere Reserve, *Laguna de Términos* Flora and Fauna Protection Area, and Los Petenes Biosphere Reserve) are located in the SWGoM, which collectively cover priority habitats such as mangroves, marshes, and swamps.

According to fisheries statistics (CONAPESCA, 2023), Campeche is the seventh largest fishery producer in the country, with 50.4 live weight tons caught in 2023 (2.4% of the national production). This state has a coastline spanning 425 km and hosts 11,600 fishers, as well as 3,330 small-scale and 133 large-scale boats. The main fishing products are horse mackerel (16%), octopus (16%), snail (11%), and fiddler crab (6%). The fisheries production for Tabasco was 28.6 live weight tons in 2023 (1.33% of the national output). The state has a coastline measuring 200 km and hosts 14,000 fishers and 4,333 small-scale boats. The main aquatic products of Tabasco are oysters (27%), mojarra (19%), flagfish (10%), and seabass (7%).

Mexico's most important fishing and hydrocarbon extraction activities originate in the resource-rich SWGoM, with the majority extracted from the *Sonda de Campeche*, which is home to the Cantarell complex, one of the world's largest oil fields. This megaproject encompasses over 304 platforms and kilometers of subsea pipelines across the operational area (PEMEX, 2024). Mexican crude oil is processed at five refineries, the most recent of which is the Olmeca refinery (2022), located on the coast of Tabasco state.

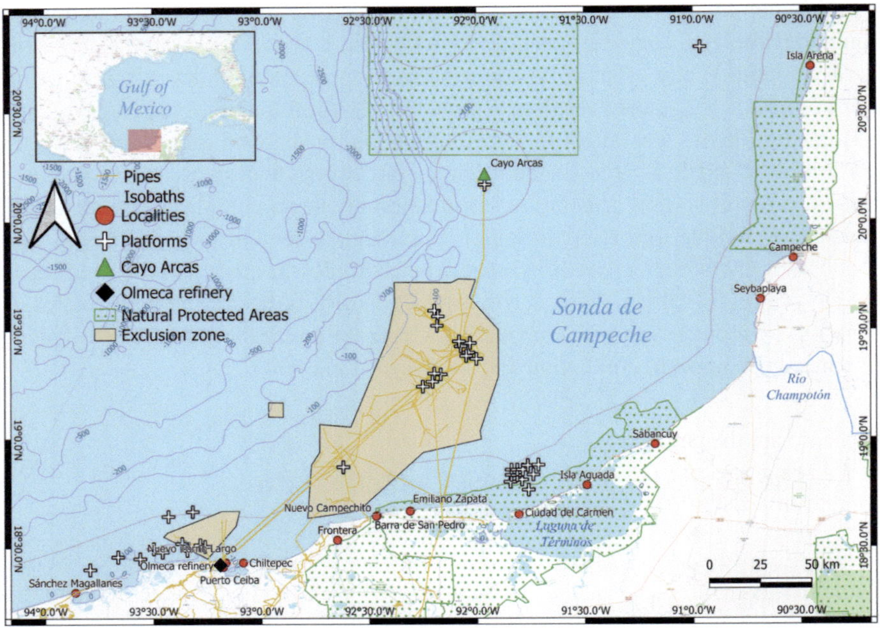

Fig. 1.2 The Southwestern Gulf of Mexico (SWGoM) includes the Coast of Campeche and Tabasco in Mexico

The oil industry in the SWGoM includes the three sectors with the supply chain divided as follows: (i) exploration and production, (ii) transport and storage, and (iii) product preparation and usage. Exploration and production involve searching for potential crude oil deposits underwater or underground by drilling exploratory wells and then bringing them to the surface. The exploitation sector operates around wells and requires a significant amount of time and substantial investments from integrated oil companies (e.g., PEMEX and ENI), as well as independent oilfield services. The second sector of transport and storage focuses mainly on pipeline transportation, marine shipping, and the transportation of refined products via pipelines. In the SWGoM, pipelines transport oil and gas from wells to coastal installations, while transportation includes both national and international movements of crude oil and gas. In the SWGoM, a bulk terminal is located on a coral reef cay system, known as *Cayo Arcas*, which is situated 180 km from the Campeche coast.

The third sector of product preparation and usage corresponds to refining oil and distributing byproducts. It is the most complex of the oil industry sectors due to the inclusion of diverse activities and the trading and marketing of derivatives. Most of the activities included in this sector are conducted in Tabasco's inland zone, resulting in critical direct revenues being concentrated among the companies established in the state. Thus, despite both Tabasco and Campeche are highly involved in the oil industry, and their economies are not equivalent. Tabasco's economy showed the greatest economic growth in the country, while Campeche exhibited the worst economic decline.

1.3.1 The Unintended Consequences of Coexistence

> When PEMEX first came here… sixty years ago, I was about ten years old. We were little. Then, the platforms came, right here in front of us, and they came in, and we saw the lights and everything: it was a wonder. "Look, let's go," my friends would tell me. And we would go at night to see the platforms on the beach. But who would have believed that we in our innocence, right? In that childhood, weren't seeing something that would affect us over time, that we would grow up and become fishers too. That was the end of the world because the sea, the water, was being polluted.

These are the perceptions regarding the arrival of the oil industry in the SWGoM and its ongoing presence, as shared by a 71-year-old retired fisherman from the community of *El Bosque*, Tabasco.

Nowadays, the oil industry in the SWGoM faces persistent scrutiny for its adverse impacts on marine resources that support critical societal needs, including food and water security. Although the industry's specific effects on the fishery resources of the SWGoM remain uncertain (Wakida-Kusunoki & Caballero-Chavez, 2009), the principal environmental risk stems from the production of light crude oil, which contains toxic, bioaccumulative monoaromatic hydrocarbons (e.g., benzene, toluene, and xylene) (Atwood & Ferguson, 1982; Soto et al., 2014). For Mexico, the 1979 *Ixtoc*-I oil spill proved particularly illuminating in revealing the oil industry's negative consequences, fundamentally transforming perceptions of this economic activity.

In the SWGoM, spatial conflicts manifest most acutely in fisheries operations, significantly impacting daily livelihoods. The fishing industry in Tabasco and Campeche operates with distinct fleets that predominantly operate in coastal waters of up to 90 meters deep—a zone contested by hydrocarbon extraction activities since the 1970s, when PEMEX began installing wells and offshore platforms at an average depth of 25 meters (Fig. 1.2). This spatial overlap has created persistent resource-use conflicts between fisheries and petroleum operations in the region.

Some of these problems create risky situations for fishers, which can result in accidents involving ships. In the words of an active fisherman (33 years old) from El Bosque, Tabasco:

> Well, out there on the high seas, 20 miles out, what worries me most are the boats, because there are some wicked boats. There are some boats that you're signaling with the spotlight to clear the boat, to clear the nets, so they don't run you over. And there are some boats that don't pay attention to you! They're like mules! And they pass you by and cut the net, or you lose the whole net. Yes, I've seen people, boat owners or skippers who have had losses. I see them sometimes come in with their boats without nets. We're talking up to 200,000 pesos per boat, carrying the nets.

Social tensions were exacerbated by the Intersectoral Agreement, which expanded the exclusion zones around the oil infrastructure in 2003 and was subsequently updated in 2014, 2017, and 2018. While empirical evidence suggests these fishing exclusion areas could function as artificial reefs, with oil extraction structures creating new ecological connectivity networks for benthic and pelagic communities, research has also demonstrated direct negative impacts on fisheries. Industry activities disrupt both the marine ecosystems supporting fisheries and the cultural dynamics of coastal communities (Quist, 2019; Quist & Nygren, 2015; Ramos-Muñoz et al., 2019).

Despite PEMEX's environmental and social commitments, along with government initiatives, competition between fisheries and hydrocarbon operations for marine space will likely intensify in the coming years. The 2013 Energy Reform fundamentally transformed this dynamic by permitting private sector participation, including 25 companies from 14 countries by 2018 (Rondas México s.f.), and opened new areas of exploitation, particularly in Tabasco's shallow waters within the SWGoM. These developments significantly expanded offshore operations, including prospecting for new oil fields via intensive seismic exploration campaigns, as well as introduced powerful new actors operating under international safety standards, significantly complicating existing resource conflicts.

A 70-year-old fisher from *Centla*, Tabasco, described this complicated situation:

> Things have been tough, or rather, things are tough! Even tougher now with these huge new fields they've found in the gulf. What are we going to do you ask? Well, we're going to have to look, look for another place… we couldn't work in this area anymore: PEMEX chased us out; the patrol car came here in Carmen; then, we had to pay a fine to clear the boats, plus the huge cost of gasoline, it's impossible. Can you imagine?
>
> Look, when I was traveling through *Antón Lizardo*, passing through *Roca Partida*, there was such an abundance of snapper that it was enough to throw out two *palandrazos* to make the *tide* and return in the same day. Now, here, they say the fish are gone, but that's not true: they've been scared away; we know there are still some, because there are places where they kill them. And there it is, on the shore, about 25–30 fathoms deep. I'll be honest with you: the fish was scared away by all the PEMEX pollution.

When describing the power asymmetries among stakeholders from the fishing and oil industries in these decision-making processes (Salazar-De la Cruz et al., 2020), he offers this testimony:

> Let me tell you. I once went to a meeting with PEMEX and CIMADE to talk about some boats that were dragging key, here on this side. I was assigned to represent the community, and you think they said something? Nothing! "PEMEX doesn't pollute, PEMEX doesn't pollute, [an attorney declared]" and I told the attorney: I'll prove to you that PEMEX pollutes in two hours. I'll go and bring you water samples, of chapo, of whatever you want… but you know the authorities, PEMEX is the government.

These cumulative adverse conditions, including coastal ecosystem degradation, overexploitation of fishing, the rising cost of gasoline, whose consumption increases as fishers are forced to travel to more distant fishing grounds, and the impacts of climate change, have forced fishers into circumstances of subsistence livelihoods and chronic uncertainty (Reyes-Grande et al., 2020) and threaten to intensify the existing asymmetrical competition between fishing and oil industries in the years to come, creating an unprecedented governance challenge for the region.

1.4 A Comprehensive, Collaborative Proposal

Effective management of complex coastal systems requires understanding their unique components and multifaceted interrelationships. Rapid coastal population growth has intensified competition for resources among diverse stakeholders, yielding both beneficial and detrimental socioeconomic and environmental impacts. Tropical nations, such as Mexico, face a particular paradox: despite possessing extraordinary natural wealth, significant economic and social disparities persist. However, documented cases of successful coexistence between competing productive activities in shared spaces remain scarce. More frequently, individualistic exploitation approaches have depleted resources and undermined collective well-being through unforeseen consequences. As the same 71-year-old retired fisherman from the community of El Bosque, sustains:

> Today, at this age, we see that what we once looked upon with joy and admiration as children, today we see as part of what ended fishing, and all those things that were happening. Well, I feel that to a certain extent, due to a lack of knowledge, or you could almost say that no one was to blame, because it's even possible that even PEMEX hadn't realized what it was doing, that this was going to happen, right?

This raises our central challenge: document how to foster sustainable coexistence between spatially overlapping yet interest-divergent productive activities, and how sustainable coexistence can be promoted among spatially co-located productive activities with conflicting interests.

We have posited that the Gulf of Mexico cannot be adequately understood through fragmented disciplinary approaches. While existing research has predominantly focused on the northern Gulf of Mexico, a systematic understanding of the SWGoM as a shared resource system is essential. This complex space supports interdependent fishing and oil and gas industries that are managed through overlapping federal governance structures and socio-political institutions. Novel methodological approaches that map and analyze industry interactions are crucial for building trust, fostering collaborative agreements, and developing sustainable solutions to resource conflicts.

Knowledge co-production has emerged as a critical framework for bridging the diverse needs of stakeholders, integrating multiple knowledge systems, and informing the improved governance of SES. As Roux notes (in Toomey et al., 2017), knowledge is something to transfer but also a process that involves negotiations and understanding between different actors. Strengthening local capacities supports sustainable development goals, while social learning processes, particularly through institutional memory and organizational development, can accelerate adaptive responses to ecosystem disturbances (Patterson et al., 2010). These approaches can enhance societal resilience in the heavily impacted marine environment of the SWGoM.

1.4.1 About this Book

This book examines the complex interplay between oil extraction and artisanal fishing in the southwestern Gulf of Mexico (SWGoM). Due to the complex challenges faced by the fishing and oil industries, our work ultimately characterizes the spatio-temporal overlaps and transformations within this coupled SES, while considering potential expansions of other economic activities, including tourism, mariculture, and renewable energy development. The four book sections integrate interdisciplinary research, providing a comprehensive assessment of regional circumstances. Each contribution presents both theoretical and methodological innovations designed to address resource-use conflicts in this contested marine space.

Through a cross-disciplinary reflection based on the lived experiences and narratives of people who live within and share the SWGoM, this narrative introduction (Chap. 1) analyzes the construction and characteristics of coupled SES in shared seascapes, where food (fisheries) and energy (oil and gas) agendas have resulted in unintended but crucial consequences in this part of the Gulf of Mexico. A glossary is included to provide key information for an international audience or students seeking a more comprehensive understanding of the SWGoM.

The first section, "A Challenging Coexistence," outlines the historical and contemporary conditions resulting from the coexistence between sectors in the SWGoM. Focusing on fishing economies, Cavieses-Núñez et al. (Chap. 2) perform a comprehensive spatiotemporal analysis of both small-scale and industrial fleets, evaluating catch trends, market values, operational costs, and the geospatial patterns of fishing effort. Coronado et al. (Chap. 3) adopt a regional perspective to systematically examine the intricate coexistence of commercial and recreational fisheries. Through official records and secondary sources, the authors characterize the sector's resource base, fleet composition, and regulatory framework. In contrast, Vite-García et al. (Chap. 4) evaluate the mariculture potential of Tabasco's coastal ecosystems, including marine waters and interconnected brackish systems—such as lagoons, estuaries, and marshes—for the sustainable cultivation of marine organisms.

Rabelo-Avalos et al. (Chap. 5) provide an in-depth analysis of the 1979 *Ixtoc*-I oil spill in the SWGoM and grassroots mobilization efforts that have redefined the region through the lens of extractive policies and media narratives. Building on this legacy of examining oil spill implications and given the lack of sufficient data for effective monitoring, Cruz-Pech et al. (Chap. 6) describe the recent presence of surface oil in the SWGoM using remote sensing techniques, distinguishing between natural seepage and anthropogenic sources.

The second section, Fishing Communities in Contested Coasts, investigates the distinct socio-ecological dynamics that emerge in fishing communities operating within oil-industrialized zones. To address data gaps in fisher demographics,

Mendoza-Carranza et al. (Chap. 7) conducted systematic surveys across coastal communities in Tabasco and Campeche, constructing a detailed socioeconomic and occupational profile of local fishers. Complementing this approach, Núñez et al. (Chap. 8) employed a well-being framework to analyze perceptions of small-scale fishers, revealing trade-offs between oil industry expansion and their livelihoods in the SWGoM. Ramírez-Pacheco (Chap. 9) further investigate the repercussions of oil industry growth through participatory and ethnographic approaches, illustrating how the sector's expansion has eroded intergenerational engagement in traditional fishing and exacerbated the ongoing crisis of livelihood succession.

Two additional chapters in this section examine the role of technology in small-scale fisheries, highlighting its transformative effects and the systemic challenges it presents. Using a mixed-methods approach (surveys, participant observation, and focus groups), Ramos-Muñoz et al. (Chap. 10) analyze how information and communication technologies (ICTs) reconfigure small-scale fisheries, delineating both adaptive opportunities and structural barriers to adoption. Martínez-Amezcua et al. (Chap. 11) present a case study of a chatbot, arguing that the interplay between traditional fishing knowledge and exogenous technological innovations—conceptualized as "technological syncretism"—is pivotal to understanding the dynamics of technology appropriation in this contested sector.

Section three, Reconfiguring Governance, scrutinizes the institutional and regulatory barriers to sustainable coexistence between industries within the SES of the SWGoM. Employing an interdisciplinary lens, the contributors dissect the legacy of environmental degradation and evolving policy regimes and explore mechanisms for promoting balanced development while enhancing system resilience, policy instruments, and territorial planning. Peña-Puch et al. (Chap. 12) trace five decades (1970–present) of coastal transformation in Campeche, synthesizing data on fisheries, oil production, legal frameworks, and stakeholder dynamics. Their analysis underscores socioenvironmental trade-offs, conservation initiatives, and spatial conflicts—and synergies—with Protected Areas, exposing patterns of contestation and collaboration. Solano and Frutos-Cortes (Chap. 13) delineate the territorial reconfiguration triggered by Mexico's 2013 energy reform, framing it within broader processes of transnationalization. By exploring the periods of these shifts, they reveal how the reform exacerbated the trends of displacement and marginalization in the fishing activities of the early twenty-first century.

Gamma-Campillo et al. (Chap. 14) critically evaluate Mexico's environmental legal framework and spatial planning instruments, including a case study from Tabasco, to elucidate the systemic hurdles to participatory decision-making and sustainable resource governance. Concluding the section and focusing on governance architectures, Soria-Gómez et al. (Chap. 15) employ network analysis and structured interviews to assess the Red Snapper Fishery Improvement Project (FIP) in the Campeche Sound, examining its effectiveness in identifying sectoral divides.

Part IV, Pathways to Sustainability, explores integrative approaches to mitigating the complex anthropogenic pressures on coastal ecosystems while enhancing their capacity to deliver sustainable well-being. Mesa-Jurado et al. (Chap. 16) and Lara-Mendoza et al. (Chap. 17) explore the provisioning ecosystem service of small-scale fisheries and the challenges for marine sustainability and territorial planning in the study area, respectively. (Chap. 19) Narro (Chap. 18) analyzes Campeche's paradoxical status as Mexico's foremost hydrocarbon producer amid its transition toward renewable energy. The study underscores how the region's underutilized solar and wind resources could catalyze a transition toward the greater use of sustainable energy while addressing historical dependencies on fossil fuels. Complementing this localized assessment, Núñez-Lara et al. (Chap. 19) propose an integrative framework that incorporates regional stakeholders and multidimensional data to guide spatial planning processes, aiming to balance economic gains, social welfare, and ecological integrity.

Acknowledgments This study was supported by the project "*La desafiante coexistencia de los sistemas socio-ecológicos acoplados; las industrias de la pesca y petróleo en la Sonda de Campeche*" (Ciencia de Frontera 2019, Project CF-2019-1564454, CONACyT). Special thanks to J.B. Cruz-Pech for map design. For quality assurance, we employed large language models (LLMs) for initial grammar and spelling checks, with all content subsequently reviewed and refined by a human editor. We are deeply grateful to A. MacTavish for her exceptional assistance with English language editing.

Competing Interests The authors have no conflicts of interest to declare that are relevant to the content of this chapter.

References

Andrews, N., & Siakwah, P. (2020). Oil and development in Ghana: Beyond the resource curse. In *Oil and development in Ghana: Beyond the Resource Curse*. Routledge. https://doi.org/10.4324/9781003013730

Appel, H., Mason, A., Watts, M., & Huber, M. T. (2015). *Subterranean estates : Life worlds of oil and gas*. Cornell University Press.

Atwood, D., & Ferguson, R. (1982). An example study of the weathering of spilled petroleum in a tropical marine environment: IXTOC-1. *Bulletin of Marine Science, 32*(1), 1–13.

Basurto, X., Gutierrez, N. L., Franz, N., del Mar Mancha-Cisneros, M., Gorelli, G., Aguión, A., Funge-Smith, S., Harper, S., Mills, D. J., Nico, G., Tilley, A., Vannuccini, S., Virdin, J., Westlund, L., Allison, E. H., Anderson, C. M., Baio, A., Cinner, J., Fabinyi, M., et al. (2025). Illuminating the multidimensional contributions of small-scale fisheries. *Nature, 637*(8047), 875–884. https://doi.org/10.1038/s41586-024-08448-z

Biggs, R., de Vos, A., Preiser, R., Clements, H., Maciejewski, K., & Schlüter, M. (2021). The routledge handbook of research methods for social-ecological systems. In *The Routledge handbook of research methods for social-ecological systems*. Taylor and Francis. https://doi.org/10.4324/9781003021339

Bobbio, N. (2013). *Democracia y secreto*. Fondo de Cultura Económica.

CONAPESCA. (2023). *Anuario Estadístico de Acuacultura y Pesca* 2023.

Cvitanovic, C., Wilson, S. K., Fulton, C. J., Almany, G. R., Anderson, P., Babcock, R. C., Ban, N. C., Beeden, R. J., Beger, M., Cinner, J., Dobbs, K., Evans, L. S., Farnham, A., Friedman, K. J., Gale, K., Gladstone, W., Grafton, Q., Graham, N. A. J., Gudge, S., et al. (2013). Critical research needs for managing coral reef marine protected areas: Perspectives of academics and managers. *Journal of Environmental Management, 114*, 84–91. https://doi.org/10.1016/j.jenvman.2012.10.051

Escamilla-Pérez, B. E., Ortiz-Lozano, L., Molina-Rosales, D., & Espinoza-Tenorio, A. (2021). Cultural importance of marine resources subject to fishing exploitation in coastal communities. *Ocean & Coastal Management, 208*(January). https://doi.org/10.1016/j.ocecoaman.2021.105605

Espinoza-Tenorio, A. (2020). ¿Mancharse las manos de negro? El dilema ético de la investigación en territorios petrolizados. *Revista Mexicana de Ciencias Políticas y Sociales, LXIV*(237), 183–210.

Espinoza-Tenorio, A., Espejel, I., Wolff, M., & Zepeda-Domínguez, J. A. (2011). Contextual factors influencing sustainable fisheries in Mexico. *Marine Policy, 35*(3), 343–350. https://doi.org/10.1016/j.marpol.2010.10.014

FAO. (2024). The State of World Fisheries and Aquaculture 2024. In *The state of world fisheries and aquaculture 2024*. Food and Agriculture Organization of the United Nations. https://doi.org/10.4060/cd0683en

García-de-Vinuesa, A., Florido, D., Vilas, C., Torres, M. Á., Delgado, M., Muñoz, I., Cabrera-Castro, R., Ramos, F., & Llope, M. (2025). Framing social systems for ecosystem-based management: The Guadalquivir estuary-Gulf of Cadiz coupled SES as case study. *Environmental Development, 55*, 101206. https://doi.org/10.1016/J.ENVDEV.2025.101206

Hilborn, R. (2007). Managing fisheries is managing people: What has been learned? *Fish and Fisheries, 8*(4), 285–296. https://doi.org/10.1111/j.1467-2979.2007.00263_2.x

Holling, C. S. (2001). Understanding the complexity of economic, ecological, and social systems. *Ecosystems, 4*(5), 390–405. https://doi.org/10.1007/s10021-001-0101-5

IEA. (2025). *Global Energy Review 2025*. www.iea.org

Kooiman, J. (1993). *Modern governance : New government-society interactions*. Sage. https://books.google.com.mx/books/about/Modern_Governance.html?id=CRXt1WgkaIMC&redir_esc=y

Levin, L. A., Amon, D. J., & Lily, H. (2020). Challenges to the sustainability of deep-seabed mining. *Nature Sustainability, 3*(October). https://doi.org/10.1038/s41893-020-0558-x

Miller, D. G. M., Slicer, N., & Sabourenkov, E. N. (2010). IUU fishing in antarctic waters: CCAMLR actions and regulations. In *Law, technology and science for oceans in globalisation: IUU fishing, oil pollution, bioprospecting, outer continental shelf* (pp. 175–196). https://doi.org/10.1163/ej.9789004180406.i-610.53

Obi, C. I. (2010). Oil extraction, dispossession, resistance, and conflict in Nigeria's oil-rich Niger Delta. *Canadian Journal of Development Studies / Revue Canadienne d'études Du Développment, 30*(1–2), 219–236. https://doi.org/10.1080/02255189.2010.9669289

Ostrom, E. (2009). A general framework for analyzing sustainability of social-ecological systems. *Science (New York, N.Y.), 325*(5939), 419–422. https://doi.org/10.1126/science.1172133

Owusu, V., Lawer, E. T., Adjei, M., & Ogbe, M. (2023). Impact of offshore petroleum extraction and "ocean grabbing" on small-scale fisheries and coastal livelihoods in Ghana. *Maritime Studies, 22*(2), 1–14. https://doi.org/10.1007/s40152-023-00307-3

Paolo, F., Kroodsma, D., Raynor, J., Hochberg, T., Davis, P., Cleary, J., Marsaglia, L., Orofino, S., Thomas, C., & Halpin, P. (2024). Satellite mapping reveals extensive industrial activity at sea. *Nature, 625*(7993), 85–91. https://doi.org/10.1038/s41586-023-06825-8

Patterson, O., Weil, F., & Patel, K. (2010). The role of community in disaster response: Conceptual models. *Population Research and Policy Review*. https://doi.org/10.1007/s11113-009-9133-x

PEMEX. (2024). *Anuario Estadístico 2024*.

Quist, L.-M. (2019). Fishers' knowledge and scientific indeterminacy: Contested oil impacts in Mexico's sacrifice zone. *Maritime Studies, 18*(1), 65–76. https://doi.org/10.1007/s40152-018-0123-7

Quist, L.-M., & Nygren, A. (2015). Contested claims over space and identity between fishers and the oil industry in Mexico. *Geoforum, 63*, 44–54. https://doi.org/10.1016/j.geoforum.2015.05.015

Ramos-Muñoz, D. E., Ramos-Reyes, R., Zamora-Cornelio, L. F., Hernández-De la Cruz, A., & Espinoza-Tenorio, A. (2019). Exclusión en el Golfo de México: una visión desde los pescadores sobre la industria petrolera en Tabasco. *Cuadernos de Geografía: Revista Colombiana de Geografía, 28*(2), 357–372. https://doi.org/10.15446/rcdg.v28n2.73511

Reyes-Grande, F., Espinoza-Tenorio, A., & Díaz-Perera, M. Á. (2020). Ya no hay pescado. Subsistencia e incertidumbre en una localidad pesquera del Golfo de México. *Intersticios Sociales, 20*, 193–223.

Rittel, H. W. J., & Webber, M. M. (1973). Dilemmas in a general theory of planning. *Policy Sciences, 4*(2), 155–169. https://doi.org/10.1007/BF01405730/METRICS

Ross, M. L. (2001). Does oil hinder democracy? *World Politics, 53160180*(3), 325–361. https://doi.org/10.1353/wp.2001.0011

Salazar-De la Cruz, C. C., Zepeda-Domínguez, J. A., Espinoza-Tenorio, A., & Ramos-Muñoz, D. E. (2020). Governance networks in marine spaces where fisheries and oil coexist: Tabasco, México. *Extractive Industries and Society, 7*(2), 676–685.

Saldívar-Moreno, A., Espinoza-Tenorio, A., Mondragón, R., Fernández-Rivera Melo, F. J., Muñoz-Ray, A., Velázquez-Castillo, M. A., Ehuan-Noh, R. G., & Oliveto-Andrade, A. (2022). *Jóvenes pescadores en la construcción de iniciativas locales y soluciones para la sostenibilidad de pesquerías de pequeña escala en México*. COBI A.C.

Soto, L. A., Botello, A. V., Licea-Durán, S., Lizárraga-Partida, M. L., & Yáñez-Arancibia, A. (2014). The environmental legacy of the Ixtoc-I oil spill in Campeche sound, southwestern Gulf of Mexico. *Frontiers in Marine Science, 1*(November), 1–9. https://doi.org/10.3389/fmars.2014.00057

Toomey, A. H., Knight, A. T., & Barlow, J. (2017). Navigating the space between research and implementation in conservation. *Conservation Letters, 10*(5), 619–625. https://doi.org/10.1111/conl.12315

Wakida-Kusunoki, A. T., & Caballero-Chavez, V. (2009). Effects of the Kab 121 well oil spill on artisanal fishing in the coast of Campeche and Tabasco, Mexico. *Ciencia Pesquera, Mayo, 17*, 12.

Part I
A Challenging Coexistence

Chapter 2
Relevance of Fisheries Dynamics in the Southwestern Gulf of Mexico

R. Cavieses-Núñez, R. I. Hirales-Cota, L. M. Uribe-Sandoval, D. G. Cupido-Santamaria, A. Espinoza-Tenorio, A. Uribe-Martínez, and E. Cuevas

2.1 Introduction

In 2023, it was noted that fish production from capture remains fundamental for human consumption in Latin America, providing approximately 12 million tons, compared to nearly five million tons from aquaculture production. In Mexico's case, fishing and aquaculture production in 2023 exceeded 1.74 million tons (FAO, 2024).

R. Cavieses-Núñez (✉)
Departamento Académico de Ingeniería en Pesquerías, Universidad Autónoma de Baja California Sur, La Paz, Mexico

Instituto de Investigaciones Oceanológicas, Universidad Autónoma de Baja California, Ensenada, Mexico
e-mail: cavieses@uabcs.mx

R. I. Hirales-Cota
Departamento Académico de Ingeniería en Pesquerías, Universidad Autónoma de Baja California Sur, La Paz, Mexico
e-mail: r.hirales@uabcs.mx

L. M. Uribe-Sandoval
Escuela Nacional de Estudios Superiores, Unidad Morelia, Universidad Nacional Autónoma de México, Morelia, Mexico
e-mail: lumus@ciencias.unam.mx

D. G. Cupido-Santamaria · A. Espinoza-Tenorio
El Colegio de la Frontera Sur, Unidad Campeche, Campeche, Mexico
e-mail: deysi.cupido@ecosur.mx; aespinoza@ecosur.mx

A. Uribe-Martínez · E. Cuevas
Instituto de Investigaciones Oceanológicas, Universidad Autónoma de Baja California, Ensenada, Mexico
e-mail: abigail.uribe@uabc.edu.mx; ecuevas@uabc.edu.mx

A. Espinoza-Tenorio et al. (eds.), *The Fishing and Oil Industries in Coexistence*,
https://doi.org/10.1007/978-3-032-02621-7_2

The Mexican states of Tabasco and Campeche, located in the southwestern Gulf of Mexico, had fishing production of 18,207 tons and 55,572 tons, respectively, in 2022. This made Tabasco the 12th national force in terms of catch volume and value, while Campeche was the seventh in catch volume and fourth in value (CONAPESCA, 2022).

On the other hand, the *Sonda de Campeche* and the coast of Tabasco are recognized as the leading oil extraction areas in Mexico, from which 81% of Mexican oil is extracted (SENER, 2017). This region also boasts high fishing productivity, particularly for small-scale fisheries, which are vital for the subsistence of coastal communities (Ramos-Muñoz et al., 2020; Espinoza-Tenorio et al., 2022). The coexistence of these sectors has generated a context of competition for resources and territory, where the environmental impacts of oil, such as spills and pollution, have affected the health of marine ecosystems and, consequently, the fishing catches (Zalik, 2009; Peña-Puch et al., 2019; Ramos-Muñoz et al., 2019). Oil exploitation in the region has significantly increased the vulnerability of local socio-ecological systems, affecting not only marine biodiversity but also the resilience of the fishing communities (Coronado et al., 2024).

The oil industry's activities exacerbate the already vulnerable socioeconomic conditions of fishing communities. These communities heavily depend on marine resources for their livelihoods and lack a robust, relevant strategy to address their losses (Espinoza-Tenorio et al., 2019). In addition, the conflicts between these two industries have been exacerbated by the displacement of fishers to other areas due to pollution or the occupation of traditional fishing grounds by oil platforms (Pérez-Sánchez & Muir, 2003).

Regarding other socioeconomic impacts, the oil industry has brought unequal development in the regions of Tabasco and Campeche; while it has generated employment and economic growth in certain areas, it has also created socioeconomic inequalities where fishing communities, particularly the smallest and most vulnerable, have been marginalized and have experienced a deterioration in their quality of life (Espinoza-Tenorio et al., 2022). In addition, the lack of access to essential services, such as health and education, is exacerbated in those communities due to the overload of local infrastructure accompanying the oil industry (Ramos-Muñoz et al., 2020).

Small-Scale Fisheries (SSF) in Mexico are characterized by their complex and diverse nature, involving multi-species and multi-gear operations that play a crucial role in local economies and food security for coastal communities (Ojeda-Ruiz et al., 2019; Coronado et al., 2020). Those fisheries typically use fiberglass vessels with lengths of less than 12 feet and engines of up to 120 horsepower, allowing them to operate in nearby coastal areas. They have significant socioeconomic importance, providing employment to numerous people in coastal communities and substantially contributing to local livelihoods (Espinosa-Romero et al., 2017; Wintergalen et al., 2022). Their operations are notably geographic and seasonally varied, which respond to local environmental and socioeconomic conditions (Pellowe & Leslie, 2017; Mendoza-Portillo et al., 2020). Along with their relevance, SSF face challenges such as overexploitation, climate change, and management limitations, requiring governance approaches that include the participation of the entire community and their traditional knowledge to ensure their long-term sustainability (Lozano et al., 2018; Finkbeiner & Basurto, 2015; Salas et al., 2007).

On the other hand, industrial and semi-industrial fishing fleets are characterized by larger vessels and greater operational capacity than small-scale fishing fleets. The semi-industrial fishing fleets (ISIF) are vessels between 20 and 25 m long that operate offshore in areas deeper than 50 m and undertake fishing trips lasting between 15 and 20 days. Those vessels have more powerful motors (120–365 HP) and larger crews (8–15 people) than the small-scale fleets, which enables them to travel between 35 and 200 nautical miles from shore. Semi-industrial vessels employ various fishing gears, including longlines, traps, jimbas, and hookah systems to target multiple species like red groupers, octopuses, and lobsters (Salas et al., 2019). Their stronger technological capacity enables them to achieve higher catch rates for specific target species, but at the cost of higher operational costs due to longer trips and increased fuel consumption required to access deeper fishing grounds (Salas et al., 2019).

In this context, our objectives were to assess the dynamics of selected fisheries variables (trends in catch volumes, production prices, operating costs based on fuel consumption per kilogram of the catch, the relationship between variations in fishing volumes and their composition, and trends in product value, among other indicators of fishing efficiency) for SSF and ISIF from 306 landing sites in Campeche and Tabasco, Mexico based on public landings reports from 2015 to 2023 to compare the overall patterns of selected variables from landings reports with a thoroughly monitored fishing fleet in Tabasco.

2.2 Methodology

2.2.1 Study Area

Our study area (Fig. 2.1) is situated in the *Sonda de Campeche*, located in the southwestern Gulf of Mexico. It represents a highly productive region rich in marine biodiversity where fishing and petroleum activities intersect, significantly shaping regional economic development for decades. This continental shelf is marked by shallow waters, with depths ranging from 30 to 200 m, featuring sandy and calcareous bottoms, and a diverse array of ecosystems, including coral reefs, seagrass meadows, and extensive mangrove zones along its shores (Rivera-Arriaga et al., 2012). Hydrologically, the area is influenced by the Yucatan Current and freshwater inflows from the Terminos fluvial-lagoon systems and the *Grijalva-Usumacinta* rivers, creating optimal conditions for the growth of commercially valuable species such as the Maya octopus (*Octopus maya*), shrimp, and various fish species (Salazar-De la Cruz et al., 2020). Climatic conditions are characterized by three distinct seasons: nortes (November–February), dry season (March–May), and rainy season (June–October), which influence both fishing practices

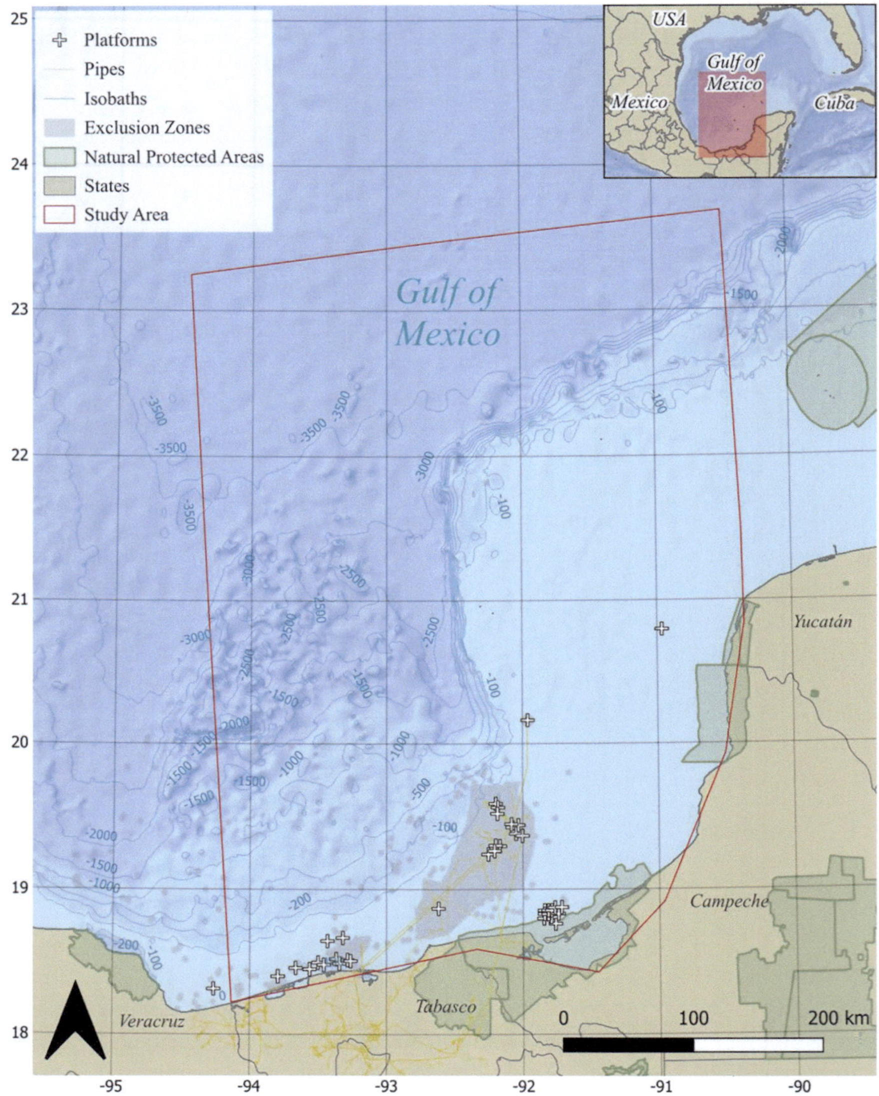

Fig. 2.1 Study Area (Red Square) in the Southwestern Gulf of Mexico

and marine ecosystem dynamics (Guarneros-Narváez et al., 2022). Socioeconomically, the region is home to significant fishing communities that primarily rely on small-scale fishing, coexisting with the petroleum industry, which has profoundly transformed the marine landscape and regional economy since the 1970s (Ramos-Muñoz et al., 2019). This transformation has created both development opportunities and environmental challenges, as well as conflicts over the use of marine space (Ramos-Muñoz et al., 2019).

2.2.2 Landing Reports Analysis

The landing report is the official document that economic units in fishing activities must complete with information about the kilograms, value, date, landing port, and species name, and submit monthly to the local fishing office. An economic unit in Mexican fisheries is an individual or collective entity that uses inputs and resources to capture, process, or market aquatic products, generating economic value and employment. It possesses specific capital assets, operates under a defined regulatory framework, and serves as the basis for implementing sectoral policies and allocating government support. Depending on the scale of operations, this could be an individual fisher, a family operation, a cooperative, or a larger enterprise (Schuhbauer et al., 2019; Wintergalen et al., 2022).

We used the landing reports of the SSF and the ISIF registered in the states of Tabasco and Campeche from 2017 to 2023 to identify the ten principal fisheries and evaluate their trends in catch volumes, average price per kilogram of catch, and frequency of species reporting. The landing reports were obtained from the National Commission of Aquaculture and Fishing (CONAPESCA, by its Spanish acronym) and contain fields such as the Fishing National Register (FNR), economic unit name, fishing area, port name, fishing office, state, species name, price per kilogram, and kilogram of catches, among other data.

The Relative Importance Index (RII) was used to determine the ten most significant fisheries, and it was calculated with the following equation:

$$RII = \frac{\%\text{Catch} + \%\text{Value} + \%\text{Frequency}}{3}$$

Where:

- %Catch is the percentage of the total landing weight (kilograms) for a particular species relative to the total landing weight of all species.
- %Value is the percentage of the total value (MXN) for a particular species relative to the total value of all species.
  %Frequency: This measure indirectly determines how aggregated a catch is in a specific period and by economic unit. This frequency represents the percentage of times a particular species is reported among the total number of reports for all species.

The foremost trends in average prices per kilogram of catch and catch volumes were identified for the ten main fisheries. Records corresponding to species associated with coastal lagoons, such as oysters and fishing resources categorized as "Others," were excluded from the arrival notices.

Economic units, such as fishing cooperatives, permit holders, or companies that record their catches in arrival notices, were identified to assess changes in fleet behavior, and comparisons were made between 2017 and 2023.

2.2.3 Vessel Monitoring System Analysis

Subsequently, data from the Vessel Monitoring System (VMS) of the Mexican industrial fishing fleet with home ports in the states of Tabasco and Campeche were analyzed from January 2018 to December 2023 with hourly resolution. This database was obtained from CONAPESCA. It contains the vessel name, economic unit names, FNR, date, geographical position, course, and velocity. The monthly distance traveled per vessel and the hours of operation were obtained by analyzing these data.

Additionally, the fishing usage areas of the industrial fleet were determined using a hexagonal binning plot created with the Matplotlib library in Python. The parameters for the hexagonal binning plot included a grid size of 40 (specifying the number of hexagons along the x-axis), a minimum count of 5 for the bins (ensuring that hexagons with fewer than 2 points are not colored), and value limits set from 0 to 500 (the minimum and maximum value for the color scale of the heatmap). This method facilitated a detailed spatial analysis of the fishing activities, highlighting the most frequented areas of the industrial fleet.

For the analysis of the small-scale fishery fleet, NEMO satellite position transmitters were installed on four small-scale fishing vessels from a cooperative society located in *Nuevo Campechito*. The data collected from the transmitters covered the period from 2022 to 2023 with hourly resolution. Using this data, the monthly distances traveled, engine operating hours, and an estimate of gasoline consumption were calculated using the following empiric equation:

$$Q \cong H \times 36.9 \frac{l}{h} \times 6000\,\mathrm{r/min}$$

Where:

- Q is the approximate consumption in liters.
- H is the number of hours the engine operates, obtained from the timestamp of the position while in motion.
- *36.9 L/Hour* is the consumption factor at 6000 revolutions per minute.

The cost per liter of gasoline was estimated using the monthly average price history in Mexico from 2022 to 2023.

2.3 Results

2.3.1 The Fishing Landing Reports Analysis

The analysis of landing reports for the small-scale fishing fleet revealed various trends and identified the principal fisheries holding fish production in Tabasco and Campeche. Table 2.1 describes the participation and importance of each reported

Table 2.1 Relative importance index ranking of small-scale fishing species

Rank	Fishery name[a]	% Capture	% Value	% Frequency	Economic units	RII
1	*Pulpo* (Octopus)	17.90	33.73	11.84	1004	21.16
2	*Robalo* (Snook)	6.78	16.26	12.40	1385	11.81
3	*Jurel* (Amberjack)	15.37	4.82	12.68	1335	10.96
4	*Bandera* (Gafftopsail catfish)	8.00	3.42	4.61	848	5.34
5	*Mojarra*	4.21	4.63	5.78	731	4.87
6	*Rubio* (Bluewing Searobin)	3.56	1.89	8.32	1094	4.59
7	*Jaiba* (blue crab)	7.15	3.35	2.61	160	4.37
8	*Guachinango* (red snapper)	2.60	6.00	3.27	707	3.96
9	*Corvina* (gulf weakfish)	2.60	2.10	6.15	1163	3.62
10	*Caracol* (snail)	6.01	2.94	1.55	174	3.50
	TOTAL	74.18	79.14	69.21	8601	74.18

[a]The local common name is kept to compare with other documents

fishery, highlighting octopus with an RII of 21.16, representing 17.90% of the catches and 33.73% of the fishing value. This data underscores the significant contribution of octopus, both in terms of highlight and economic value, with 1004 economic units participating in the fisheries from 2017 to 2023.

Additionally, the fishery species such as snook, amberjack, gafftopsail catfish, mojarra, red snapper, bluewing searobin, and gulf weakfish stand out, collectively accounting for 45.15% of the RII. These species are crucial for the economic sustainability of small-scale fishing and diversification in the catch. Snook, for example, ranks second in importance with an RII of 11.81, representing 6.78% of the catches and 16.26% of the economic value, involving 1385 economic units, surpassing the number of units participating in octopus fishing. Regarding frequency, the most reported fishery is amberjack, representing 12.68% of the records.

The analysis of the ISIF fleet's catches, as reported in arrival notices from 2017 to 2023, identified the main species caught and their trends. Table 2.2 shows that shrimp fishing stands out significantly, with an RII of 67.92, representing 73.81% of the catches and 84.81% of the total value of ISIF.

Red snapper represents an RII of 5.75, contributing 3.77% of the catches and 4.38% of the value reported by 42 economic units. Meanwhile, with an RII of 5.33, amberjack accounts for 5.44% of the catches and 3.06% of the monetary value, with 33 economic units.

Bycatch also shows a notable presence with an RII of 3.91, capturing 1.79% and representing 0.11% of the economic value, distributed among 45 economic units. Generally, this set of species is an incidental catch or bycatch.

The capture frequency also reveals valuable information. Shrimp were captured in 45.14% of the reports, followed by bycatch with 9.81% and red snapper with 9.09%.

Table 2.2 Relative importance index ranking of industrial and semi-industrial scale fisheries species

Rank	Fishery name[a]	% Capture	% Value	% Frequency	Economic units	RII
1	*Camarón* (Shrimp)	73.81	84.81	45.14	54	67.92
2	*Guachinango* (Red snapper)	3.77	4.38	9.09	42	5.75
3	*Jurel* (Amberjack)	5.44	3.06	7.50	33	5.33
4	*Fauna* (Bycatch)	1.79	0.11	9.81	45	3.91
5	*Mero* (Mere)	2.60	2.30	6.03	32	3.64
6	*Besugo* (Sea bream)	2.46	1.26	5.49	29	3.07
7	*Pargo* (Porgy)	1.22	0.96	4.44	37	2.20
8	*Bandera* (Gafftopsail catfish)	2.84	0.53	2.14	18	1.83
9	*Rubia y Villajaiba* (Yellowtail Snapper)	0.94	0.52	3.53	30	1.66
10	*Tiburon* (Shark)	1.01	0.75	1.05	12	0.94
	Total	95.88	98.68	94.22	332	96.25

[a]The local common name is kept to compare with other documents

When visualizing the time series for catches reported in the landings reports corresponding to the states of Tabasco and Campeche during the same period as the VMS data, we can observe in Fig. 2.2a, which represents small-scale fisheries, that octopus is the dominant species in terms of catch volume throughout the entire period. Starting in 2018, the catch volume of octopus experienced a significant decrease, with a brief increase in production in 2021, followed by a gradual decline until 2023, when it reported the lowest catch volumes. Other species, such as amberjack, gafftopsail catfish, and snook, also exhibit similar trends in catch volume. Notably, amberjack's catch volume increased steadily until 2021, when it peaked before decreasing to a minimum of around 5000 tons in 2023.

In Fig. 2.2b, corresponding to the catches reported by the ISIF, shrimp stand out notably, representing the most significant catch volume during the studied period. The catch volume increased until it peaked in 2018, with 2100 tons, followed by a steady decline until 2023, when it reached a minimum of 600 tons.

When comparing the average prices per kilogram of catch and the total landed weight for small-scale fishing, we observe which fisheries show the most significant changes between 2017 and 2023. In Fig. 2.3a, decreases in catches are characterized by the fisheries for blue crab, amberjack, octopus, and mojarra. The landed weight of crab decreased by 64.65%, but its price per kilogram increased by 45.99%. Amberjack exhibited a 15.25% weight reduction and a 3.99% price increase per kilogram. Octopus experienced a significant decrease in weight (62.64%) and average price (19.23%). Mojarra maintained its price relatively stable, with a 27.28% decrease in landed weight.

Regarding shrimp fishing (Fig. 2.3b), there was the most significant decrease in price per kilogram (14.23%), followed by snook, which also experienced a notable reduction in landed weight (13.02%). On the other hand, red snappers did not show significant changes in either catch or price, with only a slight reduction of 15.25%

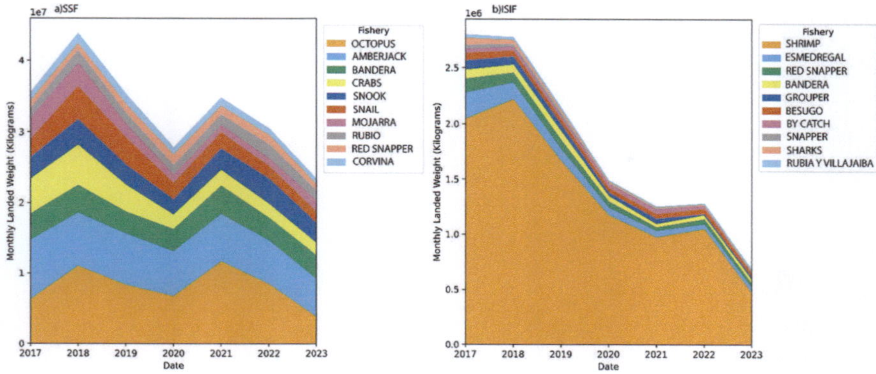

Fig. 2.2 Catch volume of the leading fisheries for 2017–2023: (**a**) Small-scale fisheries (SSF) and (**b**) Semi-industrial fishing fleets (ISIF)

in weight and a 3.99% increase per kilogram. These ten fisheries have generally seen a decline in reported catches, with a few exceptions, such as snook, red snapper, and Gulf weakfish, which remained relatively stable. For price variations, prices generally remained stable or increased.

In contrast to small-scale fishing, ISIF shows significant price increases per kilogram. In Fig. 2.3b, shrimp fishing is losing its advantage as it is the most important fishery for the price SIF by its RII. For this fishery, 2038 tons were reported in 2017, and by 2023, the catch had dropped to 466 tons, representing a 77.13% decrease. There was an approximately 83% decrease in red snapper fishing from 39 tons in 2017 to only six tons in 2023. On average, these ten fisheries saw a 72% reduction in their catches.

In contrast to the catches, the average prices per kilogram of catch showed significant increases in most fisheries, with an average increase of 42%. The snapper and shrimp fisheries were the only ones that showed decreased prices, with reductions of 11.6% and 2%, respectively.

2.3.2 *The Movements of the SSF Fleets*

In the study conducted throughout 2021 on the operation of four small vessels equipped with NEMO transmitters specialized in the red snapper fishery, significant patterns were observed in the distance navigated and fuel consumption, as illustrated in Fig. 2.4. During the year, the average distance traveled was 626.39 km, with notable fluctuations. The maximum distance reached was 1331 km, and the minimum was 240 km, with peaks in May and November, and a considerable decline in March (Fig. 2.4a). The second panel (Fig. 2.4b) displays the estimated monthly fuel consumption, measured in liters; these parameters do not consistently exhibit a linear relationship, suggesting complex operational dynamics. These

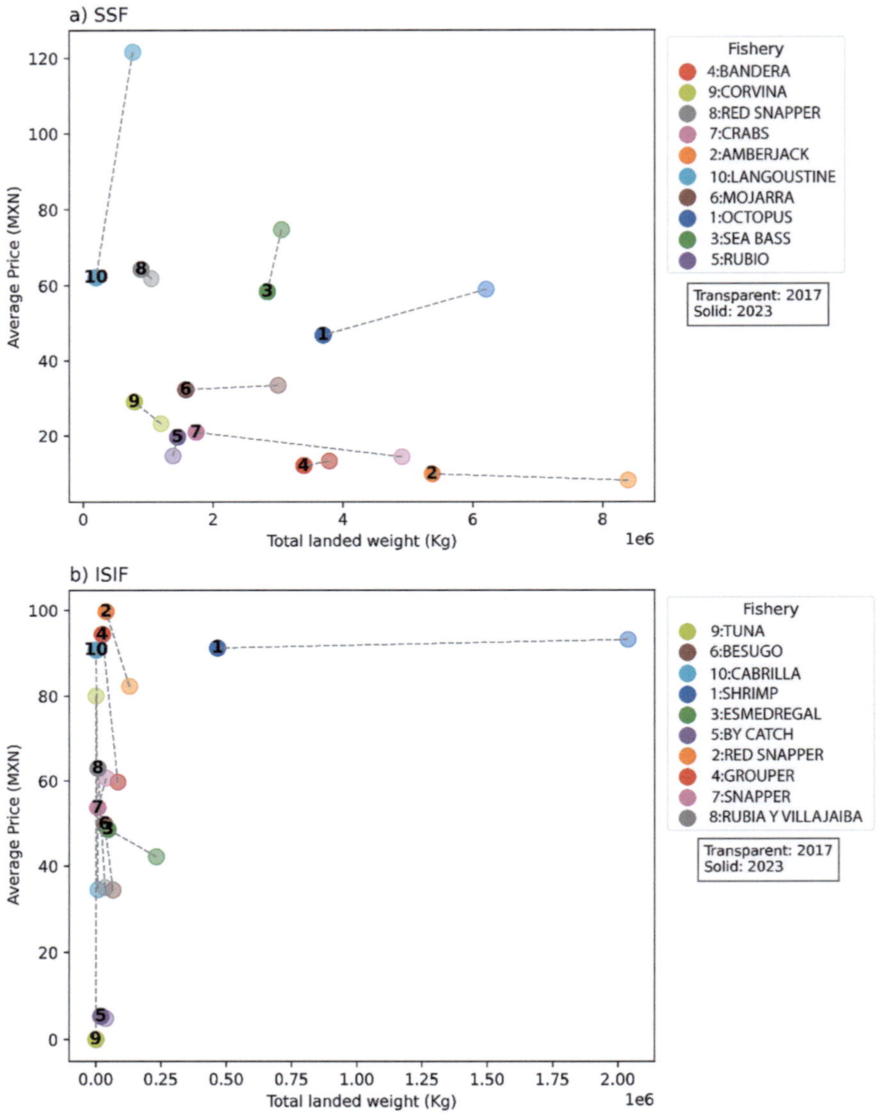

Fig. 2.3 Comparison of landed weights and average prices of the main species captured by their RII in 2017 and 2023 in the regions of Tabasco and Campeche: (**a**) small-scale fishery analysis and (**b**) industrial-scale fishery

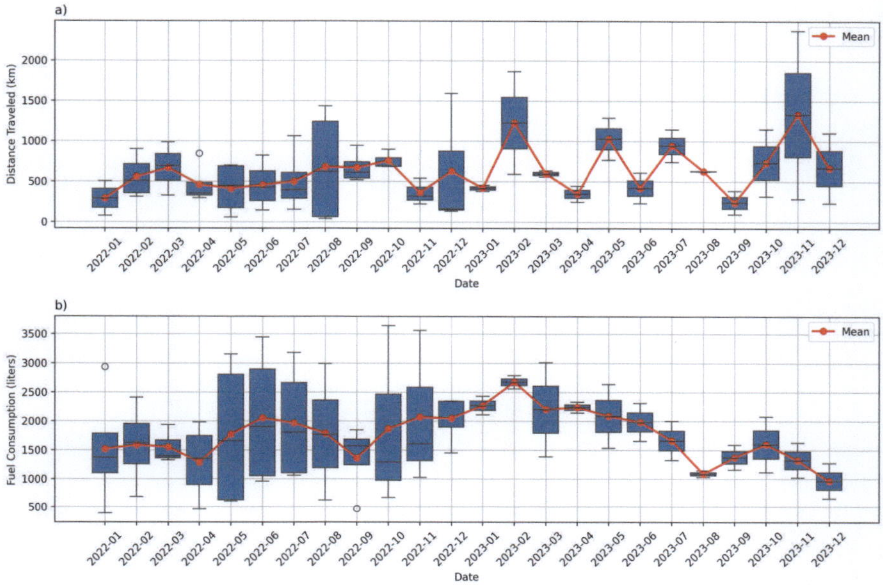

Fig. 2.4 (**a**) Monthly distance traveled and (**b**) estimated fleet fuel consumption for the smaller fleet with NEMO transmitters

parameters do not consistently exhibit a linear correlation, as evidenced by periods where increased distance does not correspond proportionally to fuel consumption. The data indicates that vessels operating for extended periods within confined geographical areas may record higher fuel consumption despite shorter navigational distances. These finding challenges conventional assumptions about distance-fuel relationships in maritime operations. However, these results should be interpreted with caution due to limitations in the estimation methodology.

In Fig. 2.5a, the average fleet fuel consumption, measured in liters and calculated from the operating hours of the vessels and their distance traveled, is shown. Throughout 2022 and up until December 2023, consumption exhibits a generally stable trend with slight fluctuations, peaking in June 2022 and reaching its highest point in February 2023 at 2054 liters. The trend then decreases toward January 2024, with a minimum of 1000 liters.

In contrast, Fig. 2.5b shows the average price of gasoline in US dollars per liter during the same period. A constant increase in gasoline price is evident, starting in 2022 at approximately USD 1.00 per liter and rising to nearly USD 1.30 per liter by early 2024.

Fig. 2.5 Time series for (**a**) estimated monthly fuel consumption in liters based on operating hours, (**b**) average price per liter of gasoline in dollars, and (**c**) monthly catches of the Red Snapper Fishery reported for the small-scale fishing fleet with NEMO

On the other hand, Fig. 2.5c shows the monthly catches of red snapper reported at the fishing office in *Frontera*, Tabasco. The catches show monthly variations throughout the study period, with notable peaks in certain months and a general decreasing trend toward the end of the period, particularly noticeable from mid-2023 onward.

2.3.3 The Movements of the ISIF Fleets

Figure 2.6 analyzes the ISIF monthly distance traveled and operating hours from January 2018 to October 2023. The graph in Fig. 2.6a depicts the monthly distance traveled in kilometers. Over the years, the average distance traveled has shown significant variations, with a notable increase in 2019, when the highest average distances were recorded. In comparison, 2023 records a considerable decrease in distance traveled. Figure 2.6b illustrates the ISIF monthly operating hours. Like the distance traveled, operating hours show fluctuations throughout the study period. The year 2019 again stands out with a peak in operating hours, suggesting intense activity during that year. In contrast, the following years show a decrease in operating hours, reaching their lowest point in 2021 with a slight rebound in 2023. Figure 2.6 illustrates how ISIF concentrates activity over shorter distances, albeit with increased activity hours.

Figure 2.7 displays heat maps of point density from VMS data for the study region across quarters of 2018 and 2023. These maps reveal distinct spatial patterns

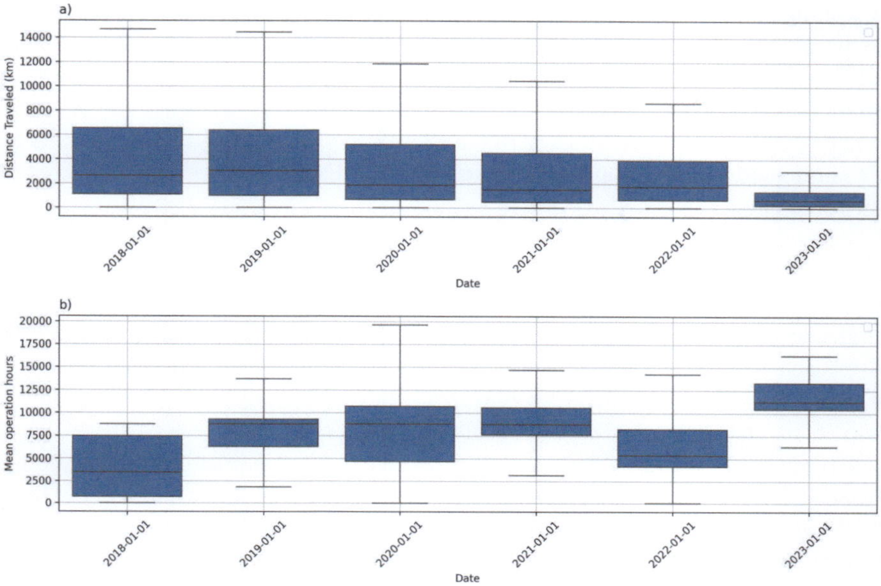

Fig. 2.6 Monthly distance traveled (**a**) and operating hours (**b**) of the Industrial and Semi-Industrial Fleets (ISIF)

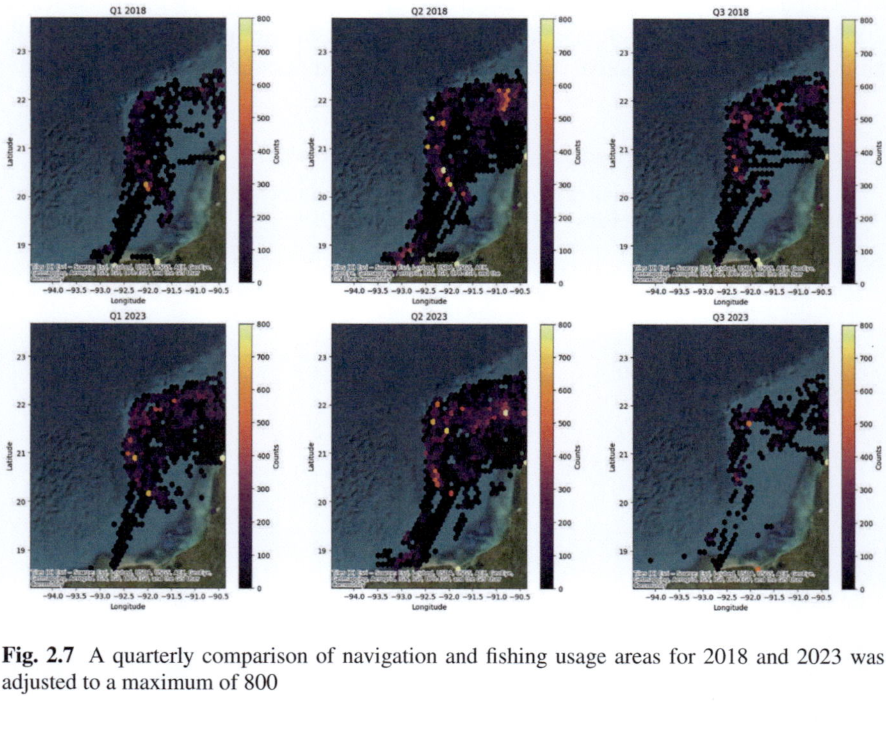

Fig. 2.7 A quarterly comparison of navigation and fishing usage areas for 2018 and 2023 was adjusted to a maximum of 800

of fishing activity, with a higher concentration of points near the coast, particularly in the second and third quarters. The heat maps for 2023 also reveal similar patterns, with a notable density of points near the coast, particularly in the second and third quarters. Comparing both years, the corresponding quarters' intensity and distribution of points are consistent. However, a significant difference is that the density in the third quarter of 2023 appears more concentrated and less dispersed compared to the same quarter in 2018.

Table 2.3 quantifies these observations, showing substantial decreases in fleet activity between 2018 and 2023. VMS records declined by 42.4% in Q1, 36.2% in Q2, and 32.3% in Q3. While the number of active vessels remained stable, with slight decreases in Q1 and slight increases in Q2 and Q3, the operational intensity decreased markedly. Days at sea were reduced significantly across all quarters: 47.3% in Q1 (from 1388.6 to 732.3 days), 43.7% in Q2 (from 1418.7 to 798.8 days), and 36.1% in Q3 (from 1386.1 to 885.1 days).

The average number of days at sea per vessel exhibited similar trends, decreasing from 22.8 to 12.8 days in Q1 (a 43.9% reduction), from 24.0 to 13.1 days in Q2 (a 45.4% decrease), and from 23.1 to 14.3 days in Q3 (a 38.1% reduction). The mean traveled distances declined across all quarters: a decrease of 29.6% in Q1 (from 1834.7 to 1292.4 km), 34.3% in Q2 (from 1859.6 to 1221.3 km), and 33.9% in Q3 (from 1882.1 to 1244.9 km). This reduction in traveled distances corresponds with the trend previously noted in Fig. 2.6, confirming a consistent decline in the spatial range of fishing operations. The spatial density comparisons indicate that while the overall pattern of fishing areas remained stable between years, the third quarter of 2023 exhibits more concentrated and less dispersed fishing effort compared to 2018, aligning with the quantitative reduction in operational parameters.

Table 2.3 Vessel monitoring system quarter description differences

Quarter	Number of points	Number of vessels	Total days at sea	Mean days at sea by vessel	Mean distance (km)
Q1 2018	24,906	61	1388.6	22.8	1834.7
Q2 2018	24,290	59	1418.7	24.0	1859.6
Q3 2018	24,396	60	1386.1	23.1	1882.1
Q1 2023	14,343	57	732.3	12.8	1292.4
Q2 2023	15,508	61	798.8	13.1	1221.3
Q3 2023	16,505	62	885.1	14.3	1244.9

2.3.4 The Fisheries Communities

Figure 2.8 compares catches (tons) and the catch value (in pesos) for the main landing sites in 2017 and 2023. The *Malecon Sansores Perez, Champotón,* and *Campeche* areas show a decrease in catches, from 6474 tons in 2017 to 4712 tons in 2023. The

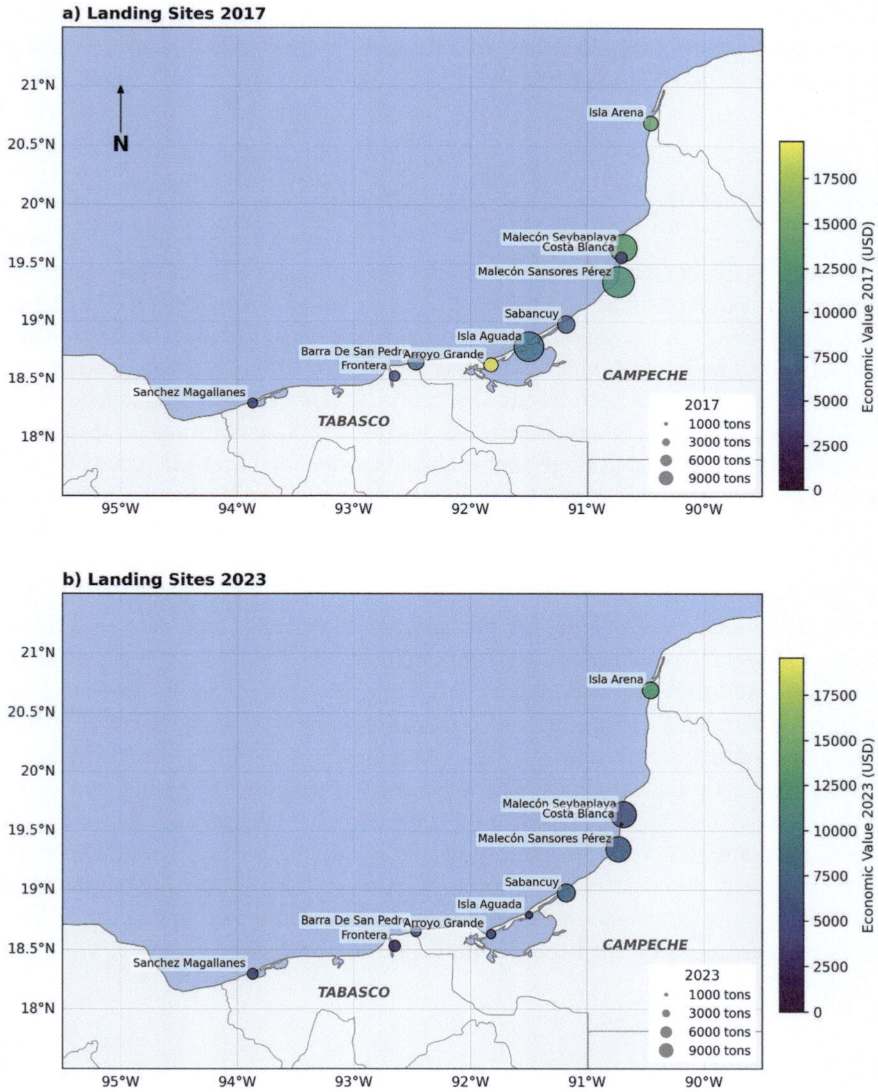

Fig. 2.8 Comparison of catches and catch value for the main SSF landing sites

value of the catches also decreased from USD 12,935 in 2017 to USD 6691 in 2023. *Isla Aguada* has shown a notable reduction in catches, from 5955 tons in 2017 to 872 tons in 2023, accompanied by a corresponding decrease in value from USD 8352 to 3276.

Malecon Seybaplaya reflects a decline in catches, from 5247 tons in 2017 to 4903 tons in 2023, and in value, from USD 13,371 to 5855. *Sabancuy* shows a slight increase in catches, from 2881 tons in 2017 to 3004 tons in 2023, while the value of the catches increased from USD 6465 to 7299. *Barra de San Pedro* has seen a decrease in catches, from 2605 tons in 2017 to 1435 tons in 2023, and a reduction in value, from USD 7334 to 6102 USD.

2.4 Discussion

The fisheries dynamics in the southwestern Gulf of Mexico exhibit significant spatiotemporal variations that demand careful examination. Our analysis reveals two concerning trends in the region: a marked reduction in fishing space utilization and a notable decline in catch volumes across several key fisheries. These patterns are evident in the observed catch trends reported in Tabasco and Campeche from 2017 to 2023. While Campeche experienced an increase in catches until 2017, reaching a peak of 69,026 tons, the average annual catch has since stabilized at around 53,800 tons over the following 5 years, according to the Statistical Yearbook of Fisheries and Aquaculture. In contrast, Tabasco has shown a clear downward trend since 2018, with a significant reduction in catches, dropping from 50,000 tons that year to 18,000 tons in 2022 (CONAPESCA, 2022).

These regional trends align with our findings at the fleet level, where we documented substantial reductions in both traveled distances and catch volumes, particularly in the industrial fleet's operations. These operations showed a 29–35% decrease in traveled distances and a nearly 40% reduction in days at sea between 2018 and 2023. The spatial contraction of fishing activities and declining catches indicate systemic changes in the region's fisheries sector that cannot be attributed to a single factor. Instead, these changes likely reflect the complex interplay of environmental shifts, economic pressures, regulatory changes, and competition with other maritime industries, particularly those involved in oil extraction and production. The following sections examine these patterns in detail for SSF and ISIF fleets, as well as their implications for the fishing communities in this region.

2.4.1 Small-Scale fisheries Landing Reports

The octopus fishery is vital for coastal communities in Campeche (Coronado et al., 2020). With a Relative Importance Index (RII) of 21.16, octopus significantly outweighs other species although its harvest is strictly seasonal (SADER, 2023). During the closure of octopus fishing, snook and amberjack fisheries became crucial alternatives that sustained these communities. In 2018, octopus catches reached

14,900 tons, primarily from Campeche State (CONAPESCA, 2022). Snook and amberjack rank second and third in importance, respectively. These species and mojarra are captured within métier fisheries, where multiple fish species classified as "Finfishery" are harvested using the same fishing gear (Monroy et al., 2010). Additionally, the observed changes in catch volumes may be influenced by environmental factors, such as variations in sea surface temperature, changes in oceanic currents, and extreme weather events, particularly in the case of red grouper (Oribe-Pérez et al., 2023; García-Rodríguez et al., 2024). Furthermore, while octopus maintains the highest economic importance, economic units that diversify their catch across multiple species demonstrate greater resilience (Wintergalen et al., 2022). This diversification strategy is a crucial adaptation mechanism, allowing fishers to maintain income stability despite seasonal closures, environmental fluctuations, and market variations (Frawley et al., 2019).

Considering the case of the red snapper fishery is a highly specialized operation with 707 economic units involved. Although it ranks eighth based on its relative importance index, this species contributes 6% of the economic value but only 2.6% of the total catch volume. This specialization renders economic units focusing solely on red snapper fishing particularly susceptible to changes in stock availability and market conditions, underscoring the risks of overspecialization in fishing operations.

When it comes to prices, the outcomes are varied. Particular species, such as the octopus, have experienced considerable declines in value, falling from USD 2.90 to 2.25 per kilogram, and snook, which decreased from USD 3.75 to 2.90 per kilogram. In contrast, other species, including crab, snail, weakfish, red snapper, and bluewing searobin, have experienced price increases. Interestingly, no scientific studies have analyzed these price variations and the potential factors influencing them in the study area. However, the observed price increases can be attributed to two main factors: first, inflation, and second, the removal of government fuel subsidies in 2018 (Revóllo-Fernández, 2024; Causa Natura, 2024), which may shift operational costs to the price per kilogram. However, alongside inflation and subsidies, fisheries like those for red snapper and octopus, which have access to national and international markets, may face additional factors such as dollar price fluctuations and international and export regulations that also impact the price.

It is essential to note that for fisheries employing catch shares—a rights-based management system that can reduce costs and increase revenues by allocating individual or group quotas—this shift enables more efficient operations, resulting in lower costs and the elimination of redundant fishing capacity, particularly when quota trading is permitted (Birkenbach et al., 2023).

2.4.2 Small-Scale Fisheries Spatial Analysis

In the context of spatial analysis, it is crucial to understand how fishing operations are conducted and to identify potential impacts on exploited stocks, particularly considering the interactions between co-occurring species and different types of fishing gear. Knowledge of the spatial distribution of fishing effort and the dynamics of gear usage, influenced by fishers' behavior, is essential for this understanding (Monroy et al., 2010; Torres-Irineo et al., 2021). The fleet equipped with NEMO transmitters showed an increase in distances traveled toward the end of the first year of study. In 2023, there were significant variations in the monthly average distances traveled, likely reflecting changes in the spatial distribution of the target species, specifically the red snapper. However, due to the limited number of boats equipped with transmitters installed, it is challenging to draw more definitive conclusions about the behavior of this fleet. The importance of using account-resolution monitoring systems to support fisheries management decisions cannot be overstated (Hernández-Álvarez et al., 2023). While the ISIF fleet benefits from satellite monitoring systems like VMS, the small-scale fleet faces significant challenges accessing such technology.

Such technology may become robust traceability mechanisms, which have become increasingly critical to addressing illegal fishing, seafood fraud, and human rights abuses in this region (Lewis & Boyle, 2017), enhancing access to monitoring technology for the small-scale fleet could be crucial for achieving more effective and equitable fisheries management.

On the other hand, VMS technology for SSF vessels may be critical to prevent catch like Hurricane Milton in 2024, which caused the disappearance of 15 fishers in Yucatán, who were unable to reach the mainland and were lost due to the extreme weather conditions and the non-communications, delayed search efforts (El Financiero, 2024).

2.4.3 *Industrial and Semi-industrial Fisheries Dynamics*

The analysis of the landing reports regarding the RII ranking highlights the shrimp fishery, which accounts for 17.38% of the total catch volume and contributes 84.8% of its economic value, being responsible for 45% of the arrival notices. This fishery supports 54 economic units and has a relative importance index of 67.92. However, it is noteworthy that, contrary to these findings, Tabasco does not consider shrimp one of the most significant fisheries (CONAPESCA, 2022). Instead, reports indicate that species from the finfish fishery are the primary focus, while Campeche ranks fourth, with maximum catches of 7182 tons in 2015, which declined to 3269 tons in 2022 (CONAPESCA, 2022).

On the other hand, for the ISIF, the red snapper ranks second in relative importance, with an index of 5.75. In this category, the species account for 7% of the total volume and 3.06% of the economic value, encompassing 42 economic units. The red snapper (*Lutjanus campechanus*) has increased production from 2015 to 2020 (Secretaría de Agricultura y Desarrollo Rural, 2023); in contrast, the red snapper time series shows a negative trend from 2018 to 2023 in catch volumes, this coincides with forecast done by the fisheries authority, that point a possible negative trend form 2018 (SADER, 2023). Another approximation of the negative trends into best practices in fisheries management implemented in recent years has influenced stock trends in small-scale multi-species fisheries in Mexico (García-Rodríguez et al., 2024). It is noted that *L. campechanus* is one of the few species that is not overexploited or overfished, with a 62% probability of not falling into these categories, which may contribute to its increase in recent catches (García-Rodríguez et al., 2024). On the other hand, the prices of species caught by ISIF have shown significant increases in the average cost per kilogram, which, as previously argued, is likely due to changes in subsidy policies. Another factor affecting prices is opening foreign markets for fishery exports (Coronado et al., 2020).

2.4.4 Industrial and Semi-industrial Fisheries Spatial Analysis

When analyzing the spatial dynamics of the ISF fleet, we observed that the average distance traveled has decreased since 2022, while sea operation hours have increased during this period. This indicates a strategic concentration of fishing efforts in specific areas that are presumably more productive. This spatial concentration aligns with the findings of Monroy et al. (2010) and Torres-Irineo et al. (2021), who documented similar patterns of fishing effort aggregation to resource availability.

Our data reveal distinct seasonal trends in the distribution of the fleet, with fishing activity primarily occurring from May to August, during the rainy season, spread across various fishing grounds. In contrast, from October to December, aligning with the onset of the "nortes" or cold front season, primary concentration zones significantly cluster off the coast of Campeche along the continental shelf slope.

This seasonal variation in spatial distribution is strongly influenced by the "nortes" phenomenon described by Ehuan-Noh et al. (2020), who found that these seasonal weather patterns significantly shape fishers' tactics. The "nortes" season (approximately October to March) brings strong northeastern winds with speeds exceeding 30 km/h and temperate water currents. According to their research with fishing captains from Seybaplaya, Campeche, this period is regarded as the most productive time for fishing, despite challenging navigation conditions. These winds drive large schools of migratory species to the region, increasing turbidity levels and enhancing nutrient concentration.

The concentration of fishing effort near the coast during the "nortes" season also represents a safety adaptation, as venturing farther offshore during this period presents higher risks due to unpredictable weather conditions. This pattern is consistent with our VMS data, which shows reduced traveled distances but increased operational hours—vessels spending more time fishing in smaller, more productive areas closer to shore.

The economic dimension of this spatial pattern cannot be overlooked. As highlighted by Avilés-Polanco et al. (2023), the shrimp fishing companies in the Gulf of Mexico were found to be more cost-efficient (0.13 inefficiency) than those operating in the Gulf of California and South Pacific (0.31 inefficiency). These efficiency differences are likely reflected in the operational strategies we observed. Our data, mostly from shrimp and red snapper fleets, show that shrimp is highly fuel-intensive (fuel price elasticity of 0.75). The significant reduction in traveled distances observed in our VMS data represents a rational economic response to increasing fuel costs and decreasing catch volumes, particularly in the shrimp fishery, which has shown a marked decline from 2100 tons in 2018 to only 600 tons in 2023.

The temporal pattern in fleet movements also correlates with the distribution of fuel subsidies documented by Avilés-Polanco et al. (2023), who found that fuel subsidies contributed to reducing cost inefficiencies, with the most significant effects observed when subsidies reached approximately 20% of total subsidized fuel. Since the states bordering the Gulf of Mexico receive different proportions of these subsidies, this may partially explain the differences in operational efficiency and spatial patterns between regions.

These findings have important implications for fisheries management. They suggest that seasonal and spatially explicit management measures may be more effective than year-round approaches, highlighting the need to consider both environmental factors and economic constraints when designing management strategies for Mexico's Gulf industrial fishing fleet.

2.4.5 Findings from Fishery Offices

When examining the temporal evolution of fishery offices where landing reports are registered, it becomes clear that most sites have experienced a significant reduction in catches from 2017 to 2023, except for *Sacunbay, Sanchez Magallanes*, and *Frontera*, which have seen an increase in catches. For example, *Malecon Sansores Perez* decreased from 6474 tons in 2017 to 4712 tons in 2023, indicating a substantial decline. This trend is also evident in locations such as *Isla Aguada* and *Malecon Seybaplaya*, where catches dropped by more than 80% and 50%, respectively. This suggests a shift in the economic focus of these localities, which face a decline in catches; it is also necessary to note that the number of economic units has decreased since 2019. A case in point is the *Isla Aguada* offices, which have reduced their reliance on fishing income while expanding their tourism sector (Peña-Puch et al., 2023).

Concurrently, the economic value of catches at some sites has also shown a downward trend. For example, at *Malecon Seybaplaya*, the total value decreased from 267,420 MXN in 2017 to 117,093 MXN in 2023, reflecting a substantial reduction. However, not all sites exhibit this trend; *Sabancuy*, for instance, shows an increase in economic value despite a modest increase in catches, which may have resulted from a shift in the species caught or an increase in market prices due to access to new value chains or exportation.

The observed decline in catch volumes from 2020 to 2023, particularly in SSF, may be linked to unreported fishing activities during the COVID-19 pandemic and complex market dynamics affecting formal landings. Castillo-Figueroa (2023) documented that many SSF operations in Puerto Madero continued uninterrupted during the pandemic, despite restrictions, with fishers stating, "*No le paramos por el* COVID" (We did not stop because of COVID). Similarly, Arreguín-Sánchez et al.

(2022) found that, although formal records indicated a decrease in fishing activity, actual fishing operations continued, with products being directed to local markets rather than formal landing sites. Bennett et al. (2020) noted this pattern globally, where fishing continued but reporting decreased as traditional market chains were disrupted. More comprehensive research is needed to fully understand these dynamics in both SSF and ISIF.

This study sheds light on the complex dynamics that affect Gulf of Mexico fisheries, which are shaped by economic, environmental, and social factors. The findings suggest a need for management approaches that strike a balance between economic viability and ecological sustainability. As marine resource pressures intensify, management policies must adapt to address these evolving conditions, thereby supporting the conservation of species and the livelihoods of fishing communities. Strengthening the use of VMS in small-scale fleets is essential for enhancing decision-making processes. Additionally, understanding the specific needs of the fishing industry is vital to ensuring food sovereignty, especially in the face of challenges such as the reduction of fishing areas and climate change.

2.5 Conclusions

The small-scale fleet is vital in the regional fishing economy, with octopus, snook, and amberjacks being the most important species. However, there has been a noticeable negative trend in the catches of these species, which could be attributed to environmental factors, shifts in fishing practices, or pressures from other economic activities. Additionally, significant price variations have been observed; while some species have experienced considerable declines in value, others have managed to increase their prices. Changes are likely to influence these price fluctuations in market demand and operational costs. Spatial analysis of the fleet's activities during the second half of the year reveals variations in cruise distances traveled, suggesting a concentration pattern in certain areas. This trend aligns with the seasonal operations of the red snapper fishery. However, despite the increase in fuel costs toward the end of the year, there was no corresponding rise in catch prices, possibly due to the limited time frame analyzed.

Moreover, the SSF is highly specialized in shrimp fishing, followed by red snapper, reflecting the economic significance of these species. While most catch volumes have remained stable, shrimp catches have varied significantly. Unlike the small-scale fleet, the ISIF catch prices have considerably increased, likely driven by inflation and access to international markets. No significant changes in the distances traveled by the fleet were observed until 2023, when the fleet began concentrating in specific areas, accompanied by an increase in operating hours. This trend toward fleet concentration at the end of the year is particularly noticeable in 2023.

Fishing communities have also experienced significant changes, with several fishery offices reporting decreased catches and a corresponding decline in income generated from fishing. This suggests a transformation in the economic activities of these localities. There is a general trend toward fewer economic units reporting fishing activities, which may be due to structural changes in the fishing industry or the migration of these units to other economic sectors. Understanding the emerging economic vocations within these communities is essential to ensuring their sustainability and the well-being of their inhabitants, particularly in the face of challenges such as changes in fishing zones and the impacts of climate change.

Expanding the use of VMS in the small-scale fleet is necessary to address these challenges. Enhanced monitoring will provide better insights into the spatial dynamics of fishing activities and supply critical data for informed decision-making in fisheries management, especially as fishing areas become increasingly restricted. Additionally, developing adaptation strategies tailored to the specific needs of fishing communities is crucial. These strategies should focus on supporting economic diversification, improving access to international markets, and strengthening the resilience of these communities against challenges such as the reduction of fishing zones and climate change. Such efforts are key to ensuring food sovereignty and the region's sustainable development.

Acknowledgments This study was supported by the project "*La desafiante coexistencia de los sistemas socio-ecológicos acoplados; las industrias de la pesca y petróleo en la Sonda de Campeche*" (Ciencia de Frontera 2019, Project CF-2019-1564454, CONACyT).

References

Arreguín-Sánchez, F., Albánez-Lucero, M. O., & Hernández-López, A. (2022). Effect of COVID-19 on the small-scale fishing in the Mexican Pacific: From poverty to subsistence and the opportunity to promote development. *SSRN Electronic Journal*. https://doi.org/10.2139/ssrn.5129920

Avilés-Polanco, G., Almendarez-Hernández, M. A., Beltrán-Morales, L. F., & Aranceta-Garza, F. (2023). Cost efficiencies of the shrimp fishery in Mexico: A stochastic frontier analysis. *Fishes, 8*, 472. https://doi.org/10.3390/fishes8090472

Bennett, N. J., Finkbeiner, E. M., Ban, N. C., Belhabib, D., Jupiter, S. D., Kittinger, J. N., Mangubhai, S., Scholtens, J., Gill, D., & Christie, P. (2020). The COVID-19 pandemic, small-scale fisheries, and coastal fishing communities. *Coastal Management, 48*(4), 336–347.

Birkenbach, A. M., Kaczan, D. J., Smith, M. D., Ardini, G., Holland, D. S., Lee, M. Y., et al. (2023). Do catch shares increase prices? Evidence from US fisheries. *Marine Resource Economics, 38*(3), 203–228.

Castillo-Figueroa, G. (2023). "No le paramos por el COVID". Narrativas sobre el trabajo pesquero en Puerto Madero (Chiapas) durante la pandemia del COVID-19. *LiminaR. Estudios Sociales y Humanísticos, 21*(2), e1004.

CONAPESCA. (2022). *Anuario Estadistico de Acuacultura y Pesca 2022*. https://nube.conapesca.gob.mx/sites/cona/dgppe/2022/ANUARIO_ESTADISTICO_DE_ACUACULTURA_Y_PESCA_2022.pdf

Coronado, E., Zepeda-Domínguez, J. A., Espinoza-Tenorio, A., Cupido Santamaria, D., Ramos Muñoz, D., & Monzón-Alvarado, C. (2024). Institutional mapping and implementation of the sustainable development goals across co-existing industries: The case of the fisheries-oil system in Mexico. *The Extractive Industries and Society, 17*, 101390. https://doi.org/10.1016/j.exis.2023.101390

Coronado, E., Salas, S., Torres-Irineo, E., & Chuenpagdee, R. (2020). Disentangling the complexity of small-scale fisheries in coastal communities through a typology approach: The Yucatan peninsula, Mexico case study. *Regional Studies in Marine Science, 36*, 101312. https://doi.org/10.1016/j.rsma.2020.101312

El Financiero. (2024, October 9). *Huracán 'Milton' en Yucatán: Desaparecen 15 pescadores; quedaron 'atrapados' en el mar.* https://www.elfinanciero.com.mx/estados/2024/10/09/huracan-milton-en-yucatan-desaparecen-15-pescadores-quedaron-atrapados-en-el-mar/

Ehuan-Noh, R. G., Mariaca-Méndez, R., Sáenz-Arroyo, A., & Espinoza-Tenorio, A. (2020). Tácticas y saberes: los capitanes de la pesca ante la variabilidad ambiental del mar. *Sociedad y Ambiente, 23*, 1–29. https://doi.org/10.31840/sya.vi23.2199

Espinoza-Tenorio, A., Millán-Vásquez, N. I., Vite-García, N., & Alcalá-Moya, M. G. (2019). People and blue carbon: Conservation and settlements in the mangrove forests of Mexico. *Human Ecology, 47*(6), 877–892. https://doi.org/10.1007/s10745-019-00123-6

Espinoza-Tenorio, A., Ehuan-Noh, R. M., Cuevas-Gómez, G. A., Narchi, N. E., Ramos-Muñoz, D. E., Fernández-Rivera-Melo, F. J., Saldívar-Moreno, A., Zepeda-Domínguez, J. A., Pérez-Jiménez, J. C., Oliveto-Andrade, A., & Torre, J. (2022). Between uncertainty and hope: Young leaders as change agents in sustainable small-scale fisheries. *Ambio, 51*(5), 1287–1301. https://doi.org/10.1007/s13280-021-01639-2

Espinosa-Romero, M., Torre, J., Zepeda, J., Solana, F., & Fulton, S. (2017). Civil society contributions to the implementation of the small-scale fisheries guidelines in Mexico. 423-449. https://doi.org/10.1007/978-3-319-55074-9_20

FAO. (2024). El estado mundial de la pesca y la acuicultura 2024. *La transformación azul en acción,*. FAO. doi:https://doi.org/10.4060/cd0683es.

Frawley, T. H., Crowder, L. B., & Broad, K. (2019). Heterogeneous perceptions of social-ecological change among small-scale fishermen in the Central Gulf of California: Implications for adaptive response. Frontiers in marine. *Science, 6*, 78. https://doi.org/10.3389/fmars.2019.00078

Finkbeiner, E., & Basurto, X. (2015). Re-defining co-management to facilitate small-scale fisheries reform: An illustration from Northwest Mexico. *Marine Policy, 51*, 433–441. https://doi.org/10.1016/J.MARPOL.2014.10.010

Torres-Irineo, E., Salas, S., Dreyfus-León, M., & Daniel, P. (2021). Factors affecting the spatial distribution of small-scale fishing effort in Campeche, Mexico. *Fisheries Management and Ecology, 28*, 24–38.

García-Rodríguez, E., Saldaña-Ruiz, L. E., Sosa-Nishizaki, O., Fajardo-Yamamoto, A., Cisneros-Soberanis, F., & Flores-Guzmán, A. (2024). From bottom to top: Effects of fishery improvement projects on the stock trends of multi-specific small-scale fisheries from Mexico. *Marine Policy, 168*, 106292. https://doi.org/10.1016/j.marpol.2024.106292

Guarneros-Narváez, P. V., Rodríguez-Canul, R., De Silva-Davila, R., Zamora-Briseño, J. A., Améndola-Pimenta, M., Souza, A. J., et al. (2022). Loliginid paralarvae from the southeastern Gulf of Mexico: Abundance, distribution, and genetic structure. *Frontiers in Marine Science, 9*, 941908.

Hernández-Álvarez, Y., Ramírez-Ortiz, G., Flores-de-Santiago, F., Amezcua-Linares, F., Muro-Torres, V., Arceo-Carranza, D., & Amezcua, F. (2023). Spatiotemporal variability in fish assemblages in a coastal and estuarine system in the tropical eastern Pacific during the Anthropause. *Diversity, 15*(8), 934.

Lewis, S. G., & Boyle, M. (2017). The expanding role of traceability in seafood: Tools and key initiatives. *Journal of Food Science, 82*(S1), A13–A21.

Lozano, A., Smith, H., & Basurto, X. (2018). Weaving governance narratives: Discourses of climate change, cooperatives, and small-scale fisheries in Mexico. *Maritime Studies, 18*, 77–89. https://doi.org/10.1007/S40152-018-0125-5

Monroy, C., Salas, S., & Bello-Pineda, J. (2010). Dynamics of fishing gear and spatial allocation of fishing effort in a multispecies fleet. *North American Journal of Fisheries Management, 30*, 1187–1202.

Mendoza-Portillo, F., Ramírez-Rodríguez, M., & Vargas-López, V. (2020). Interactions of small-scale fisheries in Mexico's Northwest Pacific. *Latin American Journal of Aquatic Research, 48*, 94–105. https://doi.org/10.3856/vol48-issue1-fulltext-2176

Ojeda-Ruiz, M., Marín-Monroy, E., La Cruz, A., & Cota-Nieto, J. (2019). *Analysis and management of multi-species fisheries: Small-scale finfish fishery at Bahía Magdalena-Almejas*. Ocean & Coastal Management. https://doi.org/10.1016/J.OCECOAMAN.2019.104857

Pellowe, K., & Leslie, H. (2017). Seasonal variability shapes the resilience of small-scale fisheries in Baja California Sur. *PLoS ONE, 12*. https://doi.org/10.1371/journal.pone.0182200

Peña-Puch, A., Rivera-Arriaga, E., & Williams-Beck, L. (2023). *Exploring governance challenges in coastal communities through key informant perceptions in Campeche*. Ocean and Coastal Management. https://doi.org/10.1016/j.ocecoaman.2023.106722

Peña-Puch, A. d. C., Pérez-Jiménez, J. C., & Espinoza-Tenorio, A. (2019). Advances in the study of Mexican fisheries with the social-ecological system (SES) perspective and its inclusion in fishery management policy. *Ocean and Coastal Management*. https://doi.org/10.1016/j.ocecoaman.2019.105065

Pérez-Sánchez, E., & Muir, J. F. (2003). Fishermen perception on resources management and aquaculture development in the Mecoacan estuary, Tabasco, Mexico. *Ocean & Coastal Management, 46*(6–7), 681–700. https://doi.org/10.1016/S0964-5691(03)00041-3

Ramos-Muñoz, D. E., Ramos-Reyes, R., Zamora-Cornelio, L. F., Hernández-De la Cruz, A., & Espinoza-Tenorio, A. (2019). Exclusión en el Golfo de México: una visión desde los pescadores sobre la industria petrolera en Tabasco. *Cuadernos de Geografía: Revista Colombiana de Geografía, 28*(2), 357–372. https://doi.org/10.15446/rcdg.v28n2.73511

Ramos-Muñoz, D. E., Zepeda-Domínguez, J. A., & Espinoza-Tenorio, A. (2020). La reforma energética mexicana en el marco de la pesca y el petróleo en el Golfo de México. *Revista Costas, 2*(1), 67–80. https://doi.org/10.26359/costas.0402

Rivera-Arriaga, E., Alpuche-Gual, L., Negrete-Cardoso, M., Nava-Fuentes, J. C., Lemus-Pablo, E., & Arriaga-Zepeda, C. (2012). *Programa de manejo costero integrado para el saneamiento de la Bahía de San Francisco de Campeche*. Universidad Autónoma de Campeche.

Salazar-De la Cruz, C. C., Zepeda-Domínguez, J. A., Espinoza-Tenorio, A., & Ramos-Munoz, D. E. (2020). Governance networks in marine spaces where fisheries and oil coexist: Tabasco, México. *The Extractive Industries and Society, 7*(2), 676–685.

SADER. (2023, July 21). Carta Nacional Pesquera 2023.

Salas, S., Chuenpagdee, R., Seijo, J., & Charles, A. (2007). Challenges in the assessment and management of small-scale fisheries in Latin America and the Caribbean. *Fisheries Research, 87*, 5–16. https://doi.org/10.1016/J.FISHRES.2007.06.015

Salas, S., Torres-Irineo, E., & Coronado, E. (2019). Towards a métier-based assessment and management approach for mixed fisheries in southeastern Mexico. *Marine Policy, 103*, 148–159. https://doi.org/10.1016/j.marpol.2019.02.040

Schuhbauer, A., Cisneros-Montemayor, A., Chuenpagdee, R., & Sumaila, U. (2019). Assessing the economic viability of small-scale fisheries: An example from Mexico. *Marine Ecology Progress Series*. https://doi.org/10.3354/MEPS12942

Wintergalen, E., Oyanedel, R., Villaseñor-Derbez, J., Fulton, S., & Molina, R. (2022). Opportunities and challenges for livelihood resilience in urban and rural Mexican small-scale fisheries. *Ecology and Society*. https://doi.org/10.5751/es-13471-270346

Zalik, A. (2009). Zones of exclusion: Offshore extraction, the contestation of space and physical displacement in the Nigerian delta and the Mexican gulf. *Antipode, 41*(3), 557–582. https://doi.org/10.1111/j.1467-8330.2009.00687.x

Chapter 3
The Complex Coexistence of Small-Scale and Recreational Fisheries in the Coastal Zone of the Southern Gulf of Mexico

E. Coronado, L. Palomo, and A. Espinoza-Tenorio

3.1 Introduction

Extractive fishing activities for commercial and recreational purposes constitute highly adaptive and complex socio-ecological systems (SES) that encompass biophysical, socioeconomic, and management components (Berkes, 2017; Charles, 2001; Ostrom, 2009). These systems involve a dynamic interaction between different types of small-scale fisheries (e.g., octopus and grouper), recreational fishing, tourism, and other sectors that impose pressures on shared marine ecosystems. However, understanding fishing SES and their operations to establish appropriate management strategies requires defining these components and comprehensively evaluating their structure and interactions across temporal and spatial scales.

Small-scale fisheries (SSF) and the potential growth of recreational fishing in coastal communities in the southern Gulf of Mexico face challenges stemming from the integration of fisheries and socio-ecological data, the increase in illegal fishing activities, intersectoral conflicts, and limited surveillance of regulatory compliance. In addition, small-scale and recreational fisheries share common interests related to

E. Coronado (✉)
Escuela Nacional de Estudios Superiores, Unidad Mérida, Universidad Nacional Autónoma de México, Mérida, Yucatán, México

ECOMORPHOSES, Consultoría, Mérida, Yucatán, México
e-mail: eva.coronado@enesmerida.unam.mx

L. Palomo
Universidad Marista de Mérida AC, Mérida, Yucatán, México
e-mail: lpalomo@marista.edu.mx

A. Espinoza-Tenorio
El Colegio de la Frontera Sur, Campeche, Lerma, Campeche, México
e-mail: aespinoza@ecosur.mx

A. Espinoza-Tenorio et al. (eds.), *The Fishing and Oil Industries in Coexistence*,
https://doi.org/10.1007/978-3-032-02621-7_3

the health of exploited populations, the degree of habitat conservation, and the delineation of operational areas (Arlinghaus et al., 2021; Schlüter et al., 2019).

A primary challenge in developing fisheries management strategies is the lack of data and detailed information on small-scale and recreational fishing. This issue is particularly critical in multi-use coastal areas, where both sectors coexist and often target the same species using different gear types, at different growth stages, and catch intensities. For example, data on catch volume, effort, and economic value from recreational fishing are often underreported or absent from official statistics, making it difficult to evaluate (Dahlet et al., 2023; Ostrom, 2007; Palomo & Hernández-Flores, 2019). It is also important to note that the approaches and regulatory policies for commercial and small-scale fishing, which are aimed at food production and trade, differ significantly from those of recreational fisheries, making them potentially incompatible.

Commercial fishing requires official permission to capture one or more species using authorized fishing gear, with the purpose of supplying food and obtaining economic compensation from the domestic and foreign exploitation of fisheries resources (DOF, 2022). In comparison, recreational fishing is an activity that contributes to the economic, social, and environmental system (Arlinghaus et al., 2021; Brownscombe et al., 2019; Perez-Cobb et al., 2014). In Mexico, recreational fishing is defined as an activity that connects humans with nature for leisure or competition (DOF, 2024) and is classified within the productive fishing sector according to the Mexican standard NOM-017-PESC-1994. Although its impact on the economy and its capacity to generate jobs has been recognized, recreational fishing and its potential remain undeveloped.

Traditionally, commercial and recreational fisheries in Mexico have not been managed in tandem, nor has their management included the participation of all stakeholders, with managerial decisions for both types of fishing resting exclusively with the authorities. However, in recent decades, community-based resource management and administration have garnered attention and momentum, encouraging the participation of public and private stakeholders in decision-making (Charles, 2013; Defeo et al., 2016). Facilitating the transition from traditional management to a co-management model requires comprehensive research that provides decision-makers with a holistic understanding of SES, including how social systems interact with and depend on fisheries resources.

This premise provides the context for this chapter, which aims to analyze the small-scale and recreational fishing system of the Mexican states of Campeche, Yucatán, and Quintana Roo, which comprise the *Yucatán Peninsula* in the southern Gulf of Mexico, using an SES approach. This chapter seeks to answer the following questions: What are the operational characteristics of small-scale and recreational fisheries? What is the magnitude of the fishing effort of the small-scale and recreational fishing fleets? Which components converge within the SES of small-scale and recreational fishing? The answers to these questions enhance our understanding of how these activities interact and provide a foundation for discussing potential strategies to promote co-management and governance. A mixed approach that integrated qualitative and quantitative information from the scientific literature and technical reports on small-scale and recreational fishing in coastal communities of Campeche, Yucatán, and Quintana Roo was employed to answer these questions.

3.2 Identification of Key SES Components in the Southern Gulf of Mexico

A mixed approach that integrated qualitative and quantitative information from the scientific literature and technical reports on small-scale and recreational fishing in the coastal communities of Campeche, Yucatán, and Quintana Roo was employed to answer the central questions of this chapter. Commercial fishing information for 2009 to 2021 was obtained from official statistics (CONAPESCA, 2021) and specialized literature reports. A list of the vessels registered for recreational and sport fishing in Campeche, Yucatán, and Quintana Roo from 1997 to early 2023 was provided by the Secretariat of the Navy and the Captaincy and Maritime Affairs Unit of Mexico (SEMARNAT, 2023). Each recorded recreational vessel was classified into one of seven categories (catamaran, kayak, fishing boat, jet ski, sailboat, yacht, other types), and the number of vessels and the percentage by category in each state were determined.

The literature baseline was used to identify key SES components in the southern Gulf of Mexico. This information enabled us to construct a hypothetical model of causal relationships using fuzzy cognitive mapping, following the methods outlined by Gray et al. (2013). Fuzzy cognitive maps are based on semi-quantitative models of positive and negative causal relationships (weighted from −1 to 1). An SES model for small-scale and recreational fisheries was generated using Mental Modeler software (http://www.mentalmodeler.org; Gray et al., 2013, 2014), which integrated the system components of users, natural resources, and governance, as well as external SES drivers such as the market and environmental factors.

3.2.1 Study Area

The southern Gulf of Mexico covers approximately 3200 km and encompasses six coastal states of Mexico. This study focuses on three of these states, Campeche, Yucatán, and Quintana Roo, which together make up the *Yucatán Peninsula* (Fig. 3.1). The marine and coastal ecosystems of this region exhibit a variety of geomorphological, hydrological, edaphic, and ecological factors, including mangroves, coastal lagoons, grassland areas, and coral reefs (Islebe et al., 2015).

The Campeche coast hosts the *Laguna de Términos*, the largest lagoon-estuarine ecosystem in Mexico. *Laguna de Términos* contains six habitat types defined by marine, estuarine, mixing zone, freshwater, lagoon inlet, and continental shelf waters. Each habitat is ideal for the recruitment of commercial species. In Yucatán, the presence of karst soil, influenced by the ring of cenotes, allows freshwater to reach the coast, creating areas of high estuarine productivity. In Quintana Roo, numerous bays provide protection and breeding areas for species associated with the Mesoamerican Reef System, the second largest barrier reef in the world (Espinoza-Tenorio et al., 2014; Islebe et al., 2015).

The characteristics of these three states create ideal conditions among habitats for a wide diversity of marine taxa, as observed in other coastal areas of tropical regions. Notably, 200 species of pelagic fish, rays, and sharks, echinoderms, crustaceans, and mollusks of commercial importance can be found in and around the *Yucatán Peninsula* (Arreguín-Sánchez & Arcos-Huitrón, 2011; Ayala-Pérez et al., 2012; Ramos-Miranda et al., 2021; Salas et al., 2022; Schmitter-Soto et al., 2009).

Figure 3.1 shows the estimated distribution of fishers in the *Yucatán Peninsula* based on fisher registry data of the *Bienpesca* program (CONAPESCA, 2021). The figure also shows the estimated percentage of the gross domestic product (GDP) for tourism in key locations based on data from the Ministry of Tourism (SECTUR, by its Spanish acronym) from 2018 to 2019. The map highlights how the coastal

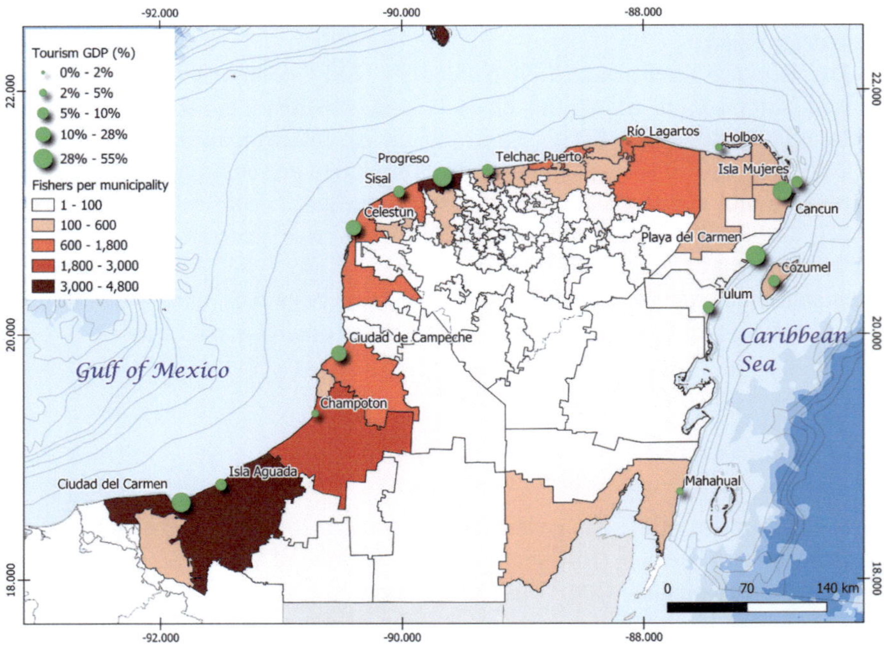

Fig. 3.1 Study area. Distribution of fishers in the *Yucatán Peninsula* based on fisher registry data of the *Bienpesca* program (CONAPESCA, 2021) and estimated percentage of the gross domestic product (GDP) for tourism in key locations based on data from the Ministry of Tourism (SECTUR) from 2018 to 2019

municipalities of Campeche, Yucatán, and Quintana Roo differ in fishing activities, fishing effort, and their contributions to tourism GDP. Notably, fishing and tourism coexist throughout the *Yucatán Peninsula*, particularly in popular tourist destinations such as *Cancún, Cozumel*, and *Playa del Carmen* in Quintana Roo, as well as *Progreso* in Yucatán, with local economies heavily reliant on both activities.

3.3 The Socio-Ecological System: Small-Scale and Recreational Fishing in the Southern Gulf of Mexico

3.3.1 Commercial Small-Scale Fishing

According to the fisheries data in the Yearbook of Fisheries and Aquaculture Statistics of the National Aquaculture and Fisheries Commission (CONAPESCA, 2021), the importance of the *Yucatán Peninsula* has become apparent in recent years. Indeed, the region is first in terms of the national production of groupers, octopus, and other high-value or exported species (Coronado et al., 2020; Salas et al., 2019). Furthermore, the volume of products from the *Yucatán Peninsula* and the value of captured resources have generated jobs, approximately 90% of which are associated with small-scale fishing, directly supporting the livelihoods of many people within coastal communities in the region.

3.3.1.1 Exploited Resources

On the *Yucatán Peninsula*, the average annual volume of commercial fishing production has fluctuated around 100,000 tons from 2009 to 2021 (Fig. 3.2). This production is valued at approximately 4 billion pesos (approximately USD 200 million) and accounts for approximately 5% of the national catch. Moreover, the average annual production of the *Yucatán Peninsula* accounts for approximately 33% of the total production on the Atlantic coast, with a value equivalent to 10% of the national total and 50% of the total for the Atlantic coast (CONAPESCA, 2021).

The distribution of target species and catch volumes varies between communities in the *Yucatán Peninsula* due to differences in the biological characteristics of exploited resources, species distributions, climate variations, habitat conditions, and changes in environmental conditions. The small-scale fleet of the *Yucatán Peninsula* comprises eight fisheries that target approximately 150 species spanning 50 families (Table 3.1). According to various studies, approximately 90% of the catch of the *Yucatán Peninsula* can be attributed to 25 species in varying proportions (Arreguín-Sánchez & Arcos-Huitrón, 2011; Coronado et al., 2020; Monroy-Garcia et al., 2019; Salas et al., 2019) (Fig. 3.3).

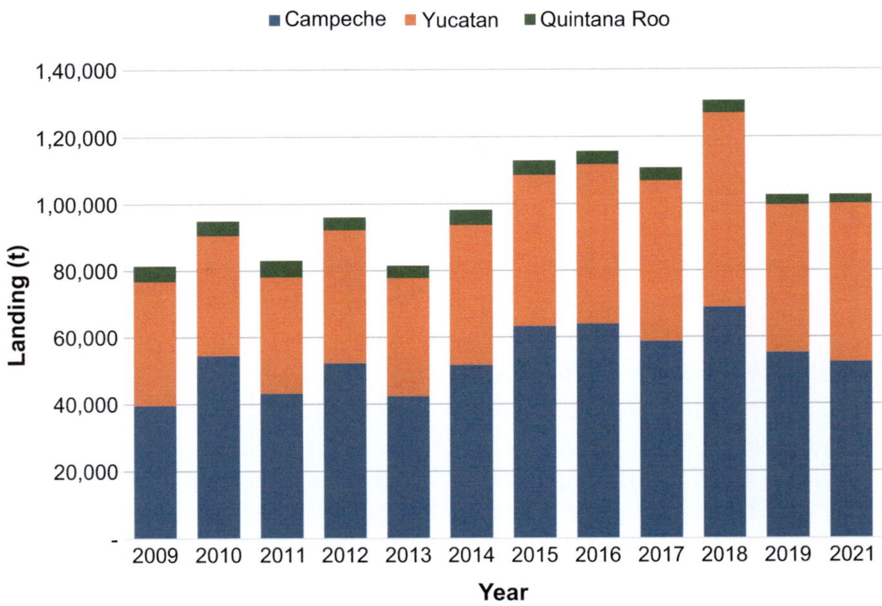

Fig. 3.2 Fishing landings in the three states of the *Yucataán Peninsula* between 2009 and 2021. (Database comes from CONAPESCA www.conapesca.gob.mx)

Table 3.1 Fisheries affected by the small-scale fleet in the *Yucatán Peninsula*, listing the number of families and associated species, the closed season periods in days, the minimum size of the target species, and specific management plans

Fishery	Family (#)	Specie (#)	Average landing 2018–2020 (t)	Closed season (Average days per year)	Minimum size[a]	Literature
Finfish	32	110	150,572	30	*Epinephelus morio* 36.3 cm (LT)	DOF (2014c)
Octopus	1	2	52,769	195	*Octopus maya* 11 cm (ML) *Octopus vulgaris* 11 cm (ML)	DOF (2014d)
Shrimps	1	7	23,760	180		DOF (2014a)
Snails	4	7	9588	210	*Triplofusus giganteus* 30 cm (LT); *Strombus costatus* 18 cm (LT): *Lobatus gigas* 20 cm (LT)	DOF (2014b)
Lobster	1	2	1068	120	*Panulirus argus* 22.3 cm (LT)	DOF (2014e)
Sharks	7	19	12,449	90		DOF (2015)
Crabs	1	2	17,037	45	*Callinectes sp.* 110 mm (LT)	DOF (1974, 2022)

LT Total Length, *ML* Mantle length, *CL* Chela length a

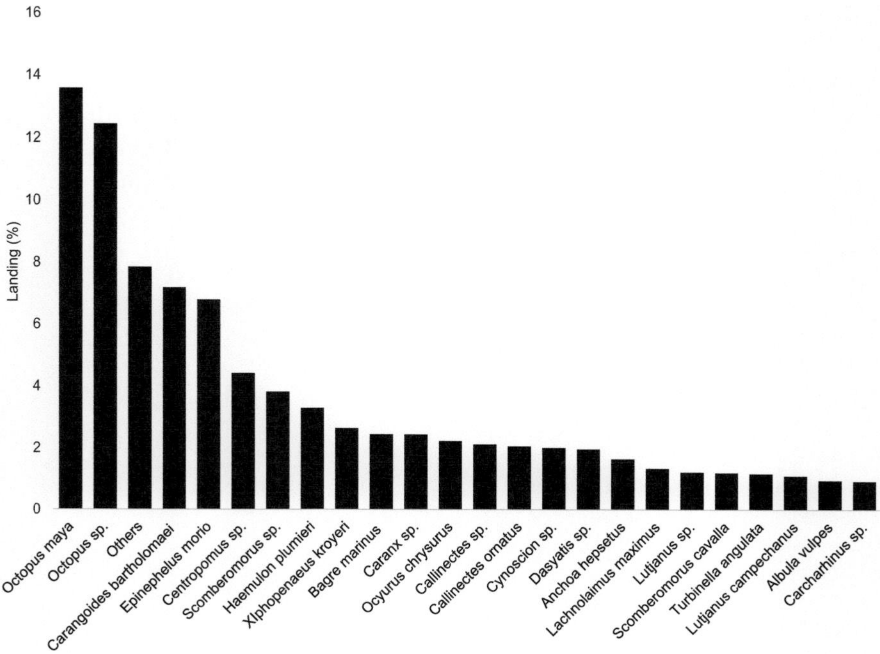

Fig. 3.3 Landing contribution per species reported by small-scale fisheries in the *Yucatán Peninsula* between 2006 and 2020. (Database comes from Coronado et al., 2022; CONAPESCA www.conapesca.gob.mx)

3.3.1.2 Fishing Effort and Fishing Permits

As of 2020, the total population of the *Yucatán Peninsula* was approximately five million inhabitants, with more than 70% residing in coastal communities or within 20 km of the coast (INEGI[1]). According to official statistics, more than 31,000 fishers and nearly 9000 vessels officially worked in the region at that time, with 93% of this workforce operating in the small-scale fishing sector (Table 3.2). It has been estimated that for every person holding a job focused on extractive fishing activities, there are at least five additional jobs associated with pre- and post-harvest activities that are held by men, women, and young people who began their working lives in the fishing sector (FAO, 2015). Therefore, an estimated 150,000 people in the *Yucatán Peninsula* could be directly or indirectly employed by the fishing sector (Coronado et al., 2022).

More than 3000 commercial fishing permits have been issued and distributed to the different fisheries operating around the *Yucatán Peninsula* (CONAPESCA, 2021; Coronado et al., 2020). Compared to the states of *Yucatán* and *Campeche*, the commercial fishing effort in *Quintana Roo* is relatively low in terms of the number of fishers, vessels, and catch volume. Instead, sport fishing is the predominant

[1] https://www.inegi.org.mx/

Table 3.2 Demographic and fishing sector characteristics of the *Yucatán Peninsula*

Characteristic	Campeche	Yucatan	Quintana Roo	Yucatan Peninsula
Coastal line length (km)	425	340	1176	1941
Population (million inhabit)	899,931	2′097,175	1′501,562	4′498,668
Total number of semi-industrial fishers[a]	645	5263	77	6165
Total number of small-scale fishers[a]	10,926	8558	2268	21,752
Total number of semi-industrial vessels (12–25 m)[a]	140	522	13	675
Total number of small-scale boats (7–12 m)[a]	5463	4279	1174	10,916
Landing of semi-industrial fisheries (t)[b]	11,606	25,520	814	37,940
Landing of small-scale fisheries (t)[b]	40,060	16,624	3605	60,289
Landing value for semi-industrial fisheries (million USD)[b]	23	56	1	80
Landing value for small-scale fisheries (million USD)[b]	57	35	11	103
Total number of seafood processing plants[a]	26	48	10	84
Total number of seafood exporting companies[a]	3	22	2	27

Population data until 2020 www.inegi.org.mx
[a]Data reported until 2020 (CONAPESCA, 2021) www.cofepris.gob.mx/
[b]Yearly average values reported between 2006 to 2020 www.conapesca.com

fishing activity in *Quintana Roo*, with sport fishing vessels operating in the same areas as those of the small-scale fishing fleet.

3.3.2 Recreational Fishing

Sport and recreational fishing are practiced extensively throughout the *Yucatán Peninsula* in numerous bays, lagoons, and estuarine systems, many of which are located within protected natural areas. In these contexts, recreational fishing has emerged as a sustainable alternative to commercial fishing, supporting new livelihood options for local communities (Ibáñez Pérez, 2011; Peña-Puch et al., 2021). These activities contribute to economic diversification and, within the framework of ecosystem management, promote coastal zone conservation (Poot-Lopez et al., 2018; Palomo-Cortés & Perez-Cobb, 2021; Quiñones-Peraza et al., 2023).

3.3.2.1 Exploited Resources

The composition of the main target species varies depending on where they are caught. The species caught in inshore areas differ from those caught offshore in open and deeper waters. Table 3.3 provides a list of key recreational fishing species by state and indicates whether each species is caught near the coast (Lt), in open waters (AA), or in both areas.

Table 3.3 Distributions of the main target species for sport and recreational fishing in coastal areas (Lt) and open waters (AA)

Specie Spanish/Scientific name	Campeche	Quintana Roo	Yucatán
Atún (*Thunnus* sp.)		AA	
Barracuda (*Sphyraena barracuda*)	Lt, AA	Lt, AA	Lt, AA
Chac chi (*Haemulon plumieri*)	Lt		Lt
Cojinuda (*Caranx crysos*)	Lt	Lt	Lt
Coronado (*Seriola dumerili*)	Lt	Lt, AA	Lt
Corvina pinta (*Cynoscion nebulosus*)	Lt	Lt	Lt
Dorado (*Coryphaena hippurus*)		AA	
Jurel (*Caranx hippos*)	Lt, AA	Lt, AA	Lt, AA
Macabí (*Albula vulpes*)	Lt	Lt	Lt
Marlin Azul (*Makaira nigricans*)		AA	
Marlin blanco (*Kajikia albida*)		AA	
Mero rojo (*Epinephelus morio*)	Lt		Lt
Mojarra (*Calamus bajonado*)	Lt		Lt
Palometa (*Trachinotus falcatus*)		Lt	
Pampano (*Trachinotus goodei*)	Lt, AA		Lt
Pargos (*Lutjanus* sp.)	Lt	Lt	Lt
Sierra, peto (*Scomberomorus* sp.)	AA	AA	AA
Vela (*Istiophorus albicans*)		AA	
Robalo (*Centropomus* sp.)	Lt	Lt	Lt
Sábalo (*Megalops atlanticus*)	Lt	Lt	Lt
Wahoo (*Acanthocybium solandri*)		AA	

Although some species are widely distributed and can be found in multiple areas, these are generally considered incidental when caught and not target species.

3.3.2.2 Size of the Recreational Fishing Fleet

Based on the information from the Secretariat of the Navy and the Captaincy and Maritime Affairs Unit of Mexico (SEMARNAT, 2023), more than 15,000 recreational and sport fishing vessels were registered from 1997 to the beginning of 2023. The vessels operating around the *Yucatán Peninsula* were divided into seven categories (Table 3.4).

The low number of boats in specific categories, such as catamarans or kayaks, suggests that the available information may be either incomplete or biased, making it difficult to accurately estimate the fishing effort. Assuming that motorboats and yachts were the main vessels involved in recreational and sport fishing activities, the fleet around the *Yucatán Peninsula* comprised 9839 vessels. Quintana Roo accounted for 46% of this fleet, reflecting its position as the primary tourist and recreational fishing destination in the Mexican Caribbean, followed by Yucatán (13%) and Campeche (4%). Given that each recreational vessel typically employed between two and four people as captains, guides, or support staff, recreational and sport fishing activities were responsible for 19,678 to 39,356 direct or indirect jobs.

Table 3.4 Types of recreational and sport fishing vessels

	Campeche		Yucatán		Quintana Roo	
Type of vessel	Number	%	Number	%	Number	%
Catamaran	4	0.006	9	0.004	20	0.002
Kayak[b]	11	0.017	94	0.037	1044	0.084
Fishing boat[b]	542	0.85	1978	0.775	6343	0.508
Jet ski	53	0.083	286	0.112	3593	0.288
Sailboat	5	0.008	52	0.02	480	0.038
Yacht[b]	19	0.03	125	0.049	831	0.067
Other types[a]	4	0.006	7	0.003	173	0.014
Total	637		2552		12,484	

Source: Prepared by the authors using data from the Secretariat of the Navy (SEMAR) and the Captaincy and Maritime Affairs Unit of Mexico

[a] Includes amphibious vehicles, barges, submersibles, among others

[b] Vessels commonly used for recreational fishing

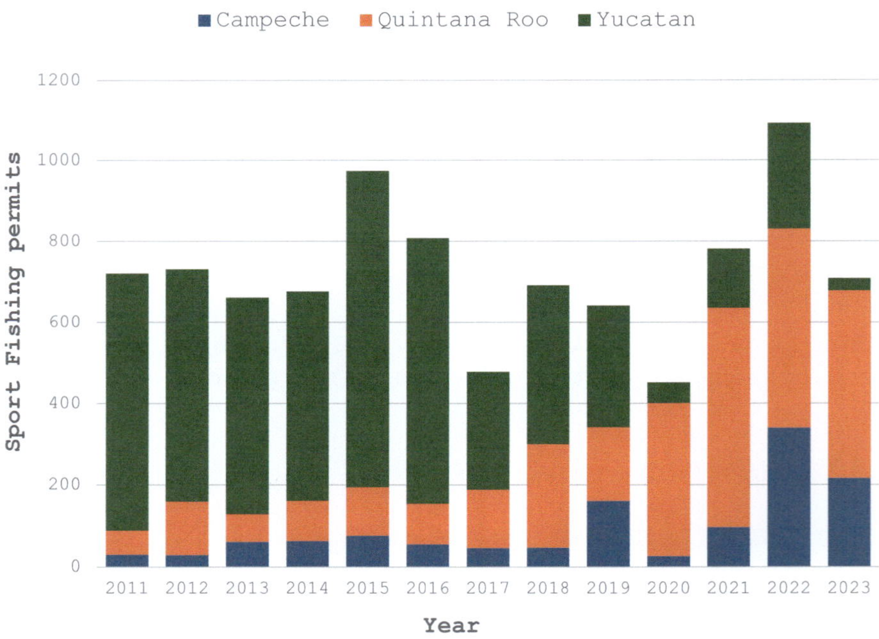

Fig. 3.4 Number of sport fishing permits issued over the past decade in the Yucatan Peninsula

3.3.2.3 Recreational and Sport Fishing Permits

A historical analysis of recreational fishing permits using data from CONAPESCA (2011–2023) revealed inconsistencies due to the lack of a reliable control and monitoring system for permit sales. Figure 3.4 shows the number of permits issued over the past decade. The data indicate an increase in the number of permits issued in Quintana Roo, accompanied by a slight rise in Campeche starting in 2021. In contrast, the number of permits issued in Yucatán decreased during the same period.

3.4 Mental Map of the Socio-Ecological System

Several studies of commercial and recreational fishing systems (Coronado et al., 2020, 2022; Kadagi et al., 2020; Palomo & Hernández-Flores, 2019, 2020; Palomo-Cortés & Perez-Cobb, 2021; Quiñones-Peraza et al., 2023; Salas et al., 2022) have resulted in the selection of key components that have been classified into social, natural resource, governance, external driver, and causal impact systems. For each component, the degree of relationship, which could be either positive or negative, was estimated, and its descriptors were identified (Table 3.5). These interactions were expressed in a cognitive map that included fishing communities, commercial and recreational fisheries, resource systems, and management components.

The mental map of small-scale and recreational fishing SES (Fig. 3.5) revealed a network of moderate complexity (degree 0.5), with the centrality of the system

Table 3.5 Descriptive table of the components and relationships pertaining to a coastal socio-ecological system (SES)

Component 1	Component 2	Interaction (+ or –)	Description
Social system (natural resource users)			
Coastal communities	Recreational fishing, small-scale fishing, or overfishing	+	Coastal communities benefit from small-scale and recreational fishing although population growth can hinder organization and increase pressure on natural resources
Natural and resource system			
Marine resources	Recreational fishing and small-scale fishing	+	Marine resource availability favors fisheries
Marine biodiversity	Nature tourism, marine resources, recreational fishing, and sustainability	+	Biodiversity favors tourism, maintains population stocks, and improves resource sustainability
Governance and legislation system			
Fisheries policies	Marine resources and small-scale fishing	+	Fisheries policies aim to protect resources and the interests of fishers
	Overfishing	–	The lack of efficient fishing policies can lead to resource overexploitation
Local governance	Sustainability of resources and fisheries policies	+	Local governance promotes sustainable fishing practices and compliance with fisheries laws
	Competition for resources	–	Failures in local governance, such as a lack of trust or collaboration, can lead to competition and conflicts in resource management
Recreational fishing regulations	Recreational fishing and resource sustainability	+	Compliance with sport fishing regulations promotes sustainable resource management
	Competition for resources	–	Lack of compliance with recreational fishing regulations leads to competition for resources and space among users

(continued)

Table 3.5 (continued)

Component 1	Component 2	Interaction (+ or –)	Description
Environmental legislation	Marine biodiversity	+	Environmental legislation favors the protection and conservation of ecosystems and marine biodiversity
External drivers of the system			
Demand for fishery resources	Small-scale fishing and overfishing	+	Demand for fishery resources affects recreational fishing and can result in overfishing
Nature tourism	Recreational fishing and coastal communities	+	Nature tourism benefits coastal communities and recreational activities
Sustainability of resources	Small-scale and recreational fishing	+	Resource sustainability is crucial for maintaining fishing activities
Impacts of system components			
Overfishing	Marine resources	–	Overfishing reduces marine resource population and affects small-scale fishing
Competition for resources	Small-scale and recreational fishing	–	Competition can produce conflicts that affect fisheries dynamics
Climate change	Biodiversity and marine resources	–	Global impact on the availability of resources and ecosystem functioning

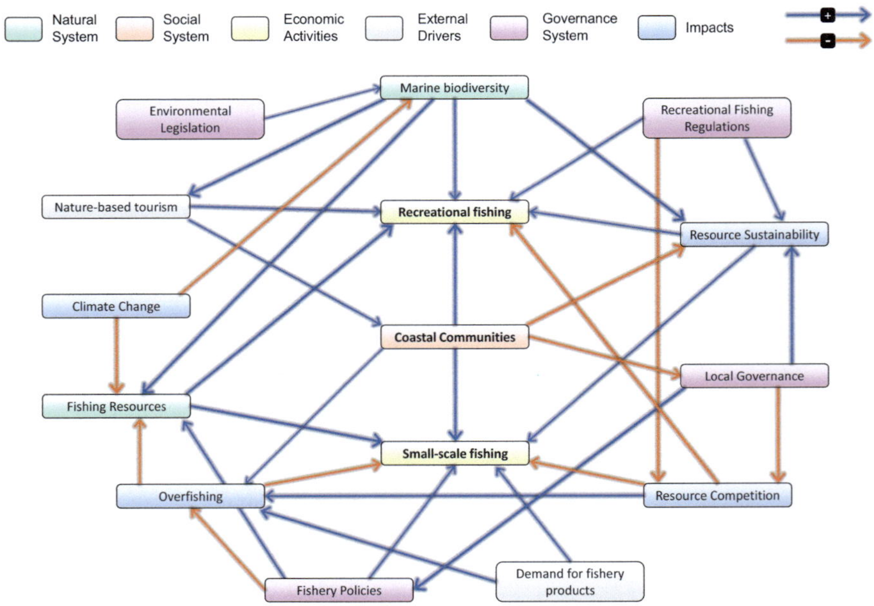

Fig. 3.5 Mental map of the structural relationships among the components of a socio-ecological system (SES) for recreational and small-scale fishing

concentrated in small-scale and recreational fishing (5.1 and 4.9). In addition, four components were identified as drivers, most notably governance and climate change. The natural system components exhibited high centrality (average 4.75), indicating their significant importance for resource availability and sustainability. These relationships were consistent with those reported in the literature and highlighted the complex coexistence of fishing activities in coastal SES.

3.5 Final Reflections: Coexistence, Conflicts, and Opportunities

In Mexico, fisheries and recreational management programs are primarily based on the biological characteristics of species, with social and economic aspects often being overlooked or neglected. As a result, systematic records of the dynamics of these activities are often lacking. In general, the regulatory strategies implemented have been unsuccessful due to resource declines, habitat degradation, uncontrolled increases in user numbers, and outdated records. Several authors have suggested that problems within fishing systems cannot necessarily be solved although they can be improved through governmental reforms, institutional redesigns, and organizational innovations that develop new tools encompassing a holistic and integrated vision for SES (Charles, 2013; Chuenpagdee & Jentoft, 2018; Ostrom, 2009).

The most common conflicts between recreational and commercial fisheries stem from competition for resources and operating spaces, or the implementation of regulations that infringe upon or otherwise affect the rights of fishers (Berkes, 2017; Charles, 2001; Kadagi et al., 2020). Spatial management tools, such as Marine Protected Areas or regulations on species, fishing gear, closed periods, and catch rates, can reduce fishing pressure on resources; however, the implementation of these regulations often excludes the participation of fishers or other actors and stakeholders (Arlinghaus et al., 2021; Espinoza-Tenorio et al., 2011; Salas et al., 2022; Voyer et al., 2017).

In Campeche, the hydrocarbon industry has restricted access to traditional fishing grounds. Nonetheless, this industry has generated pollution and introduced tensions between communities and oil companies, creating a complex institutional context (Coronado et al., 2024; Espinoza Tenorio, 2019). In Quintana Roo, the growth of recreational fishing has led to conflicts due to the increase in the number of vessels, necessitating an analysis of the system's carrying capacity to prevent ecosystem deterioration and a subsequent decline in the quality of recreational services (Palomo & Hernández-Flores, 2020).

Currently, fishing refuge zones are key tools that foster participation and co-management among users and local authorities, as seen in *Celestún, Yucatán*, and some *Sian Ka'an* communities in Quintana Roo. In these areas, the objectives have focused on recovering habitats and resources, as well as involving fishers in the design of rules for common-use areas and their management. Approaches that focus on co-responsibility strengthen social capital and integrate local knowledge into decision-making processes (Defeo et al., 2016; Schlüter et al., 2019), thereby informing the design of sustainable management policies.

The socioeconomic conditions in the states of the *Yucatán Peninsula* are diverse, complex, and dynamic, with evident differences between communities within the same state. The small-scale fishing fleet coexists with the sport and recreational fishing fleet, resulting in shared resources and stakeholders who are involved in both types of fishing. However, these forms of fishing have historically been analyzed in isolation in research and treated as such in management strategies despite their evident interconnectedness, which has generated social conflicts and environmental consequences.

This chapter has explored the socio-ecological complexity of two highly interconnected activities that are officially managed and evaluated separately. The information in this chapter provides a baseline for further studies and highlights the need to jointly assess and manage small-scale and recreational fishing in the *Yucatán Peninsula*. This chapter also opens an avenue for future reflections on the importance of comprehensive assessments that encompass the various productive activities in the southern Gulf of Mexico, namely those related to the oil industry and sport and recreational fishing.

References

Arlinghaus, R., Aas, Ø., Alós, J., Arismendi, I., Bower, S., Carle, S., Czarkowski, T., Freire, K. M. F., Hu, J., Hunt, L. M., Lyach, R., Kapusta, A., Salmi, P., Schwab, A., Tsuboi, J., Trella, M., McPhee, D., Potts, W., Wołos, A., & Yang, Z.-J. (2021). Global participation in and public attitudes toward recreational fishing: International perspectives and developments. *Reviews in Fisheries Science & Aquaculture, 29*(1), 58–95. https://doi.org/10.1080/23308249.2020.1782340

Arreguín-Sánchez, F., & Arcos-Huitrón, E. (2011). La pesca en México: Estado de la explotación y uso de los ecosistemas. *Hidrobiológica, 21*(3), UNAM.

Ayala-Pérez, L. A., Terán-González, G. J., Flores-Hernández, D., Ramos-Miranda, J., & Sosa-López, A. (2012). Variabilidad espacial y temporal de la abundancia y diversidad de la comunidad de peces en la costa de Campeche, México. *Latin American Journal of Aquatic Research, 40*(1), 63–78.

Berkes, F. (2017). Environmental governance for the Anthropocene? Social-ecological systems, resilience, and collaborative learning. *Sustainability, 9*(7), Article 7. https://doi.org/10.3390/su9071232

Brownscombe, J. W., Hyder, K., Potts, W., Wilson, K. L., Pope, K. L., Danylchuk, A. J., Cooke, S. J., Clarke, A., Arlinghaus, R., & Post, J. R. (2019). The future of recreational fisheries: Advances in science, monitoring, management, and practice. *Fisheries Research, 211*, 247–255. https://doi.org/10.1016/j.fishres.2018.10.019

Charles, A. (2013). Governance of tenure in small-scale fisheries: Key considerations. *Land Tenure Journal, 1*, 9–38.

Charles, A. T. (2001). *Sustainable fishery systems*. Wiley-Blackwell.

Chuenpagdee, R., & Jentoft, S. (2018). Transforming the governance of small-scale fisheries. *Maritime Studies, 17*(1), 101–115. https://doi.org/10.1007/s40152-018-0087-7

CONAPESCA. (2021). *Anuario Estadístico de Acuacultura y Pesca.* Gobierno de México. Secretaria de Agricultura y Desarrollo Rural.

Coronado, E., Chuenpagdee, R., Salas, S., Torres-Irineo, E., Palomo-Cortés, L., & Espinoza-Tenorio, A. (2022). Exploración socioecológica de la pesca artesanal del sur del Golfo de México por medio del marco de análisis de gobernanza interactiva. *Ciencia Pesquera, 30*(1–2), 79–95.

Coronado, E., Salas, S., Torres-Irineo, E., & Chuenpagdee, R. (2020). Disentangling the complexity of small-scale fisheries in coastal communities through a typology approach: The case study of the Yucatan Peninsula, Mexico. *Regional Studies in Marine Science, 36*, 101312. https://doi.org/10.1016/j.rsma.2020.101312

Coronado, E., Zepeda-Domínguez, J. A., Espinoza-Tenorio, A., Santamaria, D. C., Ramos-Muñoz, D., & Monzón-Alvarado, C. (2024). Institutional mapping and implementation of the sustainable development goals across co-existing industries: The case of the fisheries-oil system in Mexico. *The Extractive Industries and Society, 17*, 101390. https://doi.org/10.1016/j.exis.2023.101390

Dahlet, L. I., Selim, S. A., & van Putten, I. (2023). A review of how we study coastal and marine conflicts: Is social science taking a broad enough view? *Maritime Studies, 22*(3), 29. https://doi.org/10.1007/s40152-023-00319-z

Defeo, O., Castrejón, M., Pérez-Castañeda, R., Castilla, J. C., Gutiérrez, N. L., Essington, T. E., & Folke, C. (2016). Co-management in Latin American small-scale shellfisheries: Assessment from long-term case studies. *Fish and Fisheries, 17*(1), 176–192. https://doi.org/10.1111/faf.12101

DOF. (1974). *Acuerdo por el que se fija la talla mínima legal para la explotación de jaiba en el Golfo de México. Diario Oficial de la Federación. México 18 de abril de 1974.*

DOF. (2014a). *Acuerdo por el que se da a conocer el Plan de Manejo Pesquero de camarón siete barbas (Xiphopenaeus kroyeri) en las costas de los estados de Campeche y Tabasco. Diario Oficial de la Federación. México 31 de marzo de 2014.*

DOF. (2014b). *Acuerdo por el que se da a conocer el Plan de Manejo Pesquero de las especies de caracol pateburro o tomburro (Turbinella angulata); sacabocado o lix (Busycon perversum); rojo o chacpel (Pleuroploca gigantea); campechana (Fasciolaria tulipa); blanco o lanceta (Strombus costatus); canelo o boxeador (Strombus pugilis); molón o nolón (Melongena melongena) y chivita o noloncito (Melongena corona bispinosa) del litoral del Estado de Campeche. Diario Oficial de la Federación. México 25 de marzo de 2014.*

DOF. (2014c). *Acuerdo por el que se da a conocer el Plan de Manejo Pesquero de Mero (Epinephelus morio) y especies asociadas en la Península de Yucatán. Diario Oficial de la Federación. México 25 noviembre 2014.*

DOF. (2014d). *Acuerdo por el que se da a conocer el Plan de Manejo Pesquero de pulpo (O. Maya y O. Vulgaris) del Golfo de México y Mar Caribe. Diario Oficial de la Federación. 28 de marzo de 2014.*

DOF. (2014e). *Acuerdo por el que se da a conocer el Plan de Manejo Pesquero para la langosta espinosa (panulirus argus) de la Península de Yucatán. Diario Oficial de la Federación. México 13 de marzo de 2014.*

DOF. (2015). Proyecto de Modificación a la Norma Oficial Mexicana NOM-029-PESC-2006. *Pesca responsable de tiburones y rayas. Especificaciones para su aprovechamiento. Diario Oficial de la Federación. México febrero, 2015.*

DOF. (2022). *Carta Nacional Pesquera. Diario Oficial de la Federación, México. 26 de julio de 2022.*

DOF. (2024). *Reglamento de la Ley de Pesca.* Publicado en el Diario Oficial de la Federación el 29 de septiembre de 1999. Última reforma publicada en el DOF el 28 de enero de 2004.

Espinoza Tenorio, A. (2019). ¿Mancharse las manos de negro? El dilema ético de la investigación en territorios petrolizados. *Revista mexicana de ciencias políticas y sociales, 64*(237), 183–209. https://doi.org/10.22201/fcpys.2448492xe.2019.237.67179

Espinoza-Tenorio, A., Espejel, I., Wolff, M., & Alberto Zepeda-Domínguez, J. (2011). Contextual factors influencing sustainable fisheries in Mexico. *Marine Policy, 35*(3), 343–350. https://doi.org/10.1016/j.marpol.2010.10.014

Espinoza-Tenorio, A., Moreno-Báez, M., Pech, D., Villalobos-Zapata, G. J., Vidal-Hernández, L., Ramos-Miranda, J., Mendoza-Carranza, M., Zepeda-Domínguez, J. A., Alcalá-Moya, G., Pérez-Jiménez, J. C., Rosete, F., León, C., & Espejel, I. (2014). El ordenamiento ecológico marino en México: Un reto y una invitación al quehacer científico. *Latin American Journal of Aquatic Research, 42*(3), 386–400. http://www.scielo.cl/scielo.php?script=sci_abstract&pid=S0718-560X2014000300001&lng=es&nrm=iso&tlng=es

FAO. (2015). *Voluntary guidelines for securing sustainable small-scale fisheries in the context of food security and poverty eradication*. FAO. https://www.fao.org/documents/card/en/c/I4356EN

Gray, S. A., Gray, S., Cox, L. J., & Henly-Shepard, S. (2013). Mental modeler: A fuzzy-logic cognitive mapping modeling tool for adaptive environmental management. In *2013 46th Hawaii international conference on system sciences* (pp. 965–973). https://doi.org/10.1109/HICSS.2013.399

Gray, S. A., Zanre, E., & Gray, S. R. J. (2014). Fuzzy cognitive maps as representations of mental models and group beliefs. In E. I. Papageorgiou (Ed.), *Fuzzy cognitive maps for applied sciences and engineering: From fundamentals to extensions and learning algorithms* (pp. 29–48). Springer. https://doi.org/10.1007/978-3-642-39739-4_2

Ibáñez Pérez, R. M. (2011). Pesca deportiva-recreativa como un atractivo turístico en México: Caracterización, estimación de su demanda futura y efecto multiplicador a otros sectores. *TURYDES: Revista sobre Turismo y Desarrollo local sostenible, 4*(10), 30. https://dialnet.unirioja.es/servlet/articulo?codigo=9157091

Islebe, G. A., Calmé, S., León-Cortés, J. L., & Schmook, B. (Eds.). (2015). *Biodiversity and conservation of the Yucatán Peninsula*. Springer. https://doi.org/10.1007/978-3-319-06529-8

Kadagi, N. I., Wambiji, N., & Swisher, M. E. (2020). Potential for conflicts in recreational and artisanal billfish fisheries on the coast of Kenya. *Marine Policy, 117*, 103960. https://doi.org/10.1016/j.marpol.2020.103960

Monroy-Garcia, C., Gutiérrez-Perez, C., Medina-Quijano, H., Uribe-Cuevas, M., & Chable-Ek, F. (2019). *La actividad pesquera de la flota ribereña en el estado de Yucatán: Pesquería de escama*. Instituto Nacional de Pesca y Acuacultura.

Ostrom, E. (2007). Institutional rational choice: An assessment of the institutional analysis and development framework. In P. Sabatier (Ed.), *Theories of the policy process* (pp. 21–64). Westview Press. https://digitalcommons.usu.edu/unf_research/60

Ostrom, E. (2009). A general framework for analyzing sustainability of social-ecological systems. *Science, 325*(5939), 419–422. https://doi.org/10.1126/science.1172133

Palomo, L. E., & Hernández-Flores, A. (2019). Application of the Ostrom framework in the analysis of a social-ecological system with multiple resources in a marine protected area. *PeerJ, 7*, e7374. https://doi.org/10.7717/peerj.7374

Palomo, L. E., & Hernández-Flores, A. (2020). Integrating a spatial model and decision theory towards optimal boating density and carrying capacity in a recreational fishery. *Marine Policy, 112*, 103740. https://doi.org/10.1016/j.marpol.2019.103740

Palomo-Cortés, L., & Perez-Cobb, A. U. (2021). *Economic impact of flats fishing in Quintana Roo, México* (p. 20). Bonefish and Tarpon Trust. https://www.bonefishtarpontrust.org/downloads/research-reports/stories/economic-assess-flats-fishing-quintana-roo.pdf

Peña-Puch, A. D. C., Pérez-Jiménez, J. C., Espinoza-Tenorio, A., Méndez-Loeza, I., & Williams-Jara, G. M. (2021). The sustainability of small-scale fisheries in oil-producing sections of the Gulf of Mexico. *The Extractive Industries and Society, 8*(4), 100975. https://doi.org/10.1016/j.exis.2021.100975

Perez-Cobb, A. U., Arce-Ibarra, A. M., García-Ortega, M., Valdéz-Moreno, M., & Azueta, J. O. (2014). Artisanal recreational fisheries: Using a combined approach to fishery assessment aimed at providing insights for fishery managers. *Marine Resource Economics, 29*(2), 89–109. https://doi.org/10.1086/676838

Poot-Lopez, G., Rubio-Bueno, S., Villegas-Hernandez, H., Gonzalez-Salas, C., Guillen-Hernandez, S., & Díaz-Gamboa, E. (2018). Catch composition, effort, and selectivity of fishes of recreational fishing in Yucatán, Mexico. *Global Journal of Science Frontier Research, 18*(I1), 1–11. https://journalofscience.org/index.php/GJSFR/article/view/2293

Quiñones-Peraza, A., Villegas-Hernández, H., Guillén-Hernández, S., & Poot-López, G. R. (2023). Recreational fishing and angling tournaments in the Yucatan coast (Campeche Bank, Mexico): Social and biological dimensions. *Regional Studies in Marine Science, 61*, 102897. https://doi.org/10.1016/j.rsma.2023.102897

Ramos-Miranda, J., Cabrera, M. A., Salas, S., López-Rocha, J. A., & Flores-Hernández, D. (2021). *Especies comerciales de la pesca artesanal en la península de Yucatán. Universidad Autónoma de Campeche. Centro de Investigación y de Estudios Avanzados del IPN Unidad Mérida.* Universidad Nacional Autónoma de México.

Salas, S., Núñez, A., Cepeda-González, M. F., Ramos-Miranda, J., Cabrera, M. A., Coronado, E., López-Rocha, J. A., & Torres-Irineo, E. (2022). *Pesca Artesanal en la Península de Yucatán. Contexto socio-económico y bienestar comunitario. Cinvestav-Unam-Epomex-UAC*. Universidad Autónoma de Campeche.

Salas, S., Torres-Irineo, E., & Coronado, E. (2019). Towards a métier-based assessment and management approach for mixed fisheries in southeastern Mexico. *Marine Policy, 103*, 148–159. https://doi.org/10.1016/j.marpol.2019.02.040

SEMARNAT, 2023. Relación de embarcaciones registradas para actividades recreativas y deportivas. Recuperado el 14/04/2023 de. https://datos.gob.mx/

Schlüter, M., Haider, L., Lade, S., Lindkvist, E., Martin, R., Orach, K., Wijermans, N., & Folke, C. (2019). Capturing emergent phenomena in social-ecological systems: An analytical framework. *Ecology and Society, 24*(3). https://doi.org/10.5751/ES-11012-240311

Schmitter-Soto, J. J., Vázquez Yeomans, L., Aguilar Perera, A., Curiel Mondragón, C., & Caballero Vázquez, J. A. (2009). Lista de peces marinos del Caribe mexicano. *Anales del Instituto de Biología serie Zoología, 71*(002), Article 002. http://www.revistas.unam.mx/index.php/zoo/article/view/7237

Voyer, M., Barclay, K., McIlgorm, A., & Mazur, N. (2017). Connections or conflict? A social and economic analysis of the interconnections between the professional fishing industry, recreational fishing, and marine tourism in coastal communities in NSW, Australia. *Marine Policy, 76*, 114–121. https://doi.org/10.1016/j.marpol.2016.11.029

Chapter 4
Mariculture and the Oil Industry in the State of Tabasco, Mexico

N. Vite-García, S. J. Serrano-Guzmán, L. Rangel-López, M. A. Perera-García, and C. Ruiz-Méndez

4.1 Introduction

Marine aquaculture or mariculture is an activity that involves the production of aquatic organisms through their cultivation, within marine ecosystems with open waters (offshore or unprotected coasts), semi-enclosed (inlets and bays), or in protected waters (such as estuaries or coastal lagoons), whose main purpose, but not the only one, is food production. The oil industry is the economic sector dedicated to the exploration, extraction, refining, distribution, and commercialization of petroleum and its derivatives. This indfustry encompasses a wide range of activities, from the exploration of crude oil deposits in the subsoil to the refining of oil into products

N. Vite-García (✉)
Universidad de Salamanca, PhD Programme Education in the Knowledge Society, Salamanca, España

Universidad Juárez Autónoma de Tabasco/ División Académica de Ciencias Agropecuarias, Villahermosa, Tabasco, México
e-mail: metodio.vite@ujat.mx

S. J. Serrano-Guzmán
Universidad del Mar, Puerto Ángel, Oaxaca, México
e-mail: sserrano@angel.umar.mx

L. Rangel-López · M. A. Perera-García
Universidad Juárez Autónoma de Tabasco/ División Académica de Ciencias Agropecuarias, Villahermosa, Tabasco, México
e-mail: lenin.rangel@ujat.mx; martha.perera@ujat.mx

C. Ruiz-Méndez
Universidad de Salamanca, Salamanca, España
e-mail: camilo@usal.es

A. Espinoza-Tenorio et al. (eds.), *The Fishing and Oil Industries in Coexistence*,
https://doi.org/10.1007/978-3-032-02621-7_4

such as gasoline, diesel, and petrochemicals. The oil industry plays a crucial role in the local and global economy by supplying energy for transportation, power generation, and various industrial applications, its export, etc.

The interaction between mariculture and the oil industry can become a daily occurrence if we consider the present circumstances, both globally and locally, and the possible scenarios that humanity will face in the coming decades. Humanity faces multiple challenges, with climate change (CC) being perhaps the most pressing. Among the most evident consequences we can mention the increase in temperature, the decrease in water resources and the considerable frequency of anomalous climate changes, the detriment of biodiversity, vicissitudes and alterations in agro-ecological, and agro-food systems (food production), as well as threats to marine environments, social consequences such as the exodus of large population centers and damage to human health, among many other examples (Day et al., 2008; Cruz & Martínez, 2015; Bruno et al., 2018; Esperón-Rodríguez et al., 2019; Chapman et al., 2020; Leal-Filho et al., 2022). The effects of climate change are being felt in various parts of the world's productive environments, natural systems, and communities (Cárdenas, 2010; Porter et al., 2020; Maulu et al., 2021).

One of the main causes of climate change is the anthropogenic emission of greenhouse gases resulting from the burning of fossil fuels. Another reason is related to food production; several agricultural activities significantly contribute to the emission of greenhouse gases (GHG), which are the primary culprits of climate change (Etim et al., 2013; Panchasara et al., 2021). Approximately 40% of the planet's surface is used for food production (Crist et al., 2017). This leads to intensive use of soils and their resources, but also significant use of water and impacts on nitrogen cycles and biodiversity (Foley et al. 2011).

Climate change presents a valuable opportunity for economic, social, and environmental transformation, which is necessary for the integral development of our society and the prevention of environmental deterioration. This awareness should be promoted to drive social change and lead to the solution of the current environmental crisis (Cruz & Martínez, 2015). In this scenario, the Sustainable Development Goals (SDGs) of the UN Agenda 2030 are an initiative to transform society (Gamboa-Bernal, 2015). The SDGs that help combat CC are #1, #2, #3, #6, #7, #12, #13, and #15 (Leal-Filho et al., 2022).

Mariculture is presented as a sustainable food production activity that partially frees inland systems from the production of good-quality protein, takes pressure off fishing systems, and serves as a brake and monitor for the industry, including the oil industry. However, for this to become a reality, it is necessary that the oil industry is carried out with maximum safety measures to minimize environmental impacts, while the mariculture activity has to take into account spaces, good practices, and the management of aquatic organisms in order to obtain products that can satisfy both local and global food markets.

The objective of this paper is to review and make visible the importance of encouraging mariculture or marine aquaculture projects as an alternative and sustainable source of food production in an environment and social environment dominated by the oil industry, as is the case in the Gulf of Mexico (GoM), particularly in the state of Tabasco. On multiple occasions, the cultivation of aquatic organisms has been discussed as a way to obtain good-quality food without having a negative impact on the environment. As an example, the oyster *Crassostrea virginica* has been used as sentinel species, that is, organisms that are indicators of the impact and restoration of ecosystems (Volety et al., 2009), in addition to the fact that, like other bivalves, it can provide ecosystem services such as improving water quality and stabilizing the coastline, seston filtration and carbon fixation, measurement of parameters related to microbiological contamination, habitat services for the benefit of biodiversity, as well as social and cultural services that provide benefits and identity to the local population (Coen et al., 2007; Grabowski et al., 2012; Guimarães-Filho et al., 2022; Thomas et al., 2022).

The relationship between marine aquaculture and oil production is a delicate balance between economic benefits and environmental risks. Cooperation and careful management are essential to minimize negative impacts and explore opportunities for synergies that benefit both industries and the marine environment in general. Due to the difference in the size of the industries, mariculture can do little to compete for adequate space and conditions for sustainable development, but it is not an activity that should be abandoned. The oil industry cannot develop in a disproportionate manner without taking into account measures for environmental protection, prevention, mitigation, restoration and/or compensation, and in this context aquaculture can be a measure for this purpose, with actions such as the establishment of protection and sustainable use zones, in which mariculture with low environmental impact can be developed.

4.2 The Coastal Region of Tabasco

According to Carranza et al. (1975) (cited by Toledo-Ocampo, 2005), the coastal strip of the state of Tabasco is 192 km long and corresponds to one of the four major continental morphotectonic units of the GoM. This region is characterized by the presence of deltas, including those formed by the *Tonalá, Grijalva-Mezcalapa, San Pedro*, and *San Pablo* rivers (Toledo-Ocampo, 2005). Rainfall discharges from the Chiapas and Guatemala mountain ranges, form the coastal lagoon systems in Tabasco's coastal plain: *Laguna El Yucateco, Laguna El Carmen-Pajonal-La Machona-Redonda, Laguna Tupilco, Laguna de Mecoacán*, and the *Julivá-Santa Anita Lagoon* Complex, which cover an area of 31,902. 7 ha and 966,274 ha of wetlands in the *Chontalpa* Region and the *Pantanos de Centla* Biosphere Reserve (Bozada-Robles, 2006).

The area has a climate type classified as Am(f), characterized by warm, humid conditions with intense precipitation in summer (INEGI, 2025). From October to January–February is the "*nortes*" season, associated with the movement of "cold fronts"; the relatively dry months are from January to April (García-Cubas et al., 1990). Some effects of climate change are evident in many coastal communities along the Tabasco coast, such as temperature elevation, flooding, and shoreline erosion, which have become one of the main vulnerabilities of these communities (Castillo-Cruz & Medrano-Pérez, 2023; de los Santos Gonzalez et al., 2023; Andrade-Velázquez et al., 2024).

Along the coastline of the state of Tabasco, there are numerous lagoons, estuaries, and wetlands that do not have a direct connection to the sea, but rather through channels activated by the alternating movement of the tides. Daily tides allow the ebb and flow of water masses, nutrients, and materials from the wetlands and mangroves that border these bodies of water to the GoM and *vice versa*. These tidal movements make possible the migration of a very varied community of plant and animal species seeking refuge, food, breeding and/or reproduction areas (Toledo-Ocampo, 2005).

In addition to its exchange with the adjacent sea, Tabasco's coastal ecosystems are significantly influenced by freshwater runoff from the Chiapas and Guatemala mountain ranges, known as the *Grijalva-Usumacinta* pluvial system (Castillo-Cruz & Medrano-Pérez 2023). This system has the second-highest rainfall discharge on Mexico's Atlantic slope, with an approximate annual load of 105,000 Hm^3 of water (Bozada-Robles, 2006). In coastal lagoons, river runoff has a direct effect on salinity, nutrient content, and other factors (Hickey et al., 2010).

In relation to atmospheric and water body temperature, in the coastal zone of the state of Tabasco, oscillation intervals of this variable can occur within the same year, even in the course of days or hours. However, in tropical and subtropical latitudes such as this one, there are much narrower annual intervals of average temperature (3–5 °C) than in temperate latitudes (10–5 °C), where the seasons are marked (Thurman, 1991; Bouman et al., 2010). This favors the breeding of a large number of fish, crustacean, and mollusk species, from which spawning can be obtained more or less continuously throughout the year, allowing the culture cycles to be distributed to achieve higher production than in temperate latitudes. In recent years, there has been an increase in atmospheric temperature in the state due to climate change (Andrade-Velázquez et al., 2024), which can have a significant impact on aquaculture because many aquatic species are affected when the temperature rises above 30 °C.

Another factor that must be taken into account for aquaculture practice is the concentration of dissolved oxygen in the water. No environmental factor with as much ecological importance for coastal and estuarine ecosystems has changed as rapidly as dissolved oxygen (Díaz, 2001; Ripley & Foran, 2007; Breitburg et al., 2018; Pezner et al., 2023; Chan et al., 2025). Over the last five decades, with particular intensity in the last two, hypoxia (or low dissolved oxygen concentration in the water), has increased in coastal areas in terms of frequency, severity, and areas affected (Díaz & Rosenberg, 1995; Wu, 2002; Bianchi et al., 2010; Rabalais et al., 2010; Pezner et al., 2023), and there is sufficient evidence to think that this problem will worsen in the coming years (Wu, 2002; Vaquer-Sunyer & Duarte, 2008; Chan et al., 2025).

Sediments in coastal systems are the result of a combination of erosional, transport, and depositional (accretion) processes that, together with chemical and biological processes, determine their most important characteristics. The biological community present in sediments fulfills specific functions within the aquatic ecosystem (Valiela, 1995; Petro et al., 2017). In sediments, a large part of the processes of decomposition and remineralization of organic matter take place (Thurman, 1991; Valiela, 1995).

Stochastic events, such as floods, storms, hurricanes, and strong continental freshwater runoff, among others, affect the transport and accretion of sediments along the coast. The large transport of inland organic matter, as well as contaminants (hydrocarbons, pesticides, herbicides, and fertilizers, among many others), causes important changes in the sediment biological community, which can have repercussions on the total ecosystem capacity to metabolize organic matter (Chen et al., 2019; Di Cesare et al., 2020). Particularly in Tabasco state, sediments may be subjected to a strong impact from the oil industry since they are the deposits where an important part of the contaminants from spills and other sources finally accumulate (Atoufi & Lampert, 2020).

The ecological conditions of Tabasco's coastal ecosystems are ideal for the practice of mariculture, both in the estuaries and coastal lagoons, as well as in the marine portion. These conditions, which favor high productivity, allow the development of all stages of fish, crustaceans and mollusks cultivation throughout the year. However, the variety in the cultivation of these three main groups of aquatic organisms has different social, economic, and environmental impacts that must be considered when choosing species and sites, so that the activity can be sustainable and represent a real alternative for the coastal population of the state. Particularly in Tabasco, the coexistence of this activity with the state's largest productive activity, the oil industry, along with all its impacts and interactions, must also be considered.

4.3 Mariculture and the Oil Industry

The coastal waters of the state of Tabasco, bathed by the southern currents of the GoM, represent an ecosystem of extraordinary biological richness that has historically supported an important fishing activity and, in more recent decades, the development of marine aquaculture (CONAPESCA, 2020). However, this fragile ecological balance is now severely compromised by the cumulative impact of the oil industry, whose operations have transformed the coastal landscape and water quality to the point of jeopardizing the very viability of aquaculture production in the region (Pinkus-Rendón & Pacheco Castro, 2013). Tabasco has one of the most complex and productive hydrological systems in Mexico (CONABIO, 2018). Its network of coastal lagoons, estuaries, and mangrove areas provides the ideal habitat for numerous commercially important species, from shrimp and oysters to diverse varieties of fish (SAGARPA, 2019). According to data from CONAPESCA (2020), prior to 2000, these waters accounted for approximately 18% of the national aquaculture production, with special relevance in the cultivation of white shrimp (*Litopenaeus vannamei*) and American oysters (*C. virginica*). However, this productive vocation clashes head-on with another reality: Tabasco is the heart of the Mexican oil industry (Javier-Jiménez et al., 2025). The *Sonda de Campeche* zone, located off its coast, is home to the most important marine reservoir in the country (PEMEX, 2021), while on land, a network of more than 5,000 km of pipelines traverses aquatic ecosystems (Muñoz et al., 2019). This forced coexistence between oil and aquaculture has generated an environmental conflict whose dimensions we are only just beginning to understand in their full magnitude (Chan-Quijano et al., 2020).

Official statistics paint an alarming picture. According to records from the National Hydrocarbons Commission (CNH, 2021), between 2000 and 2015, 395 oil spills occurred in Tabasco, accounting for 40% of the national total. The volume spilled exceeded 48,000 barrels of crude oil, with direct consequences for the state's main aquaculture areas (Muñoz et al., 2019). But far from improving, the situation has worsened: recent data from PEMEX (2023) reveal that between 2015 and 2022, 2,250 oil incidents were recorded in the region (Fig. 4.1), equivalent to an average of five accidents every week (Arias, 2025). Due to the state's extensive river system,

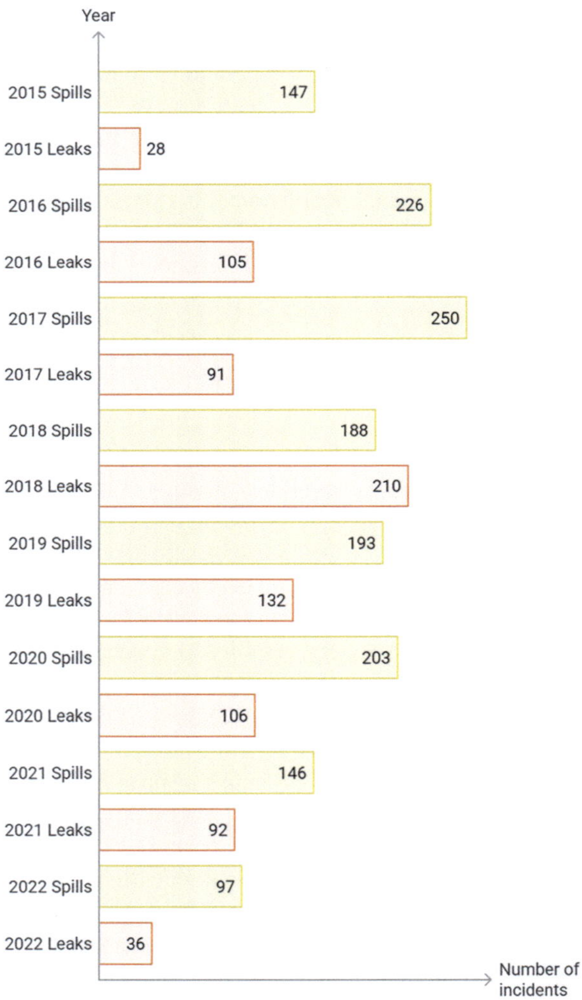

Fig. 4.1 Number of oil spills and leaks from 1915 to 1922 in the state of Tabasco. (Source: PEMEX Transparency, cited by Arias, 2025)

oil spills have a devastating effect on aquatic biotic communities, putting at risk not only all attempts at aquacultural and fishing activities in these bodies of water, but also the very integrity of the ecosystems. Among the most common causes of oil spills in the state are the following: Inadequate handling practices of crude oil and its products, clandestine outlets in pipeline areas; spills caused by road or railroad accidents; corrosion of fuel storage tanks; corrosion or rupture of surface and subway pipelines; failures due to lack of maintenance of facilities; infrastructure failure caused by extreme natural phenomena (winds, hurricanes, and earthquakes, among others); human errors in operation; abandoned underground facilities; rupture of oily drains; accidental or deliberate discharges of process wastewater into canals, canals, rivers, streams, and rivers; accidental or deliberate discharges of process wastewater into canals, rivers, streams and rivers; abandoned subway facilities; rupture of oily drains; accidental or deliberate discharges of process wastewater into canals and rivers; operation of facilities with intensive use of hydrocarbons on natural terrain without slabs or impermeable dikes; absence of infrastructure replacement programs at the end of their useful life; absence of oily drains; deficiencies in well drilling activities and, in general, lack of a culture of prevention (Saval et al., 2004; Javier-Jiménez et al., 2025).

Each of these spills triggers a chain of lethal effects for the aquaculture industry. First, there are acute effects, such as the formation of hydrocarbon films on the water surface, which block gas exchange and drastically reduce the oxygen available to cultured organisms (Bennani et al., 1995). Studies from the Autonomous University of Campeche (UAC, 2019) document mortalities greater than 70% in shrimp larvae exposed to crude oil concentrations below 1 ppm. In addition, there are chronic effects derived from the accumulation of polycyclic aromatic compounds (PAHs) in sediments, which are then incorporated by filter-feeding organisms such as oysters (Yogui & Sericano, 2009). Analyses by CINVESTAV (2020) detected concentrations of benzo(a)pyrene, a potent carcinogen, up to 15 times higher than the permitted limits in mollusks collected in *Laguna de Términos*. Added to all this are the economic and social impacts (Upton, 2011; García-Cuéllar et al., 2018).

The problem, however, transcends catastrophic events such as spills. The routine operation of the oil industry generates less obvious, but equally devastating systemic impacts (Weis & Weis, 2019). One of these is hydrological alteration: the construction of oil infrastructure, including pipelines, platforms, and docks, has modified flow patterns in coastal lagoons such as Mecoacán and Carmen-Pajonal (ECOSUR, 2021). This modification affects the nutrient cycles essential for phytoplankton, the basis of the aquaculture food chain (Garnier et al., 2021). Research by ECOSUR (2021) shows that these alterations have reduced primary productivity by 30% in areas near petrochemical complexes. Another systemic effect is the diffuse contamination derived from legal discharges of process water containing heavy metals, volatile organic compounds, and other pollutants, which, although they comply with official standards that are often obsolete, progressively accumulate in the aquatic environment (Romeo et al., 2000). López-Luna et al. (2021) found elevated levels of cadmium and led in *Crassostrea virginica*, exceeding the limits permitted by NOM-129-SSA1-195 and the FDA (United States Food and Drug Administration). Furthermore, a progressive increase in the metals was observed with decreasing distance from the coastal area of the municipality of Paraíso, Tabasco, where the main petrochemical development is located (López-Luna et al., 2021). Likewise, the loss of critical habitats such as mangroves represents a silent threat. Oil expansion has destroyed more than 12,000 hectares of mangroves in Tabasco, according to CONABIO (2018). These ecosystems function as natural "nurseries" for aquaculture species (Padmini & Usha Rani, 2009). It is estimated that the loss of each hectare of mangrove reduces potential shrimp production in extensive farming systems by 280 kg per year (Arkoosh et al., 2010).

In the face of this bleak panorama, the regulatory framework and mitigation actions implemented have been clearly insufficient (Saval et al., 2004). Although regulations such as NOM-163-SEMARNAT exist, the reality is that contingency plans are rarely activated in a timely manner. An illustrative example is the Ixachi well spill in 2019, which affected 45 shrimp farms before containment measures were implemented (Hill et al., 2010). Meanwhile, cleanup programs are often limited to the mechanical collection of visible crude oil, leaving more toxic dissolved components in the water that continue to affect organisms for months (Madhav et al., 2020). The economic sanctions imposed on the responsible companies, when applied, represent less than 1% of the market value of the production lost by aquaculturists, which discourages improvements in industrial practices (Guimarães-Filho et al., 2022). This combination of institutional negligence and inadequate redress contributes to the widespread perception of impunity and abandonment in affected communities.

Faced with such a complex and structural problem, solutions cannot be superficial. In-depth and structural measures are required, many of which have already been proposed by experts and environmental organizations (Prakash, 2021). One of the most urgent issues is the creation of oil exclusion zones. It is proposed to establish a 20-km perimeter around the main aquaculture zones, in which all new oil infrastructure would be prohibited (Dai et al., 2023). This measure would preserve key ecosystems and provide a respite for aquaculture activity. Equally important is the creation of an environmental guarantee fund, financed by a specific tax on oil production, which would automatically compensate fish farmers affected by oil-related incidents (Huang et al., 2021). In the technological field, it is urgent to implement advanced remediation systems, such as the use of indigenous hydrocarbonoclastic microorganisms that degrade toxic compounds. Tests conducted by the National Polytechnic Institute (Barros et al., 2013) have demonstrated the effectiveness of this method in restoring contaminated water bodies in significantly shorter time periods than traditional methods.

Finally, participatory monitoring should be encouraged, providing aquaculture cooperatives with real-time water quality analysis equipment connected to digital platforms for public denunciation, which would strengthen citizen vigilance and increase pressure on the authorities and companies responsible. The experience of Tabasco shows that coexistence between the oil industry and marine aquaculture is, under the current model, a chimera (Weis & Weis, 2019). Every barrel of oil extracted in the southern GoM has a hidden cost measurable in tons of shrimp that will never be harvested, in contaminated oysters that will end up in landfills, and in lost rural jobs. The choice is clear: either the oil operation in sensitive areas is radically reformulated, or the aquaculture potential of one of the richest regions in the country in terms of biodiversity is sacrificed (Pinkus-Rendón & Pacheco Castro, 2013). The time to decide is running out, while oil slicks continue to expand over waters that were once synonymous with life and prosperity for Tabasco's coastal communities (Javier-Jiménez et al., 2025).

4.4 Perspective of Mariculture in Tabasco

Under the scenario presented above, the prospects for the establishment of an industry around the cultivation of aquatic species in the coastal zone of the state of Tabasco are not very encouraging. However, adequate planning of the activity could constitute a viable production option for a sector of the state's coastal population. In addition, this activity could represent a conservation measure for coastal aquatic environments, given the great threats to which they are exposed.

Traditionally, mariculture in Tabasco has developed around the American oyster, *C. virginica*, a resource of great economic importance for both Mexican and US states bordering the GoM (Díaz et al., 2022) (Fig. 4.2). However, its abundance has decreased over the last century, particularly in recent decades, due to factors such as

Fig. 4.2 Farm for farming American oysters (*Crassostrea virginica*) in *Laguna de Mecoacán*, Tabasco. (photo by Nicolás Vite-García)

overexploitation, pollution, degradation, and habitat destruction, among others (Rothschild et al., 1994; Mackenzie Jr, 2007). Although the state of Tabasco, until the 1980s, was recognized as a site historically rich in oyster beds (Diego, 1986), the settlement of the oil industry near the mouth of the bar that connects the lagoon to the sea (Gold-Bouchot et al. 1995) and overexploitation by fishers have been factors that have caused pressure on the oyster biocenosis, resulting in the first two closed seasons in the State since 2000 (George-Zamora & Aldana-Aranda et al., 2000).

Tabasco and Veracruz are states that lead the national production of oyster, accumulating 46,803 tons in live weight between 2004 and 2013, with a peak of 50,715 tons in 2010 and a minimum of 42,250 tons in 2009. Between 2014 and 2023, production was more variable, averaging 48,851 tons per year, with its highest point in 2015 at 61,996 tons and its lowest in 2020 at 22,535 tons (Fig. 4.3). As for Tabasco, production has shown a decreasing trend: while it averaged 17,231 tons per year between 2004 and 2013, equivalent to 36.8% of the national total, the average dropped to 11,869 tons per year in the following decade, representing only 27.1% of national production. This decline has been especially pronounced since 2019, a period during which the state has not managed to exceed 8,000 tons per year (CONAPESCA, 2013, 2023).

Among mariculture activities, the cultivation of bivalve mollusks has the least negative impact on the environment and may even have beneficial effects. These organisms obtain their food by filtering the surrounding water, from which they remove the dissolved oxygen necessary for respiration and trap suspended particles, meaning that no feed needs to be administered. On the one hand, this greatly reduces production costs and, on the other, it prevents the introduction of excessive amounts of nutrients into water bodies, as is the case with fish or crustacean farming, which bases its production on the use of extruded balanced feed. In addition, bivalve

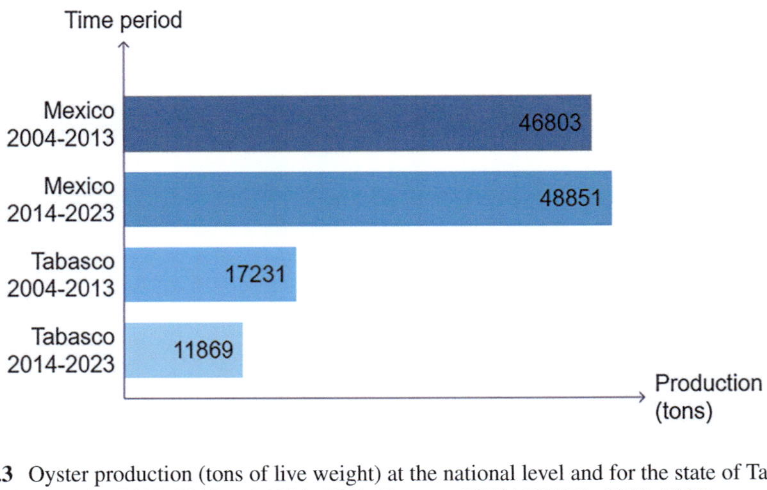

Fig. 4.3 Oyster production (tons of live weight) at the national level and for the state of Tabasco, during the decades 2004–2013 and 2014–2023. (Source: CONAPESCA, 2013, 2023)

mollusks, particularly oysters, provide other types of ecosystem services. By forming dense natural beds, they create a biocoenosis (a community heavily dependent on trophic interactions, symbionts, commensals, and parasites among coexisting populations) that improves water quality and stabilizes the coastline (van der Schatte Olivier et al., 2020; Thomas et al., 2022; Smith et al., 2023). By filtering seston (microscopic material suspended in the water column), they remove significant amounts of particulate and flocculated organic matter, promoting the creation of refuges from predation, creating a feeding habitat for juveniles and adults of mobile species and sessile stages of species that attach themselves to the shells of mollusks, among others (Coen et al., 2007; Grabowski et al., 2012). As a further point in their favor, bivalve mollusks (oysters, mussels, and clams) have been used as sentinel species, i.e., organisms that indicate environmental quality or the impact and restoration or mitigation actions of estuarine ecosystems (Volety et al., 2009; Aguilar et al., 2012; Martins et al., 2017), which can be used to measure the impact of industrial activities on coastal water bodies. Another use of bivalve species is the differentiation of biogeographical regions based on the presence or absence of bivalve species (Caballero-Herrera et al., 2022).

In order for oyster farming to continue in the region, it is necessary to establish areas where the impact of oil activity is minimal. *C. virginica* is affected not only by heavy metal concentrations, but also by the presence of polycyclic aromatic hydrocarbons (Gold-Bouchot et al., 1995; Araújo et al., 2020). The presence of particles of these elements has been shown to decrease physiological activity in this species, thereby causing alterations in its reproductive periods and development from the gonad (Baqueiro-Cárdenas et al., 2007). According to García-Cubas et al. (1990), there are around 20 species of bivalve mollusks in the Mecoacán lagoon, of which exploitation is almost exclusively focused on one species (the American oyster: *C. virginica*). Species diversification could be an important alternative for more varied production, focused on the pursuit of quality (rather than quantity).

As an example of this, recent studies carried out by the authors of this paper show that there is a considerable uptake of *Ischadium recurvum* mussel seed, a species that shares its habitat and coexists with the oyster, especially in the deeper parts of the coastal lagoons of the state of Tabasco. Mussel farming is an alternative that, in combination with oyster farming, could meet the needs of the fishing sector based in the vicinity of the lagoons. It can provide a considerable economic boost in months when the oyster ban is in force, and alternate the farming cycles between the two species.

The intense pressure exerted on semi-protected coastal ecosystems, such as lagoons, estuaries, and wetlands, suggests that the future of bivalve mollusk farming in Tabasco may need to focus on species found in the marine environment. Coastal waters that are not confined have a much faster turnover rate and are less prone to pollution and nutrient enrichment problems. To this end, it is important to first investigate all conditions, including the concentration and type of pollutants. Another option for bivalve mariculture in Tabasco, which, as it is not intended for human consumption, can be developed even in the presence of contaminants, is the cultivation of oysters to produce mother-of-pearl (cosmetics, creams, and emulsions) or mabes (half-pearls), and pearls (jewelry). There are two native species in the region: *Pteria colymbus* and *Pinctada imbricata*. The technology for very similar species (*Pteria sterna* and *Pinctada mazatlanica*) in the Mexican Pacific has already been developed and is available for transfer. Significant advances have been made in research into the cultivation of these two Atlantic species, with promising results (Lodeiros et al., 2002, 2017; Semidey et al., 2010; Velasco & Barros 2010; Saucedo et al., 2023). Pearl oysters have even been considered as tools for environmental bioremediation due to their ability to sequester contaminants through water filtration (Gifford et al., 2005).

Despite being products with high commercial value, and which can be considered a luxury, pearl farming and pearl cultivation can be carried out by social groups, such as aquaculture production cooperatives, in which economic gains are given the same priority as the benefits obtained (employment, use of natural resources, community cohesion, inclusion, gender equality, etc.), thus providing an alternative means of production for the inhabitants of the coastal area of Tabasco (Saucedo et al., 2021). In Southeast Asia, pearl farming is a significant source of income for a substantial sector of the population (Gueguen et al., 2016; Johnston et al., 2020).

For its part, crustacean farming, whose main exponent is shrimp farming, is an activity that—given the negative results due to the accelerated eutrophication processes it induces and which it has had in coastal water bodies in states such as Sinaloa and Sonora, where its management has been intensive—(Moreno, 2010), in Tabasco should be carried out with an intensity and using farming techniques that do not cause further pollution through nutrient enrichment. One alternative is Biofloc, which involves the cultivation of aquatic organisms, such as crustaceans or fish, in an environment dominated by microorganisms. These microorganisms are grouped into bioflocs, serving as the main food source (El-Sayed, 2021; Emerenciano et al., 2021). Another option would be the construction of supralittoral ponds in which producers are responsible for controlling their effluents. Shrimp farming can be a sustainable activity if it is managed with the advice of well-trained technicians and experts in scientific research and technological development. In the growth and expansion of this activity, not only the economic benefits but also, and primarily, the environmental and ecological aspects involved must be considered (Martínez-Córdova et al., 2009; Röthig et al., 2023).

With adequate training and financing, the farming of crustaceans, such as crabs, can also become a reality that helps alleviate poverty, provides sustainable employment and activity, and contributes to the conservation of the coastal, estuarine, and marine environments. Small-scale crustacean farming involves lower risks and greater flexibility, as well as a more sustainable outlook, than the same activity carried out intensively (Apine et al., 2023).

In terms of marine fish farming, Tabasco faces a number of challenges that must be taken into account (Fig. 4.4). The farming of native species should be prioritized over exotic species, firstly because the environmental conditions are those in which they have evolved, and secondly because of the environmental problems that exotic species have caused in many parts of the world and in our country (tilapia: *Tilapia*

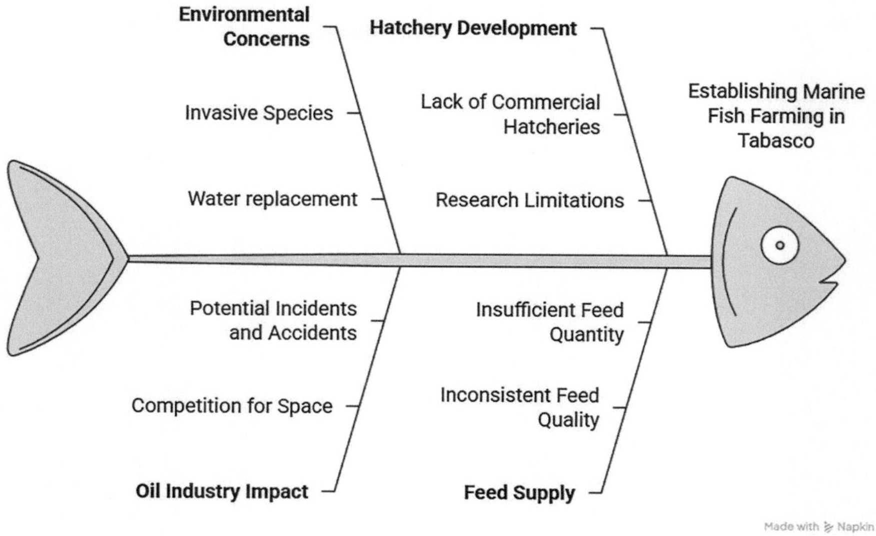

Fig. 4.4 Challenges for marine fish farming in Tabasco

sp. and *Oreochromis* sp., devil fish: *Pterygoplichthys* sp.), which in many cases have become invasive species (Lowe et al. 2000; Tricarico et al. 2016; Cuthbert et al. 2021).

There is a significant list of euryhaline fish that can be considered for farming, such as snook (Centropomidae), mullet (Mugilidae), croaker (Sciaenidae), trevally and pomfret (Carangidae), snapper and red snapper (Lutjanidae), among others. Currently, there is a global trend to move fish farming to cages, and from coastal areas to offshore areas, where interaction with human activity is lower, water exchange is greater, and therefore the impact on the environment is reduced (Benetti et al., 2006, 2021; Skladany et al., 2007; Stieglitz et al., 2021) (Figs. 4.5 and 4.6). However, in the state of Tabasco, it is crucial to consider the high level of activity in the marine area due to the oil industry, which means that the establishment of cage farms would have to be based on the establishment of exclusive zones for this purpose, avoiding the transit of vessels and possible accidents that could occur on platforms and other oil installations.

Currently, there are no commercial-scale marine fish hatcheries in Tabasco (and very few projects nationwide). The Juárez Autonomous University of Tabasco and other institutions are conducting research on sea bass hatchery production with very promising results in understanding the processes involved in reproduction, larval culture, nutrition, and other aspects (Alvarez-González et al., 2010; Ibarra-Castro et al., 2011; Contreras-García et al., 2015, 2020, 2024), but this is not enough to sustain an industry. The first step in developing marine fish farming projects is to establish production centers for high-quality fry with consistent quality. This can be done in parallel with the planning of "plots" at sea that could be concessioned for

Fig. 4.5 Cage for farming marine fish *Sciaenops ocellatus* in the *Sonda de Campeche* zone. (photo by Nicolás Vite-García)

Fig. 4.6 Cage for marine fish farming in open waters at Campeche, Mexico. At the top, divers separate and check the condition of the net, while below, the fish are confined within a net marked with a yellow buoy for partial harvesting. (photo by Nicolás Vite-García)

this activity. Another vital step to be resolved is the continuous supply of balanced feed, manufactured in quantities and of a quality that meets the nutritional and dietary requirements of these species.

There is also a more recent global trend to explore multitrophic production systems, in which different species with complementary trophic requirements are farmed within the same space in order to promote nutrient recycling between different trophic levels. The use of these systems enables the production of economically important species while mitigating some aquaculture problems through the biomitigation of aquaculture waste and the reduction of disease outbreaks (Granada et al., 2018). This may be a low-impact production alternative suitable for the coastal areas of the state of Tabasco, and it constitutes another door opening for aquaculture research.

In general, legislative measures for environmental protection exist, but in practice, little is done to protect ecosystems. The success of mariculture in the state of Tabasco could depend on compliance with regulations, laws, and other mechanisms for environmental protection, but also on generating a new awareness among producers and consumers and, subsequently, a culture of sustainable use not only of resources but also of the environment itself, water, in a way that sustains the so-called "Blue Economy" (FAO, 2014; Julian, 2014; UNEP, 2015; Ivanova et al., 2017).

4.5 Conclusion

The coastal region of Tabasco has great potential for the development of mariculture, thanks to its ecological richness and favorable natural conditions. However, the multiple negative impacts resulting from human activities, especially those of the oil industry and pollution, pose serious challenges to its sustainable implementation. Hydrocarbons alter essential biological processes: they reduce aquatic photosynthesis, generate hypoxia, and cause endocrine disruption in fish (Bodiguel et al., 2009). In addition, the destruction of 12,000 ha of mangroves (CONABIO, 2018) has eliminated key natural nurseries. These impacts have turned aquaculture into an economically high-risk activity. Policies are urgently needed to establish: (1) oil exclusion zones in critical aquaculture areas, (2) automatic compensation systems, and (3) ecological restoration programs with community participation. The survival of regional aquaculture depends on breaking the current paradigm of oil environmental impunity.

Nevertheless, with proper planning, the use of native species, clean technologies, and a strong focus on environmental conservation, mariculture can become a productive, ecological, and socially viable alternative. To achieve this, it is essential to promote effective public policies, environmental education, constant monitoring, and a transition toward a culture of sustainability that strengthens the so-called blue economy.

Acknowledgments To Dr. Alejandro Espinoza, Dr. Dora Ramos, and the entire Fishing and Oil project team for the invitation to participate in this publication. To the reviewers and editors for their corrections, comments, and suggestions to enrich the manuscript. NapkinAI was used to generate Figs. 4.1, 4.3, and 4.4.

References

Aguilar, C. A., Montalvo, C., Rodríguez, L., Cerón, J. G., & Cerón, R. M. (2012). American oyster (*Crassostrea virginica*) and sediments as a coastal zone pollution monitor by heavy metals. *International journal of Environmental Science and Technology, 9*, 579–586. https://doi.org/10.1007/s13762-012-0078-y

Alvarez-González, C. A., Gaxiola-Cortés, G., Jiménez-Martínez, L. D., Sanchez-Zamora, A., Arena-Ortiz, L., Martínez-Bruguete, T., et al. (2010). Avances en la fisiología digestiva del robalo blanco (Centropomus undecimalis) en Tabasco, México. *Avances en nutrición acuícola.*

Andrade-Velázquez, M., José Montero-Martínez, M., & Vargas Magaña, M. F. (2024). Climatic temperature indices for the coastal resilience observatory in Tabasco (Dos Bocas)-LANRESC. *Earth Sciences Research Journal, 28*(2), 213–229. https://doi.org/10.15446/esrj.v28n2.112335

Apine, E., Ramappa, P., Bhatta, R., Turner, L. M., & Rodwell, L. D. (2023). Challenges and opportunities in achieving sustainable mud crab aquaculture in tropical coastal regions. *Ocean & Coastal Management, 242*, 106711. https://doi.org/10.1016/j.ocecoaman.2023.106711

Araújo, M. E. D., Ramalho, C. W. N., & Melo, P. W. D. (2020). Artisanal fishers, consumers and the environment: Immediate consequences of the oil spill in Pernambuco, Northeast Brazil. *Cadernos de Saúde Pública, 36*, e00230319. https://doi.org/10.1590/0102-311x00230319

Arias, N. (2025). *Ocurren en Tabasco cinco accidentes petroleros a la semana, revela Pemex.* Recuperado 20 de mayo de 2025 https://petroquimex.com/ocurren-en-tabasco-cinco-accidentes-petroleros-a-la-semana-revela-pemex/.

Arkoosh, M. R., Boylen, D., Dietrich, J., Anulacion, B. F., Bravo, C. F., Johnson, L. L., et al. (2010). Disease susceptibility of salmon exposed to polybrominated diphenyl ethers (PBDEs). *Aquatic Toxicology, 98*(1), 51–59. https://doi.org/10.1016/j.aquatox.2010.01.013

Atoufi, H., & Lampert, D. J. (2020). Impacts of oil and gas production on contaminant levels in sediments. *Current Pollution Reports, 6*, 43–53. https://doi.org/10.1007/s40726-020-00137-5

Baqueiro-Cárdenas, E. R., Borabe, L., Goldaracena-Islas, C. G., & Rodríguez-Navarro, J. (2007). Los moluscos y la contaminación: una revisión. *Revista Mexicana de Biodiversidad, 78*, 1–7. https://doi.org/10.22201/ib.20078706e.2007.002.293

Barros, P., Sobral, P., Range, P., Chícharo, L., & Matias, D. (2013). Effects of sea-water acidification on fertilization and larval development of the oyster *Crassostrea gigas*. *Journal of Experimental Marine Biology and Ecology, 440*, 200–206. https://doi.org/10.1016/j.jembe.2012.12.014

Benetti, D., Brand, L., Collins, J., Orhun, R., Benetti, A., O'Hanlon, B., Danylchuk, A., Alston, D., Rivera, J., & Cabarcas, A. (2006). Can offshore aquaculture of carnivorous fish be sustainable. *World Aquaculture, 37*(1), 44–47.

Benetti, D. D., Suarez, J., Camperio, J., Hoenig, R. H., Tudela, C. E., Daugherty, Z., et al. (2021). A review on cobia, *Rachycentron canadum*, aquaculture. *Journal of the World Aquaculture Society, 52*(3), 691–709. https://doi.org/10.1111/jwas.12810

Bennani, N., Schmid-Alliana, A., & Lafaurie, M. (1995). Evaluation of phagocytic activity in a teleost fish, Dicentrarchus labrax. *Fish Shellfish Immunology, 5*, 237–246. https://doi.org/10.1016/S1050-4648(05)80017-8

Bianchi, T. S., DiMarco, S. F., Cowan, J. H., Jr., Hetland, R. D., Chapman, P., Day, J. W., & Allison, M. A. (2010). The science of hypoxia in the Northern Gulf of Mexico: A review. *Science of the Total Environment, 408*(7), 1471–1484. https://doi.org/10.1016/j.scitotenv.2009.11.047

Bodiguel, X., Loizeau, V., Le Guellec, A. M., Roupsard, F., Philippon, X., & Mellon-Duval, C. (2009). Influence of sex, maturity and reproduction on PCB and p,p′DDE concentrations and repartitions in the European hake (*Merluccius merluccius*, L.) from the Gulf of Lions (N.W. Mediterranean). *Science of the Total Environment 408*, 304–311. doi:https://doi.org/10.1016/j.scitotenv.2009.10.004.

Bouman, H. A., Nakane, T., Oka, K., Nakata, K., Kurita, K., Sathyendranath, S., & Platt, T. (2010). Environmental controls on phytoplankton production in coastal ecosystems: A case study from Tokyo Bay. *Estuarine, Coastal and Shelf Science, 87*(1), 63–72. https://doi.org/10.1016/j.ecss.2009.12.014

Bozada-Robles, L. M. (2006). Las pesquerías de la zona costera de Tabasco: una actividad amenazada. Universidad Veracruzana Intercultural, sede Selvas., Universidad Istmo-americana, A.C. 76p.

Breitburg, D., Levin, L. A., Oschlies, A., Grégoire, M., Chavez, F. P., Conley, D. J., et al. (2018). Declining oxygen in the global ocean and coastal waters. *Science, 359*(6371), 7240. https://doi.org/10.1126/science.aam7240

Bruno, J. F., Bates, A. E., Cacciapaglia, C., Pike, E. P., Amstrup, S. C., Van Hooidonk, R., et al. (2018). Climate change threatens the world's marine protected areas. *Nature Climate Change, 8*(6), 499–503. https://doi.org/10.1038/s41558-018-0149-2

Caballero-Herrera, J. A., Olivero, J., von Cosel, R., & Gofas, S. (2022). An analytically derived delineation of the west African Coastal Province based on bivalves. *Diversity and Distributions, 28*(12), 2791–2805. https://doi.org/10.1111/ddi.13454

Cárdenas, M. (2010). *México ante el cambio climático: Evidencias, impactos, vulnerabilidad y adaptación*. Greenpeace México.

Castillo-Cruz, Z. G., & Medrano-Pérez, O. R. (2023). Análisis geomorfológico de las subcuencas Usumacinta y Grijalva en el sureste de México. *Acta universitaria, 33*. https://doi.org/10.15174/au.2023.3684

Chan, F., Sokolova, I., & Vopel, K. (2025). Ocean hypoxia: The science of climate change in the sea. *Scientific Reports, 15*(1), 1–3. https://doi.org/10.1038/s41598-025-86706-4

Chan-Quijano, J. G., Torres-López, K. L., Martínez-Rabelo, F., & González-Consuelo, M. B. (2020). Soil contamination by petroleum in Tabasco, México, and its environmental repercussions. *Gaia Scientia, 14*(3), 75–91.

Chapman, E. J., Byron, C. J., Lasley-Rasher, R., Lipsky, C., Stevens, J. R., & Peters, R. (2020). Effects of climate change on coastal ecosystem food webs: Implications for aquaculture. *Marine Environmental Research, 162*, 105103. https://doi.org/10.1016/j.marenvres.2020.105103

Chen, J., McIlroy, S. E., Archana, A., Baker, D. M., & Panagiotou, G. (2019). A pollution gradient contributes to the taxonomic, functional, and resistome diversity of microbial communities in marine sediments. *Microbiome, 7*, 1–12. https://doi.org/10.1186/s40168-019-0714-6

CINVESTAV. (2020). *Evaluación de contaminantes en moluscos bivalvos de importancia comercial en el Golfo de México*. Centro de Investigación y de Estudios Avanzados del IPN.

Coen, L. D., Brumbaugh, R. D., Bushek, D., Grizzle, R., Luckenbach, M. W., Posey, M. H., Powers, S. P., & Tolley, S. G. (2007). Ecosystem services related to oyster restoration. *Marine Ecology Progress Series, 341*, 303–307. https://doi.org/10.3354/meps341303

Comisión Nacional de Hidrocarburos (CNH). (2021). *Reporte anual de incidentes operativos en la industria petrolera nacional*. https://www.gob.mx/cnh/es

CONABIO. (2018). *La biodiversidad en Tabasco: Estudio de Estado*. Comisión Nacional para el Conocimiento y Uso de la Biodiversidad.

CONAPESCA. (2013). *Anuario estadístico de acuacultura y pesca. De la comisión nacional de acuacultura y pesca*. https://nube.conapesca.gob.mx/sites/cona/dgppe/2013/ANUARIO_ESTADISTICO_DE_ACUACULTURA_Y_PESCA_2013.pdf

CONAPESCA. (2020). *Anuario estadístico de acuacultura y pesca. De la comisión nacional de acuacultura y pesca*. https://nube.conapesca.gob.mx/sites/cona/dgppe/2020/ANUARIO_ESTADISTICO_DE_ACUACULTURA_Y_PESCA_2020.pdf

CONAPESCA. (2023). *Anuario estadístico de acuacultura y pesca. De la comisión nacional de acuacultura y pesca.* https://nube.conapesca.gob.mx/sites/cona/dgppe/2023/ANUARIO_ESTADISTICO_DE_ACUACULTURA_Y_PESCA_2023.pdf

Contreras-García, M. D. J., Contreras-Sánchez, W. M., Hernández-Vidal, U., & Mcdonal-Vera, A. (2015). Induced spawning of the common Snook (*Centropomus undecimalis*) in captivity using GnRH-a implants. *Ecosistemas y recursos agropecuarios, 2*(6), 357–362.

Contreras-García, M., Contreras-Sánchez, W. M., Mcdonal-Vera, A., Hernández-Vidal, U., & Cruz-Rosado, L. (2020). Reproducción inducida del robalo chucumite (Centropomus parallelus) en Tabasco, México. *AquaTIC, 56*, 15–25.

Contreras-García, M., Contreras-Sánchez, W. M., Mcdonal-Vera, A., & Cruz-Rosado, L. (2024). Spawning induction, Larviculture and rearing of Mexican Snook *Centropomus poeyi* (Chávez, 1961). *Tropical Aquaculture, 2*(1). https://doi.org/10.19136/ta.a2n1.5729

Crist, E., Mora, C., & Engelman, R. (2017). The interaction of human population, food production, and biodiversity protection. *Science, 356*(6335), 260–264. https://doi.org/10.1126/science.aal2011

Cruz, Y. Y. P., & Martínez, P. C. C. (2015). Cambio climático: bases científicas y escepticismo. *Cultura Científica y Tecnológica, 46.*

Cuthbert, R. N., Pattison, Z., Taylor, N. G., Verbrugge, L., Diagne, C., Ahmed, D. A., et al. (2021). Global economic costs of aquatic invasive alien species. *Science of the Total Environment, 775*, 145238. https://doi.org/10.1016/j.scitotenv.2021.145238

Dai, M., Zhao, Y., Chai, F., Chen, M., Chen, N., Chen, Y., et al. (2023). Persistent eutrophication and hypoxia in the coastal ocean. *Cambridge Prisms: Coastal Futures, 1*, e19. https://doi.org/10.1017/cft.2023.7

Day, J. W., Christian, R. R., & Boesch, D. M. (2008). Consequences of climate change on the Ecogeomorphology of coastal wetlands. *Estuaries and Coasts, 31*, 477–491. https://doi.org/10.1007/s12237-008-9047-6

de los Santos Gonzalez, C. C., Perera, M. Á. D., Muñoz, D. E. R., & Campillo, L. M. G. (2023). Vulnerabilidad en la costa de Tabasco: Coronel Andrés Sánchez Magallanes. *Publicaciones e Investigación, 17*(1), 1–14. https://doi.org/10.22490/25394088.6806

Di Cesare, A., Pjevac, P., Eckert, E., Curkov, N., Šparica, M. M., Corno, G., & Orlić, S. (2020). The role of metal contamination in shaping microbial communities in heavily polluted marine sediments. *Environmental Pollution, 265*, 114823. https://doi.org/10.1016/j.envpol.2020.114823

Díaz, R. J. (2001). Overview of hypoxia around the world. *Journal of Environmental Quality, 30*, 275–281. https://doi.org/10.2134/jeq2001.302275x

Díaz, R. J., & Rosenberg, R. (1995). Marine benthic hypoxia A review of its ecological effects and the behavioural responses of benthic macrofauna. *Oceanography and Marine Biology: An Annual Review, 33*, 245–303.

Díaz, J. S. S., Brito-Manzano, N. P., Vargas-Falcón, P. M., & Bernat-Rodríguez, J. (2022). Estudio poblacional del ostión *Crassostrea virginica* en la laguna de Mecoacán Tabasco, México. *Brazilian Journal of Animal and Environmental Research., 5*(1), 469–479. https://doi.org/10.34188/bjaerv5n1-036

Diego, P. M. (1986). *Breve informe sobre la pesquería del ostión C. virginica en las lagunas de Tabasco* (61 p). Centro de Producción Acuícola de Puerto Ceiba.

ECOSUR. (2021). *Impacto de la infraestructura petrolera sobre los ecosistemas lagunares de Tabasco*. El Colegio de la Frontera Sur.

El-Sayed, A. F. M. (2021). Use of biofloc technology in shrimp aquaculture: A comprehensive review, with emphasis on the last decade. *Reviews in Aquaculture, 13*(1), 676–705. https://doi.org/10.1111/raq.12494

Emerenciano, M. G., Miranda-Baeza, A., Martínez-Porchas, M., Poli, M. A., & Vieira, F. D. N. (2021). Biofloc technology (BFT) in shrimp farming: Past and present shaping the future. *Frontiers in Marine Science, 8*, 813091. https://doi.org/10.3389/fmars.2021.813091

Esperón-Rodríguez, M., Beaumont, L. J., Lenoir, J., Baumgartner, J. B., McGowan, J., Correa-Metrio, A., & Camac, J. S. (2019). Climate change threatens the most biodiverse regions of Mexico. *Biological Conservation, 240*, 108215. https://doi.org/10.1016/j.biocon.2019.108215

Etim, N. N., Offiong, E. E., & Ukpanah, U. A. (2013). Farm animals: Culprits and victims of global warming and climate change. *AJAAR, 1*, 10–24.

FAO. (2014). *Blue growth: Unlocking the potential of seas and oceans*. Recuperado de http://www.fao.org/zhc/detail-events/en/c/233765/

Foley, J. A., Ramankutty, N., Brauman, K. A., Cassidy, E. S., Gerber, J. S., Johnston, M., et al. (2011). Solutions for a cultivated planet. *Nature, 478*(7369), 337–342. https://doi.org/10.1038/nature10452

Gamboa-Bernal, G. A. (2015). Los Objetivos de Desarrollo Sostenible: una perspectiva bioética. *Persona y Bioética, 19*(2), 175–181. https://doi.org/10.5294/pebi.2015.19.2.1

García-Cubas, A., Escobar, D. F., González, A. L., & Reguero, M. (1990). Moluscos de la laguna Mecoacán, Tabasco, México: Sistemática y ecología. In *Anales del Instituto de Ciencias del Mar y Limnología, Universidad Nacional Autónoma de México* (Vol. 17, No. 1, pp. 1–30). Instituto de Ciencias del Mar y Limnología.

García-Cuéllar, J. A., Pérez-Castañeda, R., & Defeo, O. (2018). Oil spills and their effects on shrimp fishing and aquaculture activities. *Latin American Journal of Aquatic Research, 46*(3), 467–478. http://www.lajar.cl/pdf/imar/v46n3/46-3-Garcia-Cuellar.pdf

Garnier, J., Billen, G., Lassaletta, L., Vigiak, O., Nikolaidis, N. P., & Grizzetti, B. (2021). Hydromorphology of coastal zone and structure of watershed agro-food system are main determinants of coastal eutrophication. *Environmental Research Letters, 16*(2), 023005. https://doi.org/10.1088/1748-9326/abc777

George-Zamora, A., & Aldana-Aranda, D. (2000). Producción somática de dos especies de *Crassostrea virginica* e *Ischadium recurvum* (Bivalvia) en Mecoacán, Tabasco, México. *Revista de Biología Tropical, 48*(1), 65–75.

Gifford, S., Dunstan, H., O'Connor, W., & Macfarlane, G. R. (2005). Quantification of in situ nutrient and heavy metal remediation by a small pearl oyster (*Pinctada imbricata*) farm at Port Stephens, Australia. *Marine Pollution Bulletin, 50*(4), 417–422. https://doi.org/10.1016/j.marpolbul.2004.11.024

Gold-Bouchot, G., Simá-Alvarez, R., Zapata-Pérez, O., & Güemez-Ricalde, J. (1995). Histopathological effects of petroleum hydrocarbons and heavy metals on the American oyster (*Crassostrea virginica*) from Tabasco, Mexico. *Marine Pollution Bulletin, 31*(4–12), 439–445. https://doi.org/10.1016/0025-326X(95)00171-I

Grabowski, J. H., Brumbaugh, R. D., Conrad, R. F., Keeler, A. G., Opaluch, J. J., Peterson, C. H., Piehler, M. F., Powers, S. P., & Smyth, A. R. (2012). Economic valuation of ecosystem services provided by oyster reefs. *Bioscience, 62*, 900–909. https://doi.org/10.1525/bio.2012.62.10.10

Granada, L., Lopes, S., Novais, S. C., & Lemos, M. F. (2018). Modelling integrated multi-trophic aquaculture: Optimizing a three trophic level system. *Aquaculture, 495*, 90–97. https://doi.org/10.1016/j.aquaculture.2018.05.029

Gueguen, Y., Gaertner-Mazouni, N., Lo, C., & Le Moullac, G. (2016). Integrated management of pearl culture in French Polynesia in the context of global change: Synopsis of existing results. *Estuarine, Coastal and Shelf Science, 182*, 229–234. https://doi.org/10.1016/j.ecss.2016.07.010

Guimarães-Filho, C. E. D. F., Calixto, F. A. A., Kasnowski, M. C., & Mesquita, E. D. F. M. D. (2022). Depuration of bivalve molluscs: A literature review. *Food Science and Technology, 42*, e06622. https://doi.org/10.1590/fst.06622

Hickey, B., Kudela, R., Nash, J., Bruland, K., Peterson, W., MacCready, P., Lessard, E., Jay, D., Banas, N., Baptista, A., Dever, E., Kosro, P., Kilcher, L., Horner-Devine, A., Zaron, E., McCabe, R., Peterson, J., Orton, P., Pan, J., & Lohan, M. (2010). River influences on shelf ecosystems: Introduction and synthesis. *Journal of Geophysical Research, 115*(2), 26. https://doi.org/10.1029/2009JC005452

Hill, E. M., Evans, K. L., Horwood, J., Rostkowski, P., Olumide, F., Gobson, R., Shears, J. A., & Tyler, C. R. (2010). Profiles and some initial identifications of (anti) androgenic compounds in fish exposed to wastewater treatment works effluents. *Environmental Science & Technology, 44*, 1137–1143. https://doi.org/10.1021/es901837n

Huang, M., Ding, L., Wang, J., Ding, C., & Tao, J. (2021). The impacts of climate change on fish growth: A summary of conducted studies and current knowledge. *Ecological Indicators, 121*, 106976. https://doi.org/10.1016/j.ecolind.2020.106976

Ibarra-Castro, L., Alvarez-Lajonchère, L., Rosas, C., Palomino-Albarrán, I. G., Holt, G. J., & Sanchez-Zamora, A. (2011). GnRHa-induced spawning with natural fertilization and pilot-scale juvenile mass production of common Snook, *Centropomus undecimalis* (Bloch, 1792). *Aquaculture, 319*(3–4), 479–483. https://doi.org/10.1016/j.aquaculture.2011.07.014

Instituto Nacional de Estadística Geografía e Informática (INEGI) (2025). https://www.inegi.org.mx/app/areasgeograficas/?ag=27014#collapse-Resumen

Ivanova, A., Cariño Olvera, M. M., Monteforte-Sánchez, M., Ramírez Ivanova, E. A., & Domínguez, W. (2017). La economía azul como modelo de sustentabilidad para estados costeros: el caso de Baja California Sur. *Sociedad y ambiente, 14*, 75–98. https://doi.org/10.31840/sya.v0i14.1768

Javier-Jiménez, D. G., Chávez, M. D. J. A., Morales-Bautista, C. M., Méndez-Olán, C., & Chablé-Candelero, R. (2025). Afectaciones en cultivos de Veracruz y Tabasco por derrame de petróleo en los últimos seis años. *Journal of Basic Sciences, 11*(30), 93–105.

Johnston, B., Kishore, P., Vuibeqa, G. B., Hine, D., & Southgate, P. C. (2020). Economic assessment of community-based pearl oyster spat collection and mabé pearl production in the western Pacific. *Aquaculture, 514*, 734505. https://doi.org/10.1016/j.aquaculture.2019.734505

Julian, R. (2014). *The blue economy: From concept to reality in the Caribbean Region*. Recuperado de https://www.central-bank.org.tt/sites/default/files/Caribbean_Blue_Economy.pdf.

Leal-Filho, W., Setti, A. F. F., Azeiteiro, U. M., Lokupitiya, E., Donkor, F. K., Etim, N. N., et al. (2022). An overview of the interactions between food production and climate change. *Science of the Total Environment, 838*, 156438. https://doi.org/10.1016/j.scitotenv.2022.156438

Lodeiros, C., Pico, D., Prieto, A., Narváez, N., & Guerra, A. (2002). Growth and survival of the pearl oyster *Pinctada imbricata* (Röding 1758) in supended and bottom culture in the Golfo de Cariaco, Venezuela. *Aquaculture International, 10*(4), 327–338. https://doi.org/10.1023/A:1022451608405

Lodeiros, C. J., Freites, L., Márquez, A., Glem, M. E., Guevara, M., & Saucedo, P. E. (2017). Comparative growth and survival of spat of the Caribbean pearl oyster, *Pinctada imbricata* cultivated indoor with microalgae diets and outdoor with natural diet. *Aquaculture Nutrition, 23*(3), 511–522. https://doi.org/10.1111/anu.12419

López-Luna, R., Manzano, N. P. B., Falcón, P. M. V., Vázquez, A. G., & Rodríguez, J. B. (2021). Determinación de Cadmio y Plomo en sedimento y el ostión *Crassostrea virginica* en la laguna Mecoacán. *Brazilian Journal of Animal and Environmental Research, 4*(4), 5982–5992. https://doi.org/10.34188/bjaerv4n4-091

Lowe, S., Browne, M., Boudjelas, S., & De Poorter, M. (2000). *100 of the world's worst invasive alien species: A selection from the global invasive species database* (Vol. 12). Invasive Species Specialist Group.

Mackenzie, C. L., Jr. (2007). Causes underlying the historical decline in eastern oyster (*Crassostrea virginica* Gmelin, 1791) landings. *Journal of Shellfish Research, 26*(4), 927–938. https://doi.org/10.2983/0730-8000(2007)26[927:CUTHDI]2.0.CO;2

Madhav, S., Ahamad, A., Singh, A. K., Kushawaha, J., Chauhan, J. S., Sharma, S., & Singh, P. (2020). Water pollutants: Sources and impact on the environment and human health. In *Sensors in water pollutants monitoring: Role of material* (pp. 43–62). https://doi.org/10.1007/978-981-15-0671-0_4

Martínez-Córdova, L. R., Martínez Porchas, M., & Cortés-Jacinto, E. (2009). Camaronicultura mexicana y mundial: ¿actividad sustentable o industria contaminante? *Revista internacional de contaminación ambiental, 25*(3), 181–196.

Martins, M. C., Santos, E. B., & Marques, C. R. (2017). First study on oyster-shell-based phosphorous removal in saltwater—A proxy to effluent bioremediation of marine aquaculture. *Science of the Total Environment, 574*, 605–615. https://doi.org/10.1016/j.scitotenv.2016.09.103

Maulu, S., Hasimuna, O. J., Haambiya, L. H., Monde, C., Musuka, C. G., Makorwa, T. H., et al. (2021). Climate change effects on aquaculture production: Sustainability implications, mitigation, and adaptations. *Frontiers in Sustainable Food Systems, 5*, 609097. https://doi.org/10.3389/fsufs.2021.609097

Moreno, E. F. (2010). Industria del camarón: su responsabilidad en la desaparición de los manglares y la contaminación acuática. REDVET. *Revista electrónica de Veterinaria, 11*(5), 1–20.

Muñoz, Y. P., Martínez, S. L., Luna, A. R. R., & Arcos, S. A. R. (2019). Evaluación de impacto socioambiental, por derrame de petróleo de un ducto en Comalcalco, Tabasco. *Journal of Basic Sciences, 5*(15).

Padmini, E., & Usha Rani, M. (2009). Evaluation of oxidative stress biomarkers in hepatocytes of grey mullet inhabiting natural and polluted estuaries. *Science of the Total Environment, 407*, 4533–4541. https://doi.org/10.1016/j.scitotenv.2009.04.005

Panchasara, H., Samrat, N. H., & Islam, N. (2021). Greenhouse gas emissions trends and mitigation measures in Australian agriculture sector—A review. *Agriculture, 11*(2), 85. https://doi.org/10.3390/agriculture11020085

PEMEX. (2021). *Informe anual de sustentabilidad*. Petróleos Mexicanos.

PEMEX. (2023). *Base de datos de incidentes operativos por estado (2015–2022)*. Dirección General de Seguridad Industrial y Protección Ambiental.

Petro, C., Starnawski, P., Schramm, A., & Kjeldsen, K. U. (2017). Microbial community assembly in marine sediments. *Aquatic Microbial Ecology, 79*(3), 177–195. https://doi.org/10.3354/ame01826

Pezner, A. K., Courtney, T. A., Barkley, H. C., Chou, W. C., Chu, H. C., Clements, S. M., et al. (2023). Increasing hypoxia on global coral reefs under ocean warming. *Nature Climate Change, 13*(4), 403–409. https://doi.org/10.1038/s41558-023-01619-2

Pinkus-Rendón, M. J., & Pacheco Castro, J. (2013). Expectativas sociales y deterioro ambiental por el petróleo. Caso de Cárdenas, Tabasco, México. *Cuadernos De Antropología, 22*.

Porter, L., Rickards, L., Verlie, B., Bosomworth, K., Moloney, S., Lay, B., et al. (2020). Climate justice in a climate changed world. *Planning Theory & Practice, 21*(2), 293–321. https://doi.org/10.1080/14649357.2020.1748959

Prakash, S. (2021). Impact of climate change on aquatic ecosystem and its biodiversity: An overview. *International Journal of Biological Innovations, 3*(2), 312–317.

Rabalais, N. N., Díaz, R. J., Levin, L. A., Turner, R. E., Gilbert, D., & Zhang, J. (2010). Dynamics and distribution of natural and human-caused hypoxia. *Biogeosciences, 7*, 585–619. https://doi.org/10.5194/bg-7-585-2010

Ripley, J. L., & Foran, C. M. (2007). Influence of estuarine hypoxia on feeding and sound production by two sympatric pipefish species (Syngnathidae). *Marine Environmental Research, 63*, 350–367. https://doi.org/10.1016/j.marenvres.2006.10.003

Romeo, M., Bennani, N., Gnassia-Barelli, M., Lafaurie, M., & Girard, J. P. (2000). Cadmium and copper display different responses towards oxidative stress in the kidney of the sea bass *Dicentrarchus labrax*. *Aquatic Toxicology, 48*, 185–194. https://doi.org/10.1016/S0166-445X(99)00039-9

Röthig, T., Barth, A., Tschirner, M., Schubert, P., Wenning, M., Billion, A., et al. (2023). Insect feed in sustainable crustacean aquaculture. *Journal of Insects as Food and Feed, 9*(9), 1115–1138. https://doi.org/10.3920/JIFF2022.0117

Rothschild, B. J., Ault, J. S., Goulletquer, P., & Héral, M. (1994). Decline of the Chesapeake Bay oyster population: A century of habitat destruction and overfishing. *Marine Ecology Progress Series, 11*, 29–39. https://doi.org/10.3354/meps111029

SAGARPA. (2019). *Informe técnico del sector pesquero y acuícola en el sureste mexicano. Secretaría de Agricultura y Desarrollo Rural.*

Saucedo, P. E., Acosta-Salmón, H., McLaurin-Moreno, D., Castillo-Domínguez, A., Melgar-Valdés, C. E., & Mazón-Suástegui, J. M. (2021). Freshwater pearl culture in Mexico: Historic context, present status and future perspectives. *Reviews in Aquaculture, 13*(3), 1379–1396. https://doi.org/10.1111/raq.12527

Saucedo, P. E., McLaurin, D., Lodeiros, C., Freites, L., Leon, L., Cáceres-Puig, J. I., et al. (2023). Progress towards reestablishing Latin America as a major pearl producing region: A review. *Reviews in Aquaculture, 15*(1), 242–260. https://doi.org/10.1111/raq.12714

Saval, S., Lara, F., Lesser, J. M., & Nieto, J. M. (2004). Contaminación de acuíferos con hidrocarburos: Causas, efectos, riesgos asociados y medidas de prevención. www.lesser.com.mx
Semidey, D., Marquez, A., & Lodeiros, C. (2010). Crecimiento y supervivencia de la madre perla *Pinctada imbricata* (Röding 1798), bajo condiciones de cultivo suspendido, en cuerdas y cestas perleras. *Zootecnia Tropical, 28*(4), 521–534.
Skladany, M., Clausen, R., & Belton, B. (2007). Offshore aquaculture: The frontier of redefining oceanic property. *Society & Natural Resources, 20*(2), 169–176. https://doi.org/10.1080/08941920601052453
Smith, R. S., Cheng, S. L., & Castorani, M. C. (2023). Meta-analysis of ecosystem services associated with oyster restoration. *Conservation Biology, 37*(1), e13966. https://doi.org/10.1111/cobi.13966
Stieglitz, J. D., Hoenig, R. H., Baggett, J. K., Tudela, C. E., Mathur, S. K., & Benetti, D. D. (2021). Advancing production of marine fish in the United States: Olive flounder, *Paralichthys olivaceus*, aquaculture. *Journal of the World Aquaculture Society, 52*(3), 566–581. https://doi.org/10.1111/jwas.12804
Thomas, S., Collins, K., Hauton, C., & Jensen, A. (2022). A review of the ecosystem services provided by the native oyster (*Ostrea edulis*): Implications for restoration. In *IOP conference series: Materials science and engineering* (Vol. 1245, No. 1, p. 012010). IOP Publishing. https://doi.org/10.1088/1757-899X/1245/1/012010
Thurman, H. V. (1991). *Introductory oceanography* (6th ed., 526p). Macmillan Pub. Co. https://doi.org/10.5962/bhl.title.59942
Toledo-Ocampo, A. (2005). *Marco conceptual: caracterización ambiental del Golfo de México, p* (25–52. In: A. V. Botello, J. Rendón-von Osten, G. Gold-Bouchot y C. Agraz-Hernández (Eds.). Golfo de México Contaminación e Impacto Ambiental: Diagnóstico y Tendencias, 2da Edición. Univ. Autón. De Campeche, Univ) (696 p). Nal. Autón. de México.
Tricarico, E., Junqueira, A. O., & Dudgeon, D. (2016). Alien species in aquatic environments: A selective comparison of coastal and inland waters in tropical and temperate latitudes. *Aquatic Conservation: Marine and Freshwater Ecosystems, 26*(5), 872–891. https://doi.org/10.1002/aqc.2711
UAC. (2019). *Informe sobre toxicidad aguda en larvas de camarón expuestas a hidrocarburos*. Universidad Autónoma de Campeche.
UNEP, (2015). Blue economy: Sharing stories to inspire change. Recuperado de www.unep.org/greeneconomy
Upton, H. F. (2011). *The Deepwater horizon oil spill and the Gulf of Mexico fishing industry*. Congressional Research Service. https://sgp.fas.org/crs/misc/R41640.pdf
Valiela, I. (1995). *Marine ecological processes* (2nd. edn, 686 p). Springer. https://doi.org/10.1007/978-1-4757-4125-4.
van der Schatte Olivier, A., Jones, L., Vay, L. L., Christie, M., Wilson, J., & Malham, S. K. (2020). A global review of the ecosystem services provided by bivalve aquaculture. *Reviews in Aquaculture, 12*(1), 3–25. https://doi.org/10.1111/raq.12301
Vaquer-Sunyer, R., & Duarte, C. M. (2008). Thresholds of hypoxia for marine biodiversity. *Proceedings of the National Academy of Sciences, 105*(40), 15452–15457. https://doi.org/10.1073/pnas.0803833105
Velasco, L. A., & Barros, J. (2010). Spat collection and experimental culture of the Atlantic pearl oyster, *Pinctada imbricata* (Bivalvia: Pteriidae), under suspended conditions in the Caribbean. *Journal of the World Aquaculture Society, 41*(3), 281–297. https://doi.org/10.1111/j.1749-7345.2010.00371.x
Volety, A. K., Savarese, M., Tolley, S. G., Arnold, W. S., Sime, P., Goodman, P., Chamberlain, R., & Doering, P. H. (2009). Eastern oysters (*Crassostrea virginica*) as an indicator for restoration of Everglades ecosystems. *Ecological Indicators, 9*(6), S120–S136. https://doi.org/10.1016/j.ecolind.2008.06.005

Weis, J. S., & Weis, P. (2019). Impacts of xenobiotics on estuarine ecosystems. In *Basic environmental toxicology* (pp. 385–407). CRC Press. https://doi.org/10.1201/9781315138091-14

Wu, R. S. S. (2002). Hypoxia: From molecular responses to ecosystem responses. *Marine Pollution Bulletin, 45*, 35–45. https://doi.org/10.1016/S0025-326X(02)00061-9

Yogui, G. T., & Sericano, J. L. (2009). Polybrominated diphenyl ether flame retardants in the U.S. marine environment: A review. *Environment International, 35*, 655–666. https://doi.org/10.1016/j.envint.2008.11.001

Chapter 5
The Petroleum Spill at the Ixtoc-I Well in the Gulf of Mexico: Extractivism and Media Narratives

A. V. Rabelo-Avalos, D. E. Ramos-Muñoz, A. K. Nygren, and M. A. Díaz-Perera

5.1 Introduction

The chapter examines the logic of natural resource extraction in Mexico following the 1973 global energy crisis, together with the narratives, explanations, and arguments disseminated in the regional mass media after the 1979 oil spill from the *Ixtoc*-I well. Based on documents from the oil industry and reports in regional newspapers, this study aims to elucidate how the extractivist model of that time was linked to the causes and consequences of the spill, as well as how perceptions of the event's impact on coastal areas in the states of Campeche and Tabasco were shaped by journalistic sources.

The concept of "extractivism" is employed to understand the management logic of shallow-water oil exploitation. At the same time, the methodology of microanalysis was adopted to study the spill from the perspectives of the oil industry and journalistic reports. Key topics discussed include how extractivist dynamics influenced the handling of the spill's impact, accusations and forms of resistance expressed in the media that shaped perceptions in coastal society, and how the government prioritized the benefits of large-scale, intensive extraction for the national economy over the costs of its dire effects on coastal populations.

A. V. Rabelo-Avalos (✉) · D. E. Ramos-Muñoz · M. A. Díaz-Perera
El Colegio de la Frontera Sur, Villahermosa, Tabasco, México
e-mail: angela.rabelo@ecosur.mx; dramos@ecosur.mx; mdiaz@ecosur.mx

A. K. Nygren
University of Helsinki, Helsinki, Finland
e-mail: anja.nygren@helsinki.fi

A. Espinoza-Tenorio et al. (eds.), *The Fishing and Oil Industries in Coexistence*,
https://doi.org/10.1007/978-3-032-02621-7_5

5.1.1 Conceptual Framework

It is difficult to comprehend the *Ixtoc*-I spill in the Gulf of Mexico and its impacts unless disasters are viewed as complex processes that intertwine ecological, political, and social aspects (Bebbington et al., 2019; Nygren, 2021; Quist & Nygren, 2015). A key concept in this regard is extractivism, which refers to an intensive mode of exploiting natural resources that often overlooks local modes of production characterized by low degrees of in situ processing (Klein, 2017; Nygren et al., 2022; Willow, 2018). Extractivism responds, instead, to demand in global markets, often at the cost of degrading environments and the traditional lifeways of people in the affected areas (Bebbington et al., 2019; Rodríguez-Padilla, 2018).

Here, the spill is analyzed as a case of degradation caused by accelerated extractivism, understood as "an event or series of events that cause significant damage" and impede recovery (Perry, 2005). Unlike what may be considered a "normal," miniscule seepage of oil from drilling sites, the *Ixtoc*-I spill was not a natural event, but the materialization of a failure in exploration technology, a catastrophe of anthropogenic origin, and a case of a calamity in which the "production and reproduction of conditions of vulnerability [...] define and determine the magnitude of the effects" (García-Acosta, 2005, p. 23). The technological conditions of the well's owner, Mexico's national oil company, *Petróleos Mexicanos* (PEMEX), exacerbated the magnitude of the spill and prolonged the time required to control it, though the resulting oil slick endangered the populations along nearby coastlines, the regional ecology, and people's lives and livelihoods. Examining the *Ixtoc*-I disaster, undoubtedly a watershed moment in offshore extraction, also reveals the role of the regional press as a purveyor of information to the affected society, which visibilized the event and opened spaces for debate. The media shared narratives from numerous sources, which influenced how coastal societies interpreted, evaluated, and responded to the spill, as well as how the ensuing interactions were documented (Slovic, 1987).

Ample literature exists on petroleum extraction in Mexico and Latin America (Breglia, 2013; Quist & Nygren, 2015, 2019; Salas-Landa, 2016), with analyses of how extractive industries strive to minimize the damage they perpetrate, either by ignoring it or going to great lengths to cover it up (Laastad, 2022). Various studies on the coastal area of Tabasco have examined changes in land use, local lifeways, and social movements related to oil extraction (Díaz-Perera & De los Santos-González, 2021; García-Meza, 1993; Pinkus-Rendón & Contreras-Sánchez, 2012; Ramos-Muñoz et al., 2019; Rabelo-Avalos et al., 2021; Tudela, 1989; Uribe-Iniesta, 2016; Velázquez-García, 2016; Velázquez-Guzmán, 1982), but the social effects of oil spills have not been widely studied. This chapter seeks to contribute to filling this gap.

5.2 Materials and Methodology

The methodological focus is microanalytical. Through this approach (González-González, 1971, 1997), the *Ixtoc*-I spill is viewed as an explanatory event that, based on contextual details, enables us to understand larger scale phenomena, such as extractivism and disasters resulting from petroleum exploitation. Briefly, "[...] microhistory leans toward typicity; a taste for the everyday. Events of little weight or fame; events that do not raise clouds of dust; daily life events" (González-González, 1971: 227). Therefore, we set out from a general phenomenon (extractivism), then narrowed our focus to a specific event that we deemed simultaneously representative and exceptional (the *Ixtoc*-I spill). This allowed us to recognize reactions and comprehend the logics of extractivism that impact people's daily lives by maintaining the inertias of economic usufruct to the detriment of local lifeways. The microanalytical perspective allows researchers to scrutinize the details of events closely. In Giovanni Levi's words: "Microhistory as a practice is based in essence on reducing the scale of observation [to] a microscopic analysis and the intensive study of documentary material" (1996: 122). Hence, it fosters an examination of the furtive, that which is pursued or deliberately made invisible, exemplified here by the actions of the Mexican State after the *Ixtoc*-I spill.

Documentary analysis, in turn, aims to interpret the arguments, testimonies, and narratives presented in reports, texts, treatises, and journalistic notes. González-González' microhistory provides a methodological framework for detailed documentary analyses of the ins-and-outs of events that "[...] rarely leave abundant traces" (1971: 228) since companies dedicated to extractivism purposefully apply mechanisms to silence, ignore, and minimize environmental and social impacts in the areas they exploit (Nygren and Rabelo Avalos 2025). The challenge in the specific case of the *Ixtoc*-1 spill was to investigate an event that had been covered up; indeed, shrouded in secrecy by formidable political and economic powers. Our search for information centered on libraries, newspaper archives, and regional and national archives (Table 5.1).

The materials analyzed were codified using the following keywords: technical, vulnerability, environmental, solution-remedy, contamination, transformation, coastal, corporate culture, perception, and risk. These key terms were detected using *Atlas.Ti* software to construct networks and identify connections, coincidences, and differences in the information gathered from the diverse sources consulted. We relied on journalistic narratives to gauge public opinion on the environmental and social impact of the spill by compiling reports from 1979 to 1980 that appeared mainly in one regional newspaper. Our goal was to understand the impact of the spill on coastal societies in terms of people's daily lives and the environmental and social problems that emerged in relation to the accident. The sources were triangulated to ensure the validity and reliability of findings.

Table 5.1 Collections consulted (2023)

Archive or repository	Section
Archivo General de la Nación (AGN), Mexico City	*Centro de Ecodesarrollo, A.C.*, 1980
Archivo General de la Nación, Mexico City	Political and social research, 1980
Biblioteca Universitaria, Universidad Autónoma del Carmen, Ciudad del Carmen, Campeche	Journals, *Nosotros los petroleros*, 1979–1980
Biblioteca Universitaria, Universidad Autónoma del Carmen, Ciudad del Carmen, Campeche	Books dated 1878–1983
Biblioteca Pública Electrónica, Campeche, Campeche	Government reports, 1975–1983
Biblioteca Pública Municipal, Ciudad del Carmen, Campeche	Books dated 1878–1999
Biblioteca Campeche, Campeche	Government reports, 1981–1991
Biblioteca Pública del Estado José María Pino Suárez, Villahermosa, Tabasco	Newspaper *Rumbo Nuevo* (1979–1980)
Biblioteca Pública del Estado José María Pino Suárez, Villahermosa, Tabasco	Newspaper *Rumbo Nuevo* (1979–1980) illustration/caricatures
Biblioteca Pública del Estado José María Pino Suárez, Villahermosa, Tabasco	Newspaper *Novedades* (1983)
Biblioteca Yucatanense, Mérida, Yucatan	Newspaper *Diario de Yucatán*, 1979–1980
Archivo Histórico de Petróleos Mexicanos (PEMEX), Mexico City	Film archive (1979)
Archivo Histórico de Petróleos Mexicanos (PEMEX), Mexico City	Books, yearbooks, and reports
Used bookstores, Mexico City	Topics: Tabasco and petroleum

5.3 Results

5.3.1 Petroleum Extraction and Corporate Culture

Scrutinizing PEMEX before *Ixtoc*-I requires understanding the circumstances that led to the disaster. PEMEX was created during the process of nationalizing the oil industry in 1938 in response to internal demand for petroleum and, tangentially, to generate revenue for the State. However, in the 1970s, the company experienced an economic crisis caused by excessive debt, both external and internal (Bonilla, 1976, p. 115). The government's efforts to salvage PEMEX included intense drive to expand the oil industry. As a result, extraction and exportation became key features of the nations' political economy. This reorientation toward offshore extraction was accentuated by the international energy crisis and its impact on the United States, conditions that convinced Mexico's government that increasing export-oriented extraction would be the most effective way to propel the national economy, as the US offered markets, technology, and international financing for such operations. At the same time, recently discovered deposits in Campeche and Tabasco reinforced the authorities' high expectations for the oil project, despite "growing financial dependence" (Bonilla, 1976, p. 115). In reality, however, these measures initiated the country's profound financial and technological subjection to foreign interests in oil production (Bonilla, 1976; Wionczek, 1983).

The appointment of Jorge Díaz Serrano as the Director of PEMEX in 1976 reflected strongly this new political vision. President José López Portillo of the *Partido Revolucionario Institucional* (Institutional Revolutionary Party, PRI) named that specific individual to signal a change in the State's petroleum strategy that consisted of transitioning toward a more intensive model of exploitation that prioritized exports over the domestic market (Díaz-Serrano, 1989; Bonilla, 1976). In the mid-1970s, the corresponding authorities announced: "We are not oil-dependent (*petrolizados*), but we are now, and will be for many years, an oil-producing country" (Díaz-Serrano, 1989, p. 43). This confirmed the importance of revenue from petroleum. In the view of Morales et al. (1988: 77), the new policy did not contemplate rationalization measures, but focused exclusively on increasing production.

This extractivist orientation of Mexico's petroleum industry expanded rapidly after 1973 (Morales et al., 1988, p. 47). Producing for export, extracting large volumes of petroleum to increase revenue, and doing so quickly, however, demanded intensifying the transfer of technologies. The policies adopted by the federal government and PEMEX defined the states of Campeche and Tabasco as key areas for oil extraction. Meanwhile, demand for oil in industrialized countries continued to grow, and other producing nations led the resulting market surge (Morales et al., 1988, p. 48). To avoid being left behind, Mexico, logically, had to reinvest its oil revenue in new extractivist projects. In fact, government intervention in the petroleum industry was nothing new in Mexico, but its economic weight increased with the income that entered official coffers through PEMEX. Mexico succeeded in emerging as a major oil-exporting nation (Bonilla, 1976), but that transition entailed shifting from "administering austerity" to "administering abundance" (Morales et al., 1988, p. 80), in the famous phrase uttered by President López Portillo in the aftermath of the *Ixtoc*-I spill.

During this period, relations between PEMEX and coastal communities in Campeche and Tabasco began to show signs of strain caused by incompatibility. For example, according to Moreno-Andrade (2007), the massive presence of PEMEX operators had increased the cost of living in those zones. In addition, the federal government began to grant fishing permits and introduced support programs for fishers in an attempt to regulate their activities and promote commerce in fishery products, aiming to industrialize the sector and mitigate a greater potential dependence on petroleum revenues (Pinkus-Rendón & Contreras-Sánchez, 2012).

The international petroleum market at that time was fluctuating but the fishing sector was beginning to prosper. Indeed, the increase in the volume of the captures of numerous species was so great that the value of fishery production was comparable to that of oil (Petróleos Mexicanos, 1980). In 1979, shrimp fishing in Campeche experienced a bonanza (Vázquez-Botello et al., 2005), as the Gulf of Mexico generated "36% of fish for human consumption and 39% of shrimp production" nationally. At the same time, small-scale, artisanal fishing was spreading along the coast (Petróleos Mexicanos, 1980, p. 209). However, it was precisely around that time that explorations in the Gulf of Mexico detected the enormous Cantarell reserves, a discovery that led directly to the drilling of the *Ixtoc*-I well in the *Sonda de Campeche* in the southwestern sector of the Gulf, 94 km northwest of Ciudad del Carmen (Petróleos Mexicanos, 1980 p. 3). That was the site that would be impacted by the largest oil spill in the Gulf of Mexico (Jernelöv & Lindén, 1981).

The events that were about to unfold around the Ixtoc-I platform were unprecedented: the scene of the well's structure consumed by flames was one that nobody could have imagined. Although Mexico's oil industry had certainly seen its share of spills and—according to Díaz-Serrano (1989) and Meyer and Morales (1990), quite often and on a larger scale—they had always occurred on land, not out at sea. The measures developed and employed to manage and control those accidents differed greatly in many fundamental aspects from the ones that would be required to manage the *Ixtoc*-I catastrophe. This not only complicated the authorities' response but also revealed that despite PEMEX's status as a vital leader in Mexico's economy and politics, it was plagued by vulnerabilities in key areas, such as safety, technology, and regulation.

5.3.2 Effects Mentioned in Journalistic Sources

The newspaper *Diario de Yucatán*, published in Mérida, has a predominantly regional readership as it circulates daily in three states: Campeche, Tabasco, and Yucatán (Jessica Ruiz Rubio, Diario de Yucatán, 03/05/2019). Characterized by a critical stance, analytical focus, and interest in social and political topics, by 1979, it had garnered a broad audience. Due to its proximity to the disaster zone, it was able to provide firsthand coverage of events, though in our documentary analysis, we found that some of the headlines it published were tendentious, even alarmist. The *Diario* played a significant role in disseminating information about the coastal regions of Campeche and Tabasco, often incorporating local perspectives. Its coverage from June 1979 to March 1980 totaled some 530 notes, including reports, editorials, and testimonies. From a partial, yet critical, perspective, the newspaper strove to inform the public and convey the true dimensions of the spill's impact, coupled with people's impressions of the event.

Those notes, which appeared almost daily, reflected the newspaper's journalistic values as they detailed how the disaster had occurred and documented the spread of the oil slick across the Gulf. It did not, however, publish any maps to illustrate the disaster and its reports tended to be somewhat repetitive, as specific phrases—"an out of control well," "flows of mud escaped and caused the fire," "the Ixtoc disaster," and "Díaz-Serrano on PERMARGO"—multiplied in the headlines, not only of the *Diario*, but also of *Rumbo Nuevo*, a newspaper edited in Tabasco that reported on the disaster on a lesser scale. The company mentioned in those headlines—*Perforaciones Marinas del Golfo* (PERMARGO)—was one of the first to establish operations in the Gulf of Mexico (Diario de Yucatán, 03/06/1979).

The *Diario* stated that PEMEX sought to minimize the spill's importance (23/07/1979); in fact, one public figure, the president of the National Chamber of the Fishing and Aquaculture Industry (*Cámara Nacional de la Industria Pesquera y Acuícola*) in Ciudad del Carmen, stated simply, "Nothing happened [...] there is calm, because what has been said about balls of tar washing up on beaches [does not] represent any problem of seawater contamination" (Diario de Yucatán, 01/10/1979). Operating in a context of uncertainty, political and union leaders likely opted to trust the information that PEMEX provided, which insisted that the crude oil spilled "was not toxic" (Diario de Yucatán, 01/10/1979; Petróleos Mexicanos, 1980).

Published articles also highlighted internal divisions in the fishing sector. While some fishers in the spill area reported finding "enormous amounts of dead fish," others affirmed that everything was normal (Díaz-Serrano, 1989; Petróleos Mexicanos, 1980). Dr. María Luisa Sevilla, Professor at Mexico's *Instituto Politécnico Nacional* (National Polytechnical Institute, IPN), and later the Director of the *Escuela Nacional de Ciencias Biológicas* (National School of Biological Sciences, ENCB), announced, strangely, that "those who live from exports, like fishing cooperatives, have not protested over the [danger] that many species face, though even minor contamination impedes their products from being accepted in [foreign markets]" (Diario de Yucatán, 27/06/1979). A few weeks later, a social movement launched legal demands seeking compensation for the severe damage that members of fishing cooperatives and their operations had suffered: "it emerged that members of fishing cooperatives, mainly in Ciudad del Carmen, were preparing a movement to sue PEMEX for the damage from the *Ixtoc*-I spill that affected the marine flora and fauna" (Diario de Yucatán, 23/07/1979). For some reason, however, no concrete information on their demands appeared in the ensuing months, only news of specific events.

The *Diario's* journalistic reports captured the perceptions of coastal society, including the voices of riverine fishers, residents, and cooperative leaders, who commented on how the spill had affected their fishing. Early on, the leader of one cooperative stated that "exploration by PEMEX and well-drilling have killed thousands of tons of species and irreversibly damaged the marine biology." He called for the support of "[...] the authorities of this branch" to stop what members called "this criminal negligence" (Diario de Yucatán, 08/06/1979). However, just a month later, the Head of the Department of Fisheries (*Departamento de Pesca*) reiterated that no adverse effects had occurred: "at no time has work been suspended on the three principal fishing banks in the Gulf as a consequence of the *Ixtoc*-I fire" (Diario de Yucatán, 20/07/1979). Reports like these suggest that PEMEX never presented data on the effects of fishing or the marine environment, instead remaining silent and going on the offensive to discredit dissenting voices. Weeks after PEMEX closed the well, it repeated that "fishing has not been affected" (Diario de Yucatán, 23/03/1980; Petróleos Mexicanos, 1980). In a sardonic tone, the political cartoonist Balmaceda of the *Rumbo Nuevo* newspaper drew a cartoon with the caption: "*Ixtoc*-I is not dangerous." It is worth noting that if studies on the impact on health had been conducted at that time, our research would have likely encountered them.

Despite the spread of the spill urther into the Gulf, economic activities along the coast continued. The export-oriented shrimp cooperatives "kept silent on the situation, awaiting official reports on the deterioration of phytoplankton." According to the Head of the local Fisheries Office, shrimp production had not declined but, in reality, "had increased," and "the fire at *Ixtoc*-I in the Gulf of Mexico since June third [1979] has not affected fishing" (Diario de Yucatán, 05/11/1979). Those notes gave the impression that the slick and fishing activity coexisted problem-free. Other observers also affirmed that work was proceeding "normally" with "no reports to date that shrimp boats or scaly fish [operators] are catching contaminated product." In fact, the media reported that the fleets out of Ciudad del Carmen and Campeche were operating with no difficulties (Diario de Yucatán, 05/11/1979), thereby reinforcing the view that the authorities had the situation well in hand. While the press apparently opened spaces for discussion, debate, and pressure, the expressed viewpoints suggested that the spill was "under control" in apparent efforts to minimize its effects or mitigate legal claims or social mobilizations in the region.

Other voices, however, gave more critical opinions. Felipe Villanueva, the leader of the cooperative, *Pulperos de la Sonda de Campeche* (Octopus Fishers of the Campeche Sound), expressed fear that regional fishing would suffer "irreversible damage" (Diario de Yucatán, 14/06/1979). Leaders of other fisher groups chimed in. For instance, the president of the *Federación de Cooperativas Pesqueras* (Federation of Fishing Cooperatives) in the state of Campeche stated: "we consider that fishing is more important than oil [because] the contamination caused by the spill from the *Ixtoc*-I well is affecting food for [human consumption]" (Diario de Yucatán, 09/06/1979). Whether those reports minimized or exaggerated the spill's effects, they featured mainly the voices of representatives of various cooperatives, perhaps due to the political power they wielded, as illustrated in *Rumbo Nuevo*.

Other newspaper articles revealed that the naval sector had issued several warnings about the spill; for example, the Fifth Naval Military Zone prohibited "fishing in a vast zone of the Gulf of Mexico" (Diario de Yucatán 12/06/1979), stating that "the damage caused by oil has been underestimated" (Diario de Yucatán 17/06/1979). This announcement earned mostly sarcastic responses, for at the outset, few voices sounded alarms over the spill's impact, although the information from the Navy clearly drew attention to the danger. The naval sector continued to disseminate information on the movement of the oil slick across the Gulf and on the actions that were underway to prevent contamination and other serious effects. For example, the Commander and Vice-Admiral of this Zone, said: "we do not want to be alarmists, but […] calculations by the team of researchers from the Mexican Navy [show] the spill advancing westward at a rate of 9 kilometers per day." They also expressed concern that it could reach the fishing zone of Veracruz (Diario de Yucatán, 11/07/1979).

Naval restrictions on fishing areas were being modified constantly. The Commander of the V Naval Military Zone reported that "the restriction on fishing and navigation in the area between the Tupilco and Xicalango lighthouses, to [a distance of] 12 miles from the coast has been lifted," but "fishing and navigation are prohibited between those points, from 12 to 50 miles […] between [those] lighthouses, a distance of some 65 nautical miles" (Diario de Yucatán, 17/06/1979). Announcements of that kind generated unease in the fishing sector during the

9 months that the spill lasted, especially for the shrimp fleets out of Campeche and Tabasco, where eventually some 500 boats were drydocked (Diario de Yucatán, 27/06/1979). They also directly contradicted Díaz-Serrano, who wrote that "fishing was never suspended due to causes attributable to *Ixtoc*-I" (1989: 146).

These reports explained that winds were modifying the slick (Diario de Yucatán, 03/10/1979), while contemporary descriptions suggest that the only observers who could understand the direction in which it was moving were likely fishers. During the *Diario's* 9-month coverage of the event, it never published a map or cartographic image of the slick, even though reconnaissance flights were providing PEMEX with precise information on its evolution. Figure 5.1 displays data from those flights at two moments during the spill.

Fishing, as we know, was essential in the region at that time, as production in Campeche alone generated approximately two billion pesos per year (Diario de Yucatán, 19/10/1979) and was a significant source of direct employment. As a result, the various prohibitions imposed by the naval sector had serious economic and social consequences for fishers, though the regional press did not report on the economic effects on producers, their families, the fishing cooperatives, or the guild as a whole, nor did it address the issues of damage to marine fauna and regional biodiversity.

At certain moments, some newspaper articles did raise challenges and criticized the actions that agencies had implemented to remedy the spill's effects—for example, "PEMEX does not inform anyone"—and raised their voices to demand "that PEMEX tell us the truth" (Diario de Yucatán 09/06/1979). Beyond these journalistic reports, some documents we reviewed indicated that a tense relationship existed

Fig. 5.1 Accumulated information from observation flights, July-October 1979; elaborated by Luis Uribe with data from a PEMEX report

between the federal government and PEMEX (Díaz-Serrano, 1989). The Presidency and the company jointly approved various actions intended to control the spill, including purchases of specialized materials and the hiring of national and foreign personnel and boats (Díaz-Serrano, 1989; Krauze, 1997; Petróleos Mexicanos, 1980). In his "Annual Report on Government" (*Informe del Gobierno*) of September 1st, 1979, President López Portillo stated that Mexico now had to "administrate abundance" (López-Portillo, 1979), for he was buoyed by the bonanza in oil exploitation, the recently discovered reserves, and high international prices (Diario de Yucatán, 19/10/1979). Hence, despite an uncontrolled oil spill and the fact that broad expanses of the Gulf of Mexico were in flames, petroleum revenue was still portrayed as the resource that would allow the government to fulfill, albeit not with instant success, all its proposals for the end of that year (Rabelo-Avalos, 2024).

The journalistic accounts reviewed indicate that several opposition parties—the National Action Party (*Partido de Acción Nacional*, PAN), the Authentic Party of the Mexican Revolution (*Partido Auténtico de la Revolución Mexicana*, PARM), and the Popular Socialist Party (*Partido Popular Socialista*, PPS)—questioned PEMEX's actions during the Ixtoc-I disaster. The monetary losses and outlays it caused led to its Director being called before Congress to report on the solution (Diario de Yucatán, 24/07/1979). The *Diario* recommended penal sanctions for those responsible and urged the government to take action even if "functionaries or employees of PEMEX" were involved (Diario de Yucatán 25/07/1979). To date, however, our research has not uncovered any evidence of such judicial processes.

While initial reports did not cite the voices of government officials, some functionaries later spoke of the spill as a lesson "for future years" (Diario de Yucatán, 22/07/1979). One significant spokesperson was the Senate leader, Joaquín Gamboa Pascoe (PRI), who stated: "we must see the *Ixtoc*-I problem in its true dimensions compared to the total advance that PEMEX has had in recent years," adding that "not everything is a tragedy in this affair" and "Mexico will continue to exploit its petroleum industry" despite the damage (Diario de Yucatán, 22/07/1979). With the phrase, "its true dimensions," Gamboa Pascoe expressed his view that the environmental degradation and impact of the spill on coastal societies in the Gulf were clearly of secondary importance compared to the advancement the country was achieving thanks to its oil revenue.

Officials entrusted with the task of elaborating financial reports also appeared in these notes: the congressmen Antonio Tenorio Adame (PRI), Edilio Hinojosa (PARM), and Arturo Salcido insisted that PEMEX and its "mafia" were hiding the truth from the people and that "problems of corruption" existed (Diario de Yucatán 17/07/1979). This raised suspicions about what was really happening in the *Sonda de Campeche*. Likewise, the *Diario del Sureste* discredited information published in other newspapers—the national paper *Excélsior*, for example—as is evidenced by the political cartoons it published. In summary, the regional and national journalistic media gave expression to diverse voices, sometimes as a consequence of debates, at others due to governmental pressure. In general, these newspapers succeeded in interviewing key regional actors and reporting their views during the spill, especially fishers, cooperative members, community leaders, and residents, as well as representatives from other sectors of these coastal societies. Our search for sources on the topic of claims exemplified this diversity of voices.

The *Diario de Yucatán* did not mention some important aspects that other journalistic sources did; for example, the national daily *El Universal* published notes with headlines like: "Fish dying due to the fire on the platform" ("*Mueren peces a causa del incendio de la plataforma,*" 06/06/1979), "Accusations against PEMEX are politically motivated" ("*Las denuncias contra Pemex son por motivos políticos,*" 13/06/1979), "The country's richest petroleum deposits threaten one of the largest continental fishing banks" ("*El manto de hidrocarburos más rico en el país amenaza a uno de los mayores bancos pesqueros continentales,*" 14/06/1979) and "It is said that the oil slick has not yet damaged fish farms" ("*Que la mancha de aceite aún no ha dañado criaderos,*" 14/06/1979). In summary, those sources suggested that accusations against PEMEX were merely opportunistic political attacks. However, as Velázquez-Guzmán (1982) and Beltrán (1993) observed, it took a few months for denouncements to be forged into what they called an "industry of claims" ("*industria de la reclamación*"). The fact that the *Diario de Yucatán*, unlike other papers (e.g., El Universal), did not publicize those claims and accusations indicates that distinct journalistic media adopted different critical views in response to the spill.

In some of its interviews, the *Diario de Yucatán* delved into legal repercussions for PEMEX due to "the damage caused to the marine ecology" (20/07/1979). The Head of the Department of Fisheries stated that "at no time did I think about charging PEMEX" because there was no certainty as to who exactly was responsible for the spill, adding that he could not issue a response until scientific analyses had evaluated the possible damage (Diario de Yucatán, 20/07/1979). At that time, a political cartoon was published that showed the US registering a complaint against Mexico. This triggered a response from Díaz-Serrano, who criticized the media for misinforming the public: "Each drop of misinformation creates [...] the idea that a disaster happened [one] that we caused [and] that we [combatted] with no regard for the most elemental technical and safety factors" (1989: 153). A PEMEX report dated in 1980 that we failed to identify any responsible parties and claimed, moreover, that no ecological damage had occurred (Rabelo-Avalos, 2024).

One significant difference between the versions offered by PEMEX and the *Diario de Yucatán* revolved around the issue of oil evaporation or depuration in the sea. While PEMEX firmly maintained that the spill would be cleared away by the regulatory properties and capacity of the marine ecosystem itself, the *Diario* insisted that it was necessary to "abandon the idea that the products leaked into the sea will be diluted and disappear—as if by magic—without leaving a trace" (14/06/1979). PEMEX reported that 475,000 metric tons of crude oil had spilled. According to the analysis by Sun et al. (2015), this means that only the 2010 Deepwater Horizon spill was of greater proportions. Jernelöv and Lindén (1981) stated that 50% of the spill evaporated into the atmosphere, 25% sank, and 12% was degraded biologically or photochemically. The rest was either burned off at the well—the source of the spill—or reached the shores of Mexico and Texas, where it was removed mechanically. This indicates that the amount of petroleum spilled exceeded the ecosystem's capacity to degrade it (Alcalá-Moya & Camargo, 2011).

Surprisingly, some parties went so far as to argue that certain benefits had accrued from the spill, citing data which indicated that airplane service to *Ciudad del Carmen* had increased, that it fostered a significant influx of technicians and modern machinery, and even led to the installation of a Telex system in the region (Diario de Yucatán, 04/11/1979).

As our analysis confirmed, over time, maritime currents deposited oil residues as far away as beaches in Texas (Diario de Yucatán 02/08/1979), the state that eventually presented a lawsuit against SEDCO,[1] the firm that leased the drilling equipment to the Mexican company PERMARGO, the contractor that PEMEX had hired (Diario de Yucatán, 23/10/1979). The charges proceeded, and Texas won its case against the Mexican government (Gómez-Robledo Verduzco, 2003). The spill was finally contained on March 9, 1980. According to the report by Sun et al. (2015) although the Deepwater Horizon spill in the Gulf of Mexico had a broader extension, the Ixtoc event required more time to be resolved, and the territories ultimately affected were more extensive. It is, therefore, difficult to accept the idea that *Ixtoc*-I did not have severe impacts on marine and coastal environments, as well as on the lifeways and living conditions of fishers in the Gulf and communities along the coast.

5.4 Discussion and Conclusion

Achieving this historical reconstruction of the *Ixtoc*-I spill was not an easy task, due to such factors as negligence in the maintenance of PEMEX's archives, the oil company's efforts to ignore and suppress dissident or critical narratives, and the exclusion and shrouding to which information was subjected at the time. Although it is well known that large investments were made to evaluate the damage (Toledo, 2002), in 2023, we found it impossible to access any works that might substantiate that claim. By 1980, social perceptions of the spill revealed that people were unsure of what had really happened during the Ixtoc-1 disaster and its aftermath, and that information on those events was sorely lacking (García-Acosta, 2005, p. 13). The key decisions taken to contain the slick reflected the enormous importance of oil in the national context. However, as often occurs in such cases, the actions that followed "omitted the social context" (García-Acosta, 2005, p. 13), especially at the local level. The worldwide attention that the spill drew became a political motor for incalculable monetary outlays, but PEMEX maintained a strategy of stolid silence.

[1] SECDO Inc. Conducted exploration and exploitation activities. It was involved in the accident at -I because it leased a semi-submergible platform for drilling the well to the Mexican company *Perforaciones Marítimas* (PERMARGO), whose founding shareholders included Jorge Díaz Serrano, and its board of advisers included then Vice-President of the United States, George Bush; see Márquez (1984).

The real dimensions of *Ixtoc*-I that the Senate leader, Joaquín Gamboa Pascoe, mentioned focused not on the damage from the spill, but rather on the benefits that Mexico needed and could only obtain with the revenue generated by oil extraction and exportation. But those operations, of course, obeyed the logic of the international petroleum market—outlined, for example, by Bebbington et al. (2019) and Rodríguez-Padilla (2018)—not the authorities' efforts to reap benefits for the nation. As our research demonstrates, neither environmental degradation, nor the crisis that disrupted fishing—both small scale and industrialized—in the Gulf of Mexico, nor the broader impacts on coastal societies, would be allowed to interrupt extraction. Although the *Ixtoc*-I well burned for 3 months and released thousands of barrels of crude oil, Mexico's President continued to uphold his pledge to "administrate the abundance" that petroleum brought (López-Portillo, 1979). Against the greater good of the nation, it was in PEMEX's best interest to leave the clean-up to evaporation and natural depuration and deal with some of the local social effects by implementing support programs or building infrastructure.

Coastal societies, meanwhile, were forced to readapt to the new scenario that the spill and accelerated extractivism had forged in the region. Some fishing cooperatives disappeared, and many others merged to strengthen their efforts to negotiate a fairer distribution of resources from oil operations and compensation from the government and PEMEX.

Regarding the spill itself, it is essential to note that during that period, PEMEX was highly dependent on foreign technology, which was necessary to profitably exploit offshore oil wells. This condition aligns with the extractivist criteria described by Heikkinen et al. (2023). The inability to successfully develop its own means of extracting crude oil placed PEMEX at a disadvantage due to the extreme backwardness of its technological apparatus (Slovic, 1987).

Meanwhile, the coastal societies affected could only respond to the disaster based on the information provided by media sources, such as the *Diario de Yucatán* and other newspapers, reports by the Navy (via radio), and word of mouth. However, key questions, such as how toxic the spill was for their key fishing species, were never clarified (Soto et al., 2014). At certain moments, the Navy acted as an intermediary between the spill and local people by elaborating charts to monitor the slick, triggering alerts, and imposing bans on fishing. Still, the culture of opacity that shrouded the disaster (Ramos-Muñoz et al., 2019) meant that local communities received only a minuscule amount of reliable information, so their imaginings of the real threat that the oil platforms posed to their lifeways grew (Quist & Nygren, 2019). As mentioned above, the newspaper we analyzed in depth never published maps that could have alleviated the fears of societies along the coast; instead, opting to simply narrate events through the voices of cooperative leaders and Navy personnel, transmit technical data on the control and closure of the Ixtoc-I well, and publish satirical political cartoons.

Although limited by the availability of sources and the potential biases in those that we were able to access, our study establishes a solid foundation for future research that seeks to deepen the historical analysis of documentary sources related to the *Ixtoc*-I spill. It demonstrates that, in PEMEX's view, what occurred at *Ixtoc*-I was simply an isolated event of a crude oil spill that was resolved by the natural depuration mechanisms and capacity of the marine ecosystem. Despite the scarcity of documentary information, the microhistory approach that we adopted in our work allowed us to connect previously hidden details to broader historical processes and demonstrate that the State's actions, which largely ignored the disaster in favor of emphasizing the benefits of intensive petroleum extraction for the nation, violated the coastal societies that came to perceive offshore platforms as a new, dire threat to their lifeways and livelihoods.

Acknowledgments The first author would like to thank the Secretariat of Science, Humanities, Technology, and Innovation (SECIHTI), formerly CONACYT, for the financial support she received during her Master's studies, which enabled her to carry out this research. This research is part of the project "The challenging coexistence of coupled socio-ecological systems: the fishing and oil industries in the *Sonda de Campeche,*" supported by the National Council of Humanities, Science and Technology of Mexico (CONAHCyT). Artificial intelligence (AI) was utilized to assist in selecting keywords and condensing the abstract.

References

Alcalá-Moya, G., & Camargo, A. (Eds.). (2011). *Pescadores en América Latina y el Caribe: Espacio, población, producción y política* (Vol. 1). Facultad de Ciencias, Universidad Nacional Autónoma de México.

Bebbington, A., Abdulai, A. G., Bebbington, D. H., Hinfelaar, M., & Sanborn, C. (2019). *Gobernanza de las industrias extractivas: Política, historia, ideas*. Universidad del Pacífico.

Beltrán, J. E. (1993). *Los impactos del petróleo. En Tabasco: Realidad y perspectivas* (pp. 1–10). Gobierno del Estado de Tabasco.

Bonilla, A. (1976). La política petrolera mexicana. Problemas del Desarrollo. Revista Latinoamericana de. *Economía, 7*(27), 102–117.

Breglia, L. (2013). *Living with oil: Promises, peaks, and declines on Mexico's Gulf Coast*. University of Texas Press.

Díaz-Perera, M. Á., & De los Santos-González, C. C. (2021). *La frontera olvidada: El poblamiento costero de Tabasco a través de cuatro momentos decisivos, 1518–2020*. El Colegio de la Frontera Sur.

Díaz-Serrano, J. (1989). *Yo, Jorge Díaz Serrano: La encubierta de ambición por el poder, escrita en la soledad de una celda*. Planeta.

García-Acosta, V. (2005). El riesgo como construcción social y la construcción social de riesgos. *Desacatos, 19*, 11–24.

García-Meza, N. E. (1993). *El caso del Pacto Ribereño. En Tabasco: Realidad y perspectivas II* (pp. 135–147). Gobierno de Tabasco.

Gómez-Robledo Verduzco, A. (2003). *El caso del Ixtoc-I en Derecho Internacional. En Biblioteca Jurídica Virtual (Capítulo, Colecciones [29594])*. Universidad Nacional Autónoma de México.

González-González, L. (1971). Microhistoria para Multiméxico. *Historia Mexicana, 21*(2), 225–241.

González-González, L. (1997). *Otra invitación a la microhistoria*. Fondo de Cultura Económica.

Heikkinen, A., Nygren, A., & Custodio, M. (2023). The slow violence of mining and environmental suffering in the Andean waterscapes. *The Extractive Industries and Society, 14*, 1–10.

Jernelöv, A., & Lindén, O. (1981). Ixtoc-I: A case study of the world's largest oil spill. *Ambio*, 299–306.

Klein, A. (2017). Pioneering extractive sector transparency: A PWYP perspective on 15 years of EITI. *The Extractive Industries and Society, 4*(4), 771–774.
Krauze, E. (1997). *La presidencia imperial: Ascenso y caída del sistema político mexicano, 1940–1996*. Tusquets Editores.
Laastad, S. G. (2022). Mapping terrains of struggle: State, space and the spatiality of oil mobilization in Ecuador and Peru. *Geoforum, 137*, 12–21.
Levi, G. (1996). "Capítulo 5. Sobre microhistoria", en Burke, P. (Ed). Formas de hacer historia, Alianza Universidad, 119–143.
López-Portillo, J. (1979). *Tercer informe presidencial*. Presidencia de la República.
Márquez, M. (1984). Las lecciones del accidente del pozo Ixtoc-I. *Cuadernos sobre prospectiva energética, 48*, 38–40.
Meyer, L., & Morales, I. (1990). *Petróleo y nación: La política petrolera en México (1900–1987)*. Petróleos Mexicanos, Secretaría de Energía Minas e Industria Paraestatal, Colegio de México, Fondo de Cultura Económica.
Morales, I., Escalante, C., & Vargas, R. (1988). *La formación de la política petrolera en México, 1970–1986*. El Colegio de México.
Moreno-Andrade, S. H. (2007). *Dilemas petroleros. Cultura, poder y trabajo en el Golfo de México*. Centro de Investigaciones y Estudios Superiores en Antropología Social, Publicaciones de la Casa Chata.
Nygren, A. (2021). Water and power, water's power: State-making and socionature shaping volatile rivers and riverine people in Mexico. *World Development, 146*, 105615.
Nygren, A., Kröger, M., & Gills, B. (2022). Global extractivism and transformative alternatives. *The Journal of Peasant Studies, 49*(4), 734–759.
Nygren, A., Rabelo Avalos, A. V. (2025). Oil extraction and indirect dispossession: Responsibility and resistance in Southern sacrifice zones. *The Journal of Peasant Studies*. https://doi.org/10.1080/03066150.2025.2480200
Perry, R. W. (2005). Disaster exercise outcomes for professional emergency personnel and citizen volunteers. *Journal of Contingencies and Crisis Management, 13*(2), 47–57.
Petróleos Mexicanos (PEMEX). (1980). *Combate del derrame del pozo Ixtoc 1: Informe de los trabajos realizados para el control del pozo Ixtoc 1, el combate del derrame de petróleo y determinación de sus efectos sobre el ambiente marino*. PEMEX.
Pinkus-Rendón, M. J., & Contreras-Sánchez, A. (2012). Impacto socioambiental de la industria petrolera en Tabasco: el caso de la Chontalpa. *Limina R, 10*(2), 122–144.
Quist, L. M., & Nygren, A. (2015). Contested claims over space and identity between fishers and the oil industry in Mexico. *Geoforum, 63*, 44–54.
Quist, L. M., & Nygren, A. (2019). Debating the unknowns of marine oil exploration in Mexico. *The Extractive Industries and Society, 6*, 855–862.
Rabelo-Avalos, A. V. (2024). *¿Historia olvidada? El derrame del Ixtoc-I, Pemex y la sociedad costeña en México* (Master- thesis). El Colegio de la Frontera Sur.
Rabelo-Avalos, A. V., Ramos Muñoz, D. E., Díaz-Perera, M. Á., & Mesa-Jurado, M. A. (2021). El petróleo en Tabasco: Propuesta para una periodización. *Revista Mexicana de Sociología, 83*(1), 129–155.
Ramos-Muñoz, D. E., Ramos-Reyes, R., Zamora-Cornelio, L. F., Hernández-De la Cruz, A., & Espinoza-Tenorio, A. (2019). Exclusión en el Golfo de México: Una visión desde los pescadores sobre la industria petrolera. *Ciencia Pesquera, 31*(1), 83–101.
Rodríguez-Padilla, V. (2018). Petróleo y minería: México en la senda del extractivismo. *Estudios Mexicanos, 34*(3), 279–304.
Salas-Landa, M. (2016). Crude residues: The workings of failing oil infrastructure in Poza Rica, Veracruz. *Mexico. Environment and Planning A, 48*(4), 718–735.
Slovic, P. (1987). Perception of risk. *Science, 236*(4799), 280–285.
Soto, L. A., Vázquez-Botello, A., Licea-Durán, S., Lizárraga-Partida, M. L., & Yáñez-Arancibia, A. (2014). The environmental legacy of the Ixtoc-I oil spill in Campeche Sound, southwestern Gulf of Mexico. *Frontiers in Marine Science*.

Sun, S., Hu, C., & Tunnell, J. W. (2015). Surface oil footprint and trajectory of the Ixtoc-I oil spill determined from Landsat/MSS and CZCS observations. *Marine Pollution Bulletin, 101*(2), 632–641.

Toledo, V. M. (2002). *Ecología, espiritualidad y conocimiento: De la sociedad del riesgo a la sociedad sustentable*. Siglo XXI Editores.

Tudela, F. (1989). *La modernización forzada del trópico: El caso de Tabasco*. El Colegio de México.

Uribe-Iniesta, R. (2016). *Tiempos y procesos en la constitución de un espacio regional: El caso de Tabasco*. CRIM-UNAM.

Vázquez-Botello, A., Rendón von-Osten, J., Gold-Bouchot, G., & Agraz-Hernández, C. (Eds.). (2005). *Golfo de México, contaminación e impacto ambiental: diagnóstico y tendencias* (2ª ed.). Universidad Autónoma de Campeche et al.

Velázquez-García, M. A. (2016). Recursos del estado mexicano contra los movimientos sociales: Las distintas manos de leviatán. *Estudios Sociológicos, 34*(101), 247–272.

Velázquez-Guzmán, M. G. (1982). Afectaciones petroleras en Tabasco: El movimiento del Pacto Ribereño. *Revista Mexicana de Sociología, 44*(1), 167–187.

Willow, A. J. (2018). *Understanding extrACTIVISM: Culture and power in natural resources disputes*. Routledge.

Wionczek, M. S. (1983). Algunas reflexiones sobre la futura política petrolera de México. *Desarrollo Económico, 23*(89), 59–78.

Chapter 6
Detection and Classification of Surface Oil in the Southwestern Gulf of Mexico

J. B. Cruz-Pech, A. Uribe-Martínez, A. Espinoza-Tenorio, and E. Cuevas

6.1 Introduction

The oil industry contributes approximately 2.5% to the global Gross Domestic Product (GDP) and provides nearly a third of the world's energy consumption (Quiroa, 2019). However, oil industry operations entail significant environmental risks throughout the production chain, including exploration, drilling, exploitation, and transportation. Accidental spills or routine discharges of petroleum products into coastal and marine ecosystems have severe, long-term consequences for ecological processes (e.g. Deepwater Horizon Natural Resource Damage Assessment Trustees, 2016; Marzialetti, 2012) and other economic activities, such as fisheries (Andrews et al., 2021). The recovery time of affected ecosystems depends on the quantity and type of oil spilled (Hidalgo, 2009).

The Gulf of Mexico is recognized as one of the world's Large Marine Ecosystems (Sherman, 1994; Sherman & Hempel, 2009) and ranks among the most productive petroleum regions globally, containing over 80% of Mexico's total crude oil reserves (Brooks, 1990; Ponce-Vélez & Botello, 2005). The *Sonda de Campeche* represents the primary oil exploration zone in the southwestern Gulf of Mexico (SWGoM).

J. B. Cruz-Pech · A. Espinoza-Tenorio
El Colegio de la Frontera Sur, Lerma, Campeche, México
e-mail: johnny.cruz@posgrado.ecosur.mx; aespinoza@ecosur.mx

A. Uribe-Martínez (✉) · E. Cuevas
Instituto de Investigaciones Oceanológicas, Universidad Autónoma de Baja California, Ensenada, Baja California, México
e-mail: abigail.uribe@uabc.edu.mx; eduardo.cuevas@uabc.edu.mx

A. Espinoza-Tenorio et al. (eds.), *The Fishing and Oil Industries in Coexistence*,
https://doi.org/10.1007/978-3-032-02621-7_6

The presence of oil in this region from both natural sources, such as biogenic emissions, and anthropogenic sources, including industrial and municipal wastewater discharges, as well as vessel emissions. Notably, locally referred to as "*chapopoteras*," natural seeps are geographically concentrated near oil platforms (Machain-Castillo et al., 2019; Torres et al., 2018).

The *Sonda de Campeche* has experienced numerous spills over 50 years of oil exploitation. While the 1979 Ixtoc spill remains the most severe incident to date (Soto et al., 2014), the ongoing frequency of spills continues to threaten coastal ecosystems and biodiversity across all SWGoM (García-Cuéllar et al., 2004; Uribe-Martínez et al., 2024). These events have a significant impact on marine ecosystems, fisheries, and public health (Wakida-Kusunoki & Caballero-Chávez, 2009). Additionally, the oil industry has historically generated conflicts with the fishing sector, a vital economic activity for local communities (Quist, 2016, 2019). Due to complex economic and political factors that impede effective monitoring of Mexico's oil industry (Checa-Artasu, 2014; Espinoza Tenorio, 2019; Salazar-De la Cruz et al., 2020), the timing, frequency, and impacts of spills in the SWGoM remain poorly documented. This lack of information hinders a comprehensive assessment of both immediate and long-term effects on marine biodiversity and human populations. Thus, rapid response measures are crucial for containing spills and mitigating environmental damage (Velaochaga & Xu, 2019).

Remote sensing is a valuable tool for detecting, identifying, and assessing the effects of hydrocarbons on the sea surface (Fingas & Brown, 2018; Kolokoussis & Karathanassi, 2018), particularly through Synthetic Aperture Radar (SAR) imagery. Unlike optical sensors, which rely on sunlight and are hindered by cloud cover, SAR provides an effective means to differentiate oil and oiled water from clear ocean water (Caruso et al., 2013). SAR satellites enable operational surveillance of ocean areas, delivering data at resolutions ranging from medium (~20–40 m) to high (<10 m) under all weather conditions, day or night. As a result, SAR has become the most widely used technique for detecting sea surface anomalies associated with hydrocarbons (Redondo-Apraiz & Platónov, 2001; Sun et al., 2016). Beyond facilitating early warning of oil spills in remote marine areas, this tool enhances the capacity to assess potential impacts on conservation priority zones and ecologically sensitive regions (Skrunes et al., 2019).

This study utilizes SAR satellite imagery to detect and analyze the presence of oil in marine waters of SWGoM, encompassing both natural seeps and anthropogenic spills. The spatial analysis presented in this chapter contributes to strategic planning for oil spill response by identifying high-risk zones and providing critical data for containment efforts. These findings offer valuable insights for the sustainable management of Mexico's marine territories and support evidence-based decision-making for coastal protection.

6.2 Methods

6.2.1 Study Area

The area analyzed using satellite imagery in this study is located between the coordinates: 20° 27.001′ N, 93° 12.626′ W and 18° 24.517′ N, 90° 39.817′ W (Fig. 6.1).

6.2.2 Satellite Image Processing

Sentinel-1 Synthetic Aperture Radar (SAR) imagery from the Copernicus database was processed to detect oil spills and natural seeps. Images covering the study area during the 2018–2021 period were analyzed. We performed geometric correction on the Vertical transmit-Vertical receive polarization band, implementing bilinear interpolation to rectify geometric distortions, including scale deformations, rotations, and misalignments.

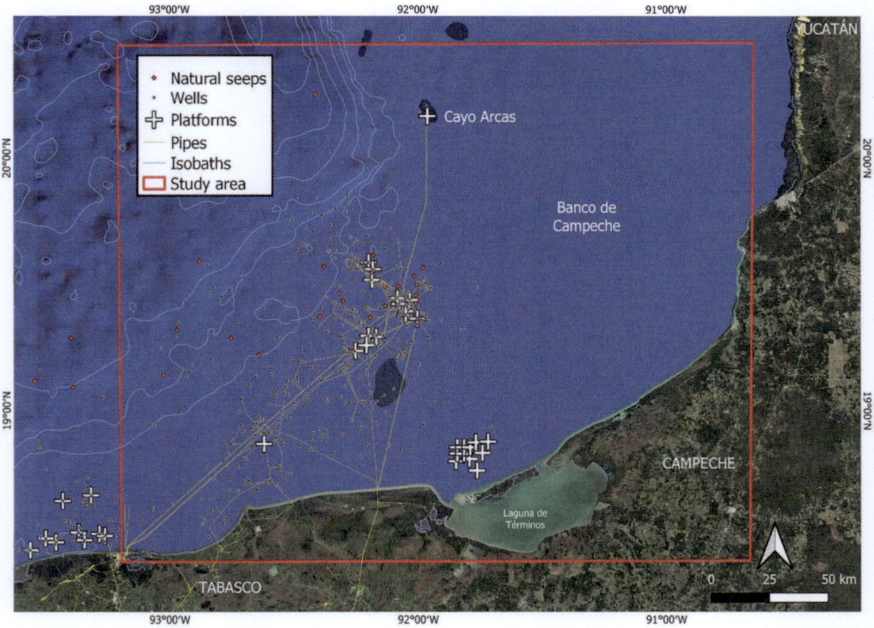

Fig. 6.1 Study area (red) for evaluating oil spills and natural seeps in the southwestern Gulf of Mexico

Each image underwent visual inspection to detect radiometric anomalies. These anomalies are areas exhibiting flattened backscatter signals that produce homogeneous textures distinct from the speckle patterns characteristic of unaffected SAR imagery. A delimitation process was performed based on the identification of signal-flattening thresholds to extract regions with low radiometric values, characteristic of possible hydrocarbon spills. The anomaly extension was estimated.

Potential hydrocarbon detections were classified by origin into four categories: (1) spills from natural seeps, (2) anthropogenic oil spills, (3) undetermined spills, and (4) combined-origin spills. This classification followed the criteria outlined in Fig. 6.2.

6.2.3 Statistical Analysis of Recurrence, Size, and Temporality

The resulting detections were converted to polygons in a vector format. A hexagonal grid with a 7 km vertex-to-vertex diameter was generated. Terrestrial areas were deliberately excluded from the hexagonal grid. We quantified the frequency of

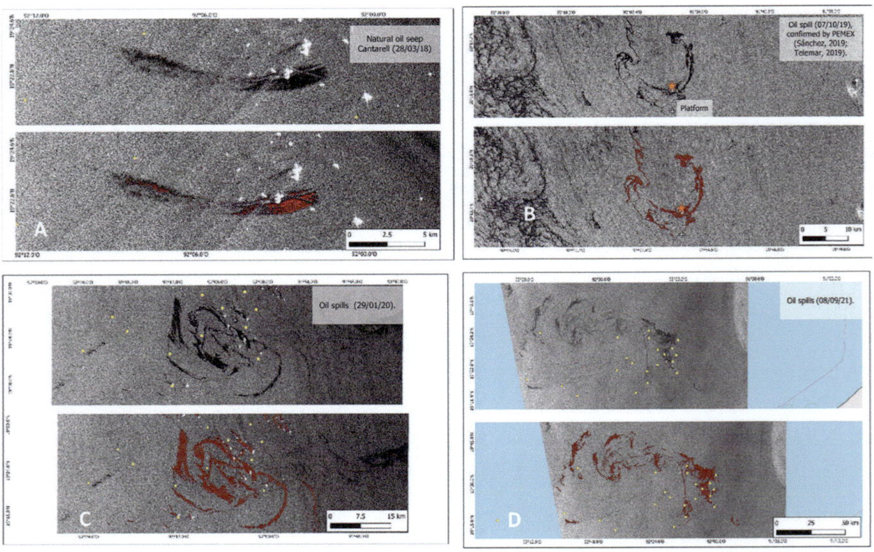

A. Natural oil seeps. They feature a feather-like shape with clearly defined, contrasting edges on one side and progressively blurred values on the opposite side, creating a characteristic sweeping effect. These seeps appear shallow and thin in profile, and their corresponding identification layers maintain minimal file sizes and spatial extents. The observed radiometric anomalies show no spatial association with offshore oil infrastructure (platforms, vessels, single point mooring buoys, or pipelines). A consistent parallel dispersion pattern is evident, with orientation typically aligned with prevailing wind or surface current directions.

B. Anthropogenic oil spills. They are characterized by identifiable sources, typically oil facilities, based on spatial proximity or direct observational evidence. These surface slicks often exhibit greater spatial extents compared to natural seeps. In some cases, supplemental evidence from news reports or social media platforms helps verify spill origins.

C. Undetermined spills. They represent cases where neither the source has been identified nor confirmation obtained from relevant authorities or operating companies. These events show no evident association with oil infrastructure, while exhibiting spatial extents exceeding those characteristics of natural oil seeps.

D. Combined-origin spills. They exhibit characteristics of both natural seep and anthropogenic spills. These events typically present morphological and radiometric signatures intermediate between the two source types, while generally exceeding the spatial scale of purely natural oil seeps.

Fig. 6.2 Oil spill classification evaluation criteria, showing: four SAR satellite image examples without and with overlaid oil spill detection polygons anomaly occurrence and the total affected area per grid cell. The derived metrics enabled computation of annual summations per spatial analysis unit (hexagon). Those were subsequently visualized through thematic maps displaying both cumulative affected areas and event frequency distributions.

The hexagonal grid outputs facilitated annual detection area analysis for each origin category. Normality assessment was performed using the Kolmogorov-Smirnov test ($N > 50$) and the Shapiro-Wilk test ($N < 50$). All cases demonstrated non-normal distributions ($p < 0.05$). Consequently, nonparametric statistics were employed, utilizing medians (as a measure of central tendency) and interquartile ranges (Q1–Q3; as a measure of dispersion).

6.3 Results

A total of 336 Sentinel-1 satellite images from the RAS sensor were downloaded and processed for the 2018–2021 study period. Radiometric anomalies potentially associated with the presence of hydrocarbons on the sea surface were found in 208 of the images (61.9%).

Overall, 118 images showed oil from natural seeps (35%), **while** 14 **contained** possible anthropogenic spills (4.16%), 42 **featured** detections with combined origins (12.5%), and 34 **had** hydrocarbons of unknown origin (10%). The analysis revealed that oil from natural seeps exhibited significantly smaller spatial extents than other oil occurrences (Fig. 6.3).

The region with the highest incidence of events is located approximately 85 km north of Atasta, Campeche, in the complex known as Cantarell, covering an area of approximately 200 km^2 (Fig. 6.4). Anthropogenic spill events were confirmed to be larger in extent and showed broader dispersion patterns, often occurring considerably closer to the coast compared to natural seep events. Among anthropogenic detections, the *Cayo Arcas* reef system emerged as particularly affected by these spills (Telemar, 2019).

The anomalies classified as indeterminate covered the largest cumulative area. They were detected up to 2 km offshore from Campeche, while the combined anomalies showed limited occurrences near the coasts of Tabasco and Campeche. Both indeterminate and combined anomalies exhibited greater spatial extents than other detection types, suggesting that some of these detections may represent unconfirmed spills.

6.3.1 Accumulated Area and Recurrence Maps

The accumulated area (sum of all identified potential oil polygons) and recurrence (number of events) parameters were calculated for different types of oil presence using a hexagonal grid (Fig. 6.5). Natural events showed higher accumulations

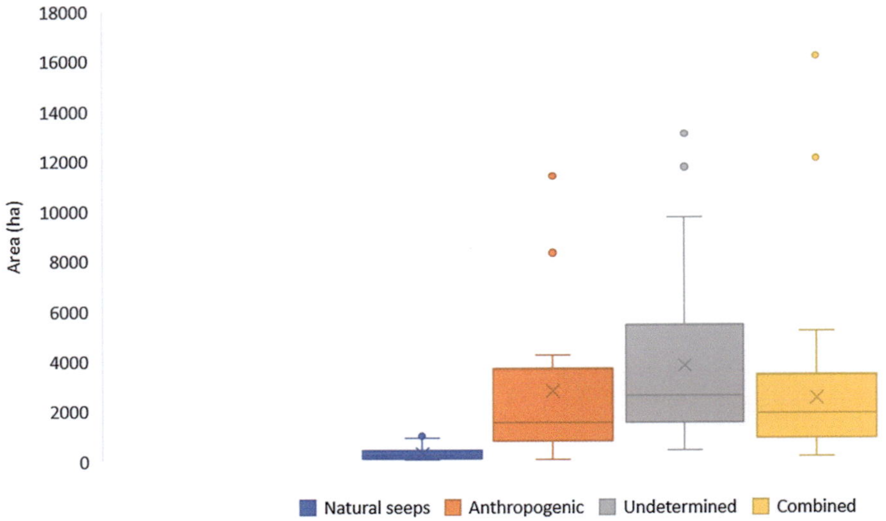

Fig. 6.3 Spill extent per hexagon: natural seeps (n = 118; median = 175.48; Q1 = 82.41; Q3 = 415.78); anthropogenic (n = 14; median = 1,514.07; Q1 = 784.57; Q3 = 3651.19); undetermined (n = 34; median = 2,584.43; Q1 = 1505.04; Q3 = 5405.11); combined (n = 42; median = 1907.88; Q1 = 880.92; Q3 = 3420.48)

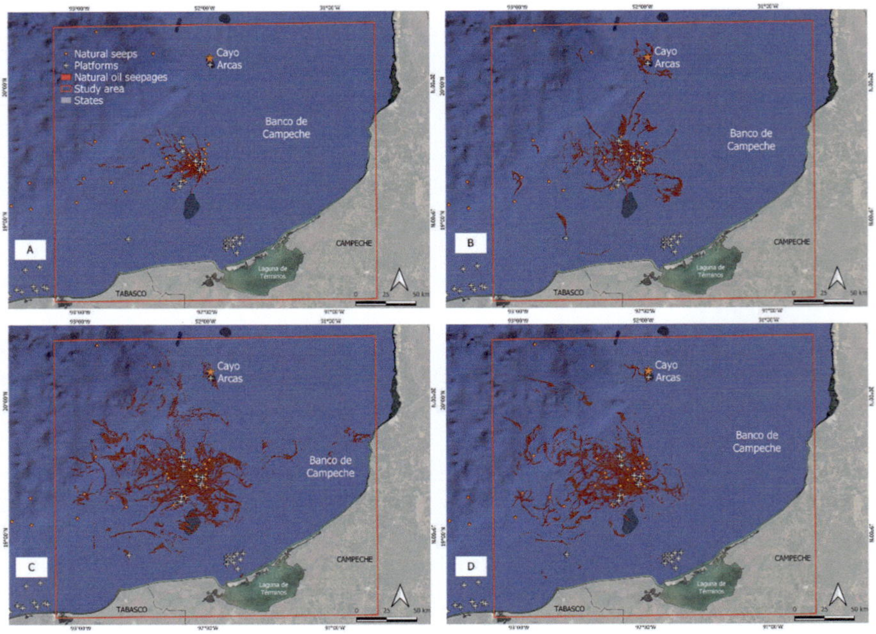

Fig. 6.4 Geographic distribution of detections during the analyzed period, classified by origin: (**a**) natural seeps; (**b**) anthropogenic spills; (**c**) undetermined oil spills; and (**d**) combined-origin events

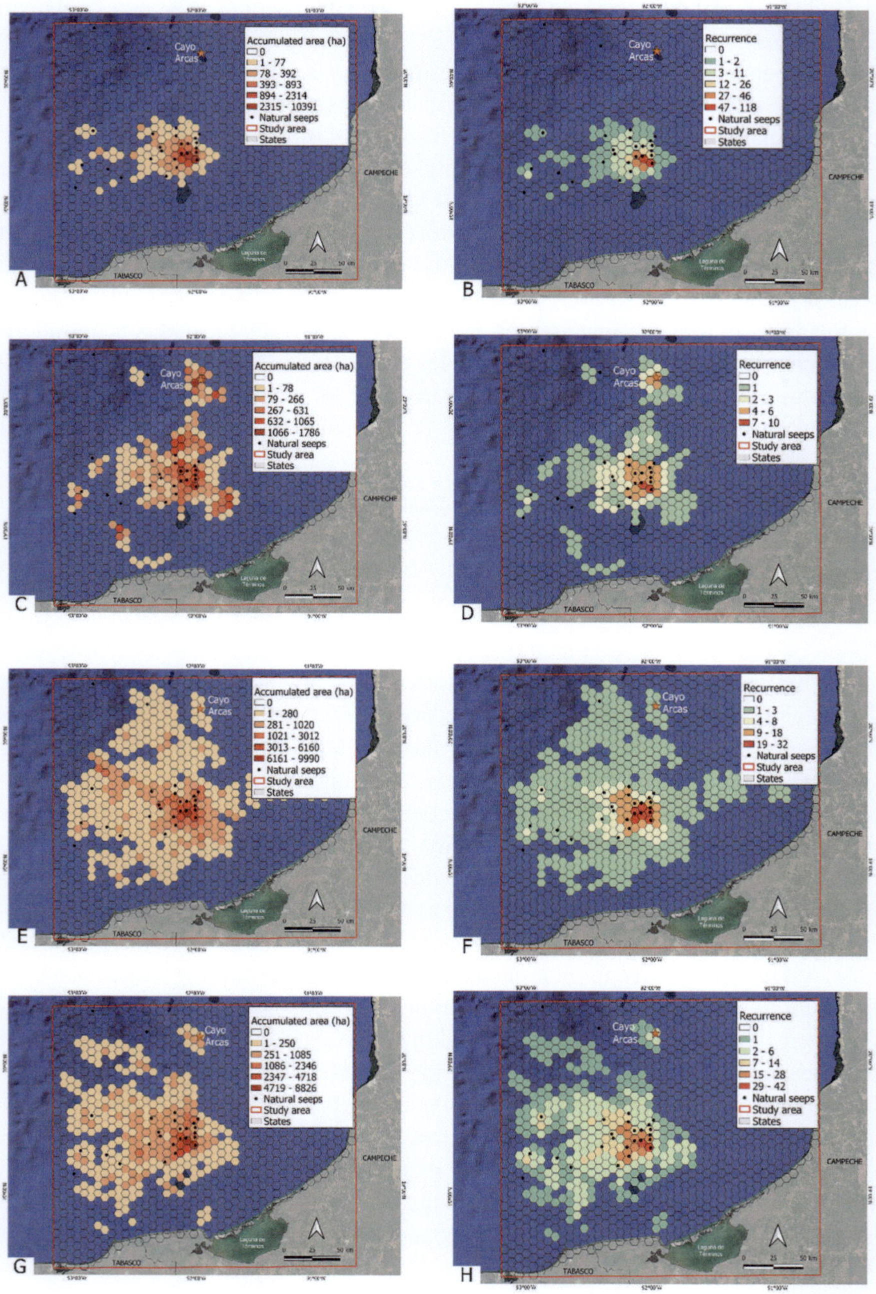

Fig. 6.5 Maps showing the accumulated area (left) and recurrence (right) of spills grouped based on their origin (2018–2021). (**a**) and (**b**) Natural; (**c**) and (**d**) Anthropogenic; (**e**) and (**f**) Undetermined; (**g**) and (**h**) Combine

around the Cantarell seep. Analysis units (hexagons) within this area contained accumulated oil coverage ranging from 2315 to 10,391 ha (Fig. 6.5a). Up to 118 natural seeps events were observed (Fig. 6.5b), extending approximately 15 km across the study area. These natural emanations exhibited year-round recurrence, representing the most frequently observed source of oil.

Anthropogenic spills exhibited the highest cumulative areas occurring in the platform zone of the Cantarell complex (Fig. 6.5c). We recorded one spill near the coast, and a significant recurrence area near *Cayo Arcas* that may be associated with the local oil infrastructure (Fig. 6.5d), with multiple hexagons around this area with at least one spill event.

For spills of undetermined origin, the accumulated area revealed concentrations in the central platform zone near natural seeps (Fig. 6.5e). The highest spill concentration was found within a 70 km area, containing 19–32 detections (Fig. 6.5f).

Combined-origin spills exhibited two focal accumulation areas, with up to 42 events at their central area (Fig. 6.5g, h).

6.4 Discussion

Based on the spill detection methodological protocol, an overview of hydrocarbon spills that occurred over a 4-year period (2018–2021) in the SWGoM was obtained. This region was identified as highly susceptible to the presence of oil, which poses a threat to local biodiversity and the well-being of human communities. Spills were recorded primarily at sites that coincide with the migration routes of protected species, such as sea turtles (Uribe-Martínez et al., 2021) and marine mammals, thereby increasing the ecological risk for this region, as some studies have documented. The information obtained in this work contributes to quantifying and ultimately preventing the effects of anthropogenic spills on natural sensitive resources.

During the study period, the highest concentrations of anthropogenic spills were observed in areas of oil exploitation and transportation, confirming a direct association with oil industry activities. These findings are valuable for both environmental management and institutional policymaking. The data offers critical monitoring parameters for key stakeholders, including the Mexican Navy (SEMAR, by its Spanish acronym) and environmental agencies, while facilitating integrated management approaches with productive sectors such as fisheries and aquaculture.

For fisheries, spill detection enables precise identification of impacted zones, which is critical for developing effective management and protection strategies for both small-scale and industrial fishing grounds (Andrews et al., 2021; Quist, 2019). This spatial data also helps mediate sectoral conflicts between fisheries and hydrocarbon industries by providing objective evidence of spill trajectories.

The detection of anthropogenic oil spills necessitates a revision of Mexico's technical and regulatory frameworks. These findings should inform the development of stronger environmental governance in the region. From a policy perspective, remote sensing enhances maritime management transparency while enabling evidence-based decision-making and intersectoral collaboration (Checa-Artasu, 2014; Espinoza Tenorio, 2019). The technology also supports adaptive governance systems for emergency response and risk mitigation (Salazar-De la Cruz et al., 2020).

6.4.1 Remote Sensing Spill Monitoring

SAR imagery has proven its capability for detecting remote oil spills, where VV polarization revealed spills as distinct dark patches (Caruso et al., 2013). However, several atmospheric conditions presented significant limitations, especially strong winds and frequent rainfall, which often alter the backscatter signal, limiting spill detection.

Source differentiation represents a significant advancement in environmental monitoring, as the existing literature is limited in its systematic characterization of spatial distinctions between natural seeps and anthropogenic spills (Brekke & Solberg, 2005; National Oceanic and Atmospheric Administration, 2021). While some studies have documented natural seep patterns (MacDonald et al., 2015), others have developed specialized detection methodologies, including algorithmic approaches and texture-based neural networks (Caruso et al., 2013; Marzialetti, 2012). Detection reliability is enhanced through multiple discriminative factors, including reflectance contrast between the ocean surface and produced water discharges, as well as spatial proximity to offshore platforms, when confirming events. Such classification systems are vital for refining remote sensing applications in support of marine spatial management within oil production zones.

Natural seeps exhibit distinct characteristics that facilitate monitoring and differentiation from anthropogenic spills or discharges. A key monitoring challenge arises when natural seep occurs directly beneath oil infrastructure, as potential facility-sourced spills in these areas would be obscured by background seeps, creating dual potential contamination sources (Machain-Castillo et al., 2019; Torres et al., 2018). Nevertheless, our study demonstrates, in agreement with MacDonald et al. (2015), that discrimination between these sources remains achievable through analysis of basin-specific oil generation patterns and spatial association with previous knowledge. This methodology successfully differentiated the Deepwater Horizon spill (NE sector) from natural seep (NW/SW sectors) in prior research, enabling effective monitoring despite spatial overlap between infrastructure and natural tar deposits in our study area.

6.4.2 Oil Spills in the Southern Gulf of Mexico

Persistent natural seeps were observed throughout the study period, consistent with documented patterns in the Gulf of Mexico (Joye et al., 2016; MacDonald et al., 2015). In contrast, several anthropogenic spills were identified within the study area, all of which were directly associated with oil industry operations. These operational releases exhibited significant variability in both extention and duration, with individual events discharging hydrocarbons intermittently over time. Quantification revealed marine-origin crude oil spills averaged 4.31 barrels per incident, with discharge events continuing for periods up to 30 days (Rosano-Hernández, 2013).

The *Sonda de Campeche* has been recognized since 1974 as one of approximately 190 global marine sites with active oil seeps (Wilson et al., 1974). Comparative examples include California's Coal Oil Point Seep, which continuously releases 30–100 barrels of oil per km^2, and Gulf of Mexico seeps contributing an estimated 10,000 barrels annually (Rosano-Hernández, 2013). At documented seep sites, sediment hydrocarbon concentrations can reach 10^4–10^5 mg kg^{-1}. Despite these substantial natural inputs, anthropogenic and unidentified spills collectively impact larger marine surface areas than natural seeps.

6.5 Conclusions

The RAS monitoring technique successfully enabled the detection and classification of oil spills on the marine surface. This approach established a foundational framework for spatiotemporal characterization of different spill types. While a significant proportion of spills could not be conclusively classified due to verification constraints, the study's results provide a robust baseline for understanding occurrence patterns. The methodology demonstrates strong potential for implementation in early warning systems and integration into regional environmental management frameworks.

The benefits of implementing an appropriate oil detection and classification technique include improved spill response, the protection of fishing grounds, the conservation of sensitive species and habitats (e.g., protected natural areas, coastal wetlands, coral reefs), the strengthening of environmental governance, the standardization of monitoring protocols, and integration with predictive dispersion models. The creation of a spill monitoring system with adequate meteorological information, including temperature, wind, and current data, in addition to data from the existing infrastructure, is essential for identifying hotspots where these spills are concentrated. This will aid and improve decision-making, thereby reducing risks to sensitive areas. All the above would allow for a more accurate assessment of the impacts and risks to coastal areas.

Acknowledgments This research is part of the project "The challenging coexistence of coupled socio-ecological systems: the fishing and oil industries in the *Sonda de Campeche*," supported by the National Council of Humanities, Science and Technology of Mexico (CONAHCyT). Large Language Model was used to verify some sentences and paragraphs for English grammar.

References

Andrews, N., Bennett, N. J., Le Billon, P., Green, S. J., Cisneros-Montemayor, A. M., Amongin, S., Gray, N. J., & Sumaila, U. R. (2021). Oil, fisheries and coastal communities: A review of impacts on the environment, livelihoods, space and governance. *Energy Research & Social Science, 75*, 102009. https://doi.org/10.1016/j.erss.2021.102009

Brekke, C., & Solberg, A. H. S. (2005). Feature extraction for oil spill detection based on SAR images. In H. Kalviainen, J. Parkkinen, & A. Kaarna (Eds.), *Image analysis* (pp. 75–84). Springer. https://doi.org/10.1007/11499145_9

Brooks, J. (1990). Classic petroleum provinces. *Geological Society Special Publication, 50*, 1–8. https://doi.org/10.1144/GSL.SP.1990.050.01.01

Caruso, M., Migliaccio, M., Hargrove, J., Garcia-Pineda, O., & Graber, H. (2013). Oil spills and spills imaged by synthetic aperture radar. *Oceanography, 26*(2), 112–123. https://doi.org/10.5670/oceanog.2013.34

Checa-Artasu, M. M. C. (2014). Geografía, poder y petróleo en México. Algunos ejemplos. *Scripta Nova. Revista Electrónica de Geografía y Ciencias Sociales, 18*. https://revistes.ub.edu/index.php/ScriptaNova/article/view/15040

Deepwater Horizon Natural Resource Damage Assessment Trustees. (2016). Deepwater horizon oil spill: Final programmatic damage assessment and restoration plan and final programmatic environmental impact statement. *NOAA (National Oceanic and Atmospheric Administration)*. http://www.gulfspillrestoration.noaa.gov/restoration-planning/gulf-plan

Espinoza Tenorio, A. (2019). Blackening your hands? The ethical dilemma of research in Petrolized territories. *Revista Mexicana de Ciencias Políticas y Sociales, 64*(237), 183–209. https://doi.org/10.22201/fcpys.2448492xe.2019.237.67179

Fingas, M., & Brown, C. E. (2018, January). A review of oil spill remote sensing. In *Sensors (Switzerland)* (Vol. 18, Issue 1). MDPI AG. https://doi.org/10.3390/s18010091

García-Cuéllar, J. Á., Arreguín-Sánchez, F., Vázquez, S. H., & Lluch-Cota, D. B. (2004). Impacto ecológico de la industria petrolera en la sonda de Campeche, México, tras tres décadas de actividad: Una revisión. *Interciencia, 29*(6), 311–319.

Hidalgo, J. C. (2009). Efectos de los derrames de petróleo sobre los hábitats marinos. *Ciencia Ahora, 24*, 22–30.

Joye, S. B., Bracco, A., Özgökmen, T. M., Chanton, J. P., Grosell, M., MacDonald, I. R., Cordes, E. E., Montoya, J. P., & Passow, U. (2016). The Gulf of Mexico ecosystem, six years after the Macondo oil well blowout. *Deep-Sea Research Part II: Topical Studies in Oceanography, 129*, 4–19. https://doi.org/10.1016/j.dsr2.2016.04.018

Kolokoussis, P., & Karathanassi, V. (2018). Oil spill detection and mapping using sentinel 2 imagery. *Journal of Marine Science and Engineering, 6*(1), 1–12. https://doi.org/10.3390/jmse6010004

MacDonald, I. R., Garcia-Pineda, O., Beet, A., Daneshgar Asl, S., Feng, L., Graettinger, G., French-McCay, D., Holmes, J., Hu, C., Huffer, F., Leifer, I., Muller-Karger, F., Solow, A., Silva, M., & Swayze, G. (2015). Natural and unnatural oil spills in the Gulf of Mexico. *Journal of Geophysical Research: Oceans, 120*(12), 8364–8380. https://doi.org/10.1002/2015JC011062

Machain-Castillo, M. L., Ruiz-Fernández, A. C., Gracia, A., Sanchez-Cabeza, J. A., Rodríguez-Ramírez, A., Alexander-Valdés, H. M., Pérez-Bernal, L. H., Nava-Fernández, X. A., Gómez-Lizárraga, L. E., Almaraz-Ruiz, L., Schwing, P. T., & Hollander, D. J. (2019). Natural and anthropogenic oil impacts on benthic foraminifera in the southern Gulf of Mexico. *Marine Environmental Research, 149*, 111–125. https://doi.org/10.1016/J.MARENVRES.2019.06.006

Marzialetti, P. (2012). *Monitoreo de derrames de hidrocarburos en cuerpos de agua mediante técnicas de sensado remoto* [Tesis de maestría]. Universidad Nacional de Córdoba, provincia de Córdoba.

National Oceanic and Atmospheric Administration. (2021). *Programa de Vigilancia de Contaminación Marina: Guía de Interpretación de Derrames de Petróleo en SAR* [Guía técnica]. https://www.noaa.gov/.

Ponce-Vélez, G., & Botello, A. V. (2005). Niveles de hidrocarburos en el Golfo de México. In A. V. Botello, J. R. Osten, G. Gold-Bouchot, & C. Agraz-Hernández (Eds.), *Golfo de México contaminación e impacto ambiental: Diagnóstico y tendencias* (2da Edición, pp. 269–298). Universidad Autónoma de Campeche, Universidad Nacional Autónoma de México, Instituto Nacional de Ecología. Ciudad de México. https://doi.org/10.13140/RG.2.1.2567.9206

Quiroa, M. (2019). La importancia del petróleo: ¿Por qué es determinante en la economía?. https://economipedia.com/actual/por-que-el-precio-del-petroleo-es-determinante-en-la-economia-mundial.html.

Quist, L.-M. (2016). In Álvaro's house: Fisher leaders, family LIFE and the ethnographer at Mexico's oil frontier. *Suomen Antropologi: Journal of the Finnish Anthropological Society, 41*(3), Article 3.

Quist, L.-M. (2019). Fishers' knowledge and scientific indeterminacy: Contested oil impacts in Mexico's sacrifice zone. *Maritime Studies, 18*(1), 65–76. https://doi.org/10.1007/s40152-018-0123-7

Redondo-Apraiz, J. M., & Platónov, A. K. (2001). Aplicación de las Imágenes SAR en el estudio de la dinámica de las aguas y de la polución del mar Mediterráneo cerca de Barcelona. *Ingeniería Del Agua, 8*(1), 15–23. https://doi.org/10.4995/IA.2001.2856

Rosano-Hernández, M. C. (2013). *El chapopote y las chapopoteras del Golfo de México: Reseña sobre el descubrimiento y usos del petróleo* (1st ed.). Instituto Mexicano del petróleo.

Salazar-De la Cruz, C. C., Alberto Zepeda-Domínguez, J., Espinoza-Tenorio, A., & E Ramos-Muñoz, D. (2020). Governance networks in marine spaces where fisheries and oil coexist: Tabasco, México. *The Extractive Industries and Society, 7*(2), 676–685. https://doi.org/10.1016/j.exis.2020.03.012

Sherman, K. (1994). Sustainability, biomass yields, and health of coastal ecosystems: An ecological perspective. *Marine Ecology Progress Series, 112*, 277–301. https://doi.org/10.3354/meps112277

Sherman, K., & Hempel, G. (2009). *The UNEP large marine ecosystem report: A perspective on changing conditions in LMEs of the world's regional seas* (182; UNEP Regional Seas Report and Studies). United Nations Environment Programme.

Skrunes, S., Johansson, A. M., & Brekke, C. (2019). Synthetic aperture radar remote sensing of operational platform produced water releases. *Remote Sensing, 11*(23). https://doi.org/10.3390/rs11232882

Soto, L. A., Botello, A. V., Licea-Durán, S., Lizárraga-Partida, M. L., & Yáñez-Arancibia, A. (2014). The environmental legacy of the Ixtoc-I oil spill in Campeche sound, southwestern Gulf of Mexico. *Frontiers in Marine Science, 1*. https://doi.org/10.3389/fmars.2014.00057

Sun, S., Hu, C., Feng, L., Swayze, G. A., Holmes, J., Graettinger, G., MacDonald, I., Garcia, O., & Leifer, I. (2016). Oil slick morphology derived from AVIRIS measurements of the Deepwater horizon oil spill: Implications for spatial resolution requirements of remote sensors. *Marine Pollution Bulletin, 103*(1–2), 276–285. https://doi.org/10.1016/j.marpolbul.2015.12.003

Telemar. (2019). Investigadores de la UNACAR revelaron presunto derrame de combustible en Cayo Arcas. *Telemar*. https://telemarcampeche.com/2019/11/12/investigadores-de-la-unacar-revelaron-presunto-derrame-de-combustible-en-cayo-arcas/

Torres, S. B., Bevilacqua, L., de Miranda, F. P., & Beisl, C. H. (2018). Detection of seep spills in Cantarell complex (Gulf of Mexico) using polarimetric SAR Radarsat-2. *Brazilian Journal of Geophysics, 36*(2), 137–150. https://doi.org/10.22564/RBGF.V36I2.928

Uribe-Martínez, A., Espinoza-Tenorio, A., Cruz-Pech, J. B., Cupido-Santamaría, D. G., Trujillo-Córdova, J. A., García-Nava, H., Flores-Vidal, X., Gudiño-Elizondo, N., Herguera, J. C., Appendini, C. M., & Cuevas, E. (2024). An affordable operational oil spill monitoring system in action: A diachronic multiplatform analysis of recent incidents in the southern Gulf of Mexico. *Environmental Monitoring and Assessment, 196*(11), 1069. https://doi.org/10.1007/s10661-024-13161-5

Uribe-Martínez, A., Liceaga-Correa, M. d. l. A., & Cuevas, E. (2021). Critical in-water habitats for post-nesting sea turtles from the southern Gulf of Mexico. *Journal of Marine Science and Engineering, 9*(8), 1–24. https://doi.org/10.3390/jmse9080793

Velaochaga, G., & Xu, H. (2019). Detección de posible derrame de petróleo en el mar peruano frente a Talara usando imágenes satelitales, febrero 2017. *Instituto Del Mar Del Perú - IMARPE, 34*(1), 265–276.

Wakida-Kusunoki, A. T., & Caballero-Chávez, V. (2009). Efectos del derrame de hidrocarburos del pozo Kab sobre la pesca ribereña en el litoral de Campeche y Tabasco, México. *Ciencias Pesqueras, 17*, 66.

Wilson, R. D., Monaghan, P. H., Osanik, A., Price, L. C., & Rogers, M. A. (1974). Natural marine oil seep: Petroleum seep into the marine environment is on the order of 0.6× 106 metric tons per year. *Science, 184*(4139), 857–865. https://doi.org/10.1126/SCIENCE.184.4139.857

Part II
Fishing Communities on Contested Coasts

Chapter 7
Socioeconomic Profile of Small-Scale Fishers in Tabasco and Campeche: Implications for Sustainable Fisheries Management

M. Mendoza-Carranza, A. Espinoza-Tenorio, M. A. Díaz-Perera, D. G. Cupido-Santamaria, and R. E. Lara-Mendoza

7.1 Introduction

Small-scale marine fisheries (SSFs) in Mexico play a vital economic and social role, supporting thousands of families and significantly influencing the entire seafood production chain. Additionally, SSFs have a profound influence on cultural identity and community cohesion in coastal regions (Finkbeiner & Basurto, 2015).

Fishing practices continue to evolve over time, reflecting fishers' adaptations to regional ecological conditions. At a national level, Campeche and Tabasco rank seventh and twelfth in fisheries production, respectively (Comisión Nacional de Acuacultura y Pesca [CONAPESCA], 2023). In these states, SSFs predominantly involve vessels under 8 m in length. This sector is particularly important due to the high number of boats and their considerable socioeconomic influence at both local and regional levels (Mendoza-Carranza et al., 2018, Peréz-Jiménez et al., 2022). In the coastal waters of Tabasco and Campeche, fisheries mainly employ bottom longlines and gill nets, characterized by multispecies catches. However, specific fisheries targeting sharks and the Mayan octopus (*Octopus maya*) in eastern Campeche have a long-standing tradition, significantly contributing to the uniqueness of regional fisheries (Markaida et al., 2017; Pérez-Jiménez et al., 2012).

M. Mendoza-Carranza (✉) · M. A. Díaz-Perera
El Colegio de la Frontera Sur–Unidad Villahermosa, Villahermosa, Tabasco, México
e-mail: mcarranza@ecosur.mx; mdiaz@ecosur.mx

A. Espinoza-Tenorio · D. G. Cupido-Santamaria
El Colegio de la Frontera Sur–Unidad Campeche, Campeche, Campeche, México
e-mail: aespinoza@ecosur.mx; deysi.cupido@ecosur.mx

R. E. Lara-Mendoza
Instituto Mexicano de Investigación en Pesca y Acuacultura Sustentables (IMIPAS), Mexico City, México
e-mail: enrique.lara@imipas.gob.mx

A. Espinoza-Tenorio et al. (eds.), *The Fishing and Oil Industries in Coexistence*,
https://doi.org/10.1007/978-3-032-02621-7_7

In Campeche, the most significant species contributing to SSFs are octopus (*Octopus maya*), crevalle jack (*Caranx hippos*), and various snail species. In Tabasco, marine fisheries are primarily supported by gafftopsail catfish (*Bagre marinus*), snook (*Centropomus* spp.), and crevalle jack (*C. hippos*) (CONAPESCA, 2023). In both states, most fisheries are operating at near maximum sustainable yield (CONAPESCA, 2022b, 2023).

Campeche has 12 primary fishing ports, whereas Tabasco has nine (Pérez-Jiménez & Mendez-Loeza, 2015). These locations host small-scale fishing fleets and the largest communities associated with fishing activities. In both states, fisher populations consist of residents as well as individuals from other states, primarily Veracruz (Pérez-Sánchez et al., 2005). The number of registered fishers in Campeche and Tabasco is 11,599 and 14,074, respectively (CONAPESCA, 2023).

The coastal and marine zones of Campeche and Tabasco, in addition to their rich fisheries resources, contain significant petroleum reserves, with the *Sonda de Campeche* hosting Mexico's largest offshore oil field (Shann, 2021). Since 2001, the Mexican government has implemented protective measures for this infrastructure, establishing large, restricted navigation zones around offshore oil platforms to ensure their security. Although the extent of these restricted zones has varied over time, they continue to displace fishers from their traditional fishing grounds, intensifying tensions between the fishing and petroleum sectors to the present day (Salazar-de la Cruz et al., 2020; Zalik, 2009). Despite these challenges, fishers have continued their activities by adapting to new circumstances. Nevertheless, they face numerous difficulties, including reduced safety, declining catches, longer fishing distances, and consequently lower profits due to higher expenses related to fuel, supplies, and an elevated risk of maritime accidents (Peña-Puch et al., 2021; Reyes Grande et al., 2020).

SSFs in Mexico and other Latin American countries lack detailed demographic and socioeconomic studies despite their economic and cultural importance. This study aims to characterize and compare socioeconomic and fishing-related data among marine small-scale fishers, using the coasts of Campeche and Tabasco as a model, considering potential differences related to environmental gradients, species diversity, and cultural variations in marine resource utilization. A total of 218 surveys were conducted, with fishers' socioeconomic and fishing-related characteristics systematically documented and analyzed. The information presented is based on fishers' perspectives, offering valuable insights that complement official data sources and sampling methods, thereby providing a more comprehensive and representative understanding of this sector. This research fills important information gaps concerning demographic and socioeconomic profiles within this sector, contributing to a more holistic understanding necessary for effective management. These data can enhance current fisheries management strategies by providing a more comprehensive and representative view of the fishing community's actual conditions and challenges.

7.2 Materials and Methods

The surveys used in this study (Appendix A) are part of a longitudinal research project aimed at understanding interactions between fisheries and the petroleum industry, titled: "The challenging coexistence of coupled socio-ecological systems: fisheries and oil industries in the *Sonda de Campeche*." The localities surveyed in Tabasco included *Sánchez Magallanes, Nuevo Torno Largo, Puerto Ceiba, Puerto Chiltepec, Frontera*, and *Barra de San Pedro*. In Campeche, the surveyed locations were *Nuevo Campechito, Colonia Emiliano Zapata, Ciudad del Carmen, Isla Aguada, Sabancuy*, and Isla Arena (Fig. 7.1). These localities are among the 12 primary small-scale fishing ports in Campeche and the 9 in Tabasco (Pérez-Jiménez & Mendez-Loeza, 2015).

To conduct the surveys, the city block sampling geo-statistical framework from Mexico's National Institute of Statistics and Geography (INEGI) was applied, as it is employed in official censuses, such as the 2020 Population and Housing Census. The number of Basic Geo-Statistical Areas (AGEBs) and polygons defining localities and populations was identified. In urban localities (e.g., *Frontera, Ciudad del Carmen*), AGEBs with fishing activities were specifically selected, whereas in rural localities (e.g., *Barra San Pedro, Isla Arena*), the entire polygon was considered for

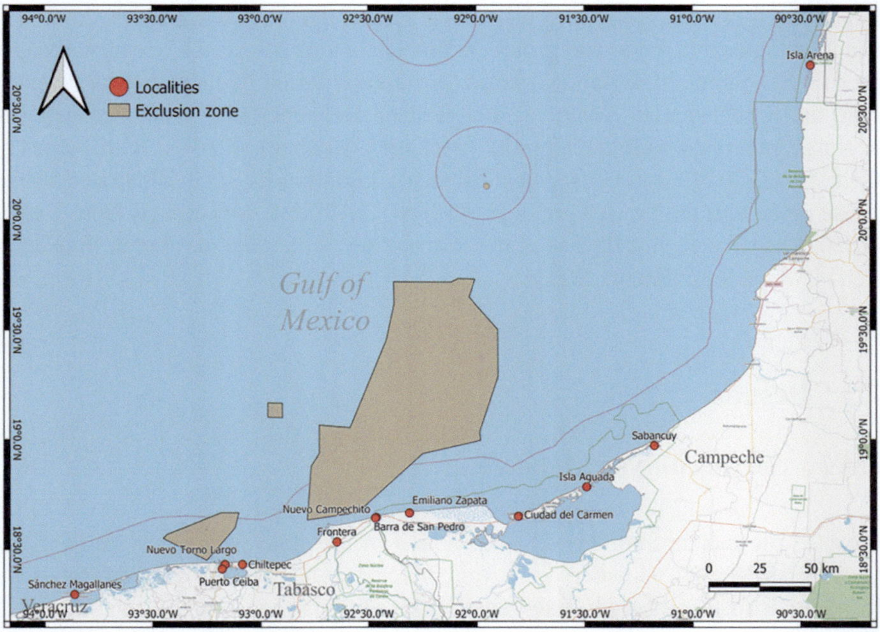

Fig. 7.1 Study area with the localities surveyed on the coast of Tabasco and Campeche

analysis. Subsequently, the number of city blocks was analyzed, and 30% were randomly selected. If the selected block lacked fishing households, it was replaced with another block from the remaining 70% or an immediately adjacent block. Surveys continued until either 20% of the inhabited homes in each community, based on INEGI's 2020 census data (INEGI, 2021), or all available fishing households had been reached. Following these procedures, a total of 218 surveys (90% men, 10% women) were conducted between September and December 2021.

The socioeconomic and fisheries-related aspects of small-scale marine fisher households were described and analyzed at the state level for Tabasco and Campeche. The specific aspects analyzed included: (a) fishers' place of origin, presented in proportions; (b) age structure and experience, with the relationship between fishers' age and their fishing experience analyzed through quantile regression (Wei et al., 2006) using the "quantreg" package (Koenker, 2005) in R software (R Core Team, 2019); (c) descriptive analysis of the number of economic dependents by gender, expressed in percentages; (d) listing other activities performed by fishers, also reported as percentages; (e) educational attainment; (f) percentage of fishers holding official fishing permits; (g) analysis of fishers' monthly income presented through frequency distribution graphs in USD (exchange rate: 1 USD = 24.0 MXN); (h) fishing type categorized by fishers as either "high-seas" and "coastal-inshore" or "freshwater." It is important to note that the local designation "high-seas" does not correspond to the FAO definition, which typically applies to fisheries conducted by large vessels, generally beyond 200 nautical miles offshore (Cunningham, 2004). (i) Catch volumes by species and groups, estimated in tons per season, are presented as mean ± SD (standard deviation), based on values reported by fishers in surveys; (j) seasonality of fisheries resources, analyzed by frequency and visualized using a heatmap generated with the "ggplot2" package (Wickham, 2016) in R software (R Core Team, 2019); and (k) species prices per kilogram by state, illustrated using pirate plots with the "yarrr" package (Phillips, 2017) in R software (R Core Team, 2019). The density plot illustrates the distribution of the data in relation to overall trends (Venables & Ripley, 2002).

7.3 Results

7.3.1 *Socioeconomic Aspects*

Most of the fishers interviewed were native to the state where they resided. In Campeche ($N = 94$), 91 fishers were from this State, while three were from Tabasco. In addition to local fishers, individuals from six other states (Tabasco, Chiapas, Oaxaca, Quintana Roo, Veracruz, and Yucatán) were also identified in Campeche. In Tabasco, besides local fishers ($n = 65$), individuals from Campeche and Veracruz were also recorded.

Fishers' ages ranged from 15 to 85 years, with 63% of the surveyed population between 45 and 65 years old. The most frequent age range for men was between 50 and 55 years, whereas for women, it was between 30 and 40 years. The average age was 52 ± 15 years (SD) for male fishers and 47 ± 13 years for female fishers.

Men fishers had significantly more years of fishing experience than female fishers (32 ± 15 vs 23 ± 9 years, respectively, $p < 0.001$). Most fishers began fishing at an early age, as evidenced by the quantile regressions at the 50th and 80th percentiles, which showed minimal differences between them (Fig. 7.2). At the tenth percentile and below, fishers were observed who had recently entered the fishing sector, regardless of age. The most extreme case was a 59-year-old fisher with only 3 years of experience.

The number of dependents per fisher was similar in both states, ranging from 1 to 9 individuals. Most fishers had between 1 and 4 economic dependents (84% in Tabasco and 86% in Campeche), with a relatively even distribution across this range (Fig. 7.3).

Fishers from both communities reported engaging in 18 different occupations (Table 7.1). The majority (52%) relied solely on fishing as their primary income source, while others supplemented their earnings with one or two additional activities (20% and 16%, respectively). Only one individual reported having seven occupations. In contrast, most women in Tabasco identified as employees of permit holders or cooperatives.

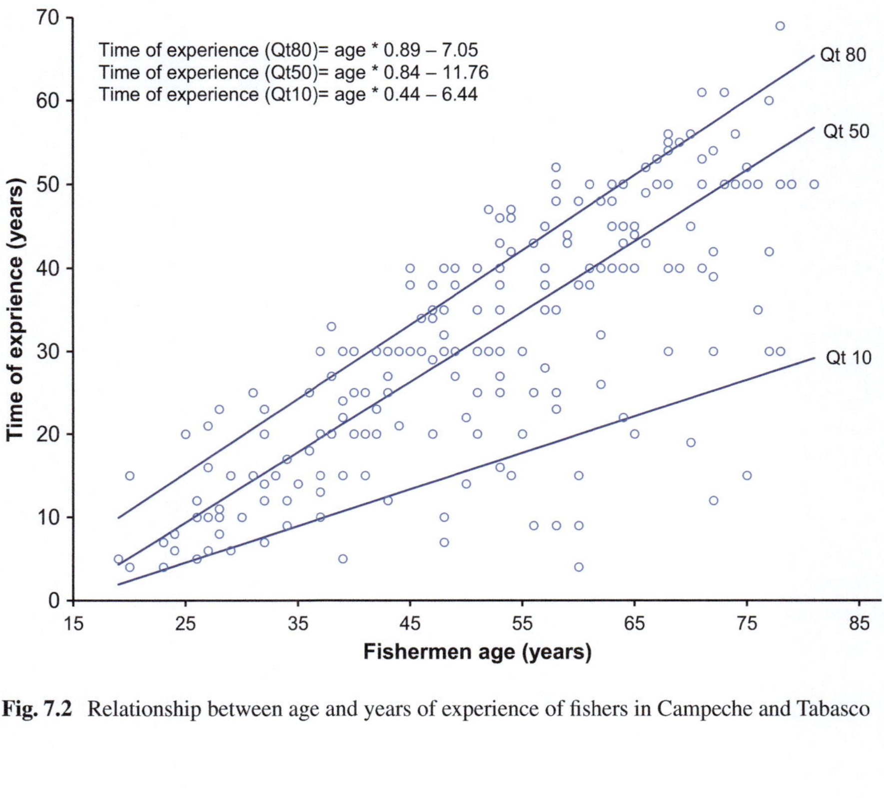

Fig. 7.2 Relationship between age and years of experience of fishers in Campeche and Tabasco

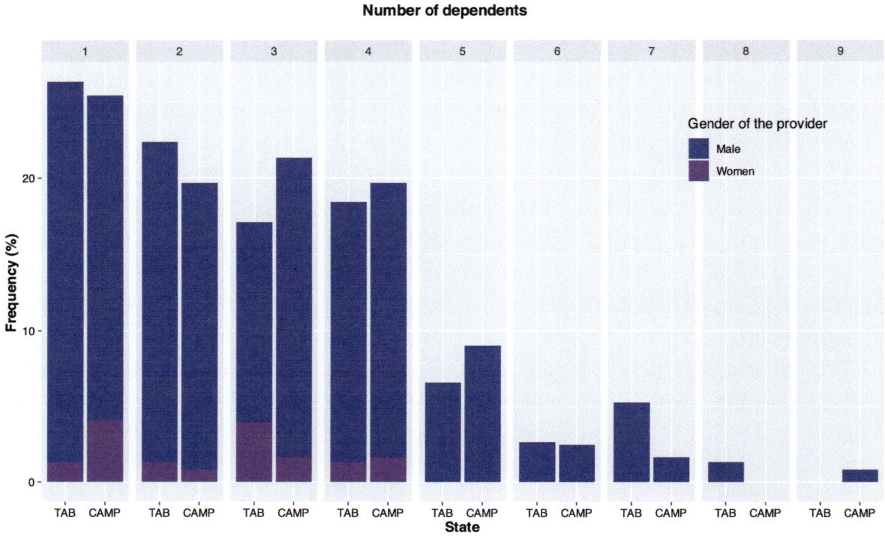

Fig. 7.3 Frequency of the number of economic dependents of fishers in Tabasco (TAB) and Campeche (CAMP)

Table 7.1 Proportion of activities performed by fishers in Tabasco and Campeche

Activity	Female	Male	Campeche	Female	Male	Tabasco	Total
Free fisher	23.08	35.86	34.38	5.00	24.72	22.73	28.91
Subsistence fisher	7.69	13.64	12.95	10.00	15.17	14.65	13.74
Fisher employed by permit holders and cooperatives	–	11.11	9.82	**25.00**	15.73	16.67	13.03
Cooperative fisher	7.69	13.64	12.95	5.00	11.24	10.61	11.85
Equipment and supply maintenance	15.38	7.07	8.04	10.00	5.06	5.56	6.87
Cooperative representative	7.69	3.54	4.02	–	5.62	5.05	4.50
Administrator	7.69	2.02	2.68	–	6.18	5.56	4.03
Paid processing worker	3.85	3.03	3.13	20.00	2.81	4.55	3.79
Fishing permit holder	–	4.04	3.57	–	3.37	3.03	3.32
Local middleman	7.69	2.53	3.13	5.00	2.81	3.03	3.08
Regional middleman	3.85	1.01	1.34	5.00	3.37	3.54	2.37
Unpaid processing worker	–	1.01	0.89	5.00	2.25	2.53	1.66
Fish evisceration worker ("Garrulera")	7.69	–	0.89	10.00	–	1.01	0.95
Retired fisher	–	0.51	0.45	–	1.12	1.01	0.71
Warehouse keeper	–	0.51	0.45	–	0.56	0.51	0.47
Driver	3.85	–	0.45	–	–	–	0.24
International middleman	–	0.51	0.45	–	–	–	0.24
Cooperative secretary	3.85	–	0.45	–	–	–	0.24

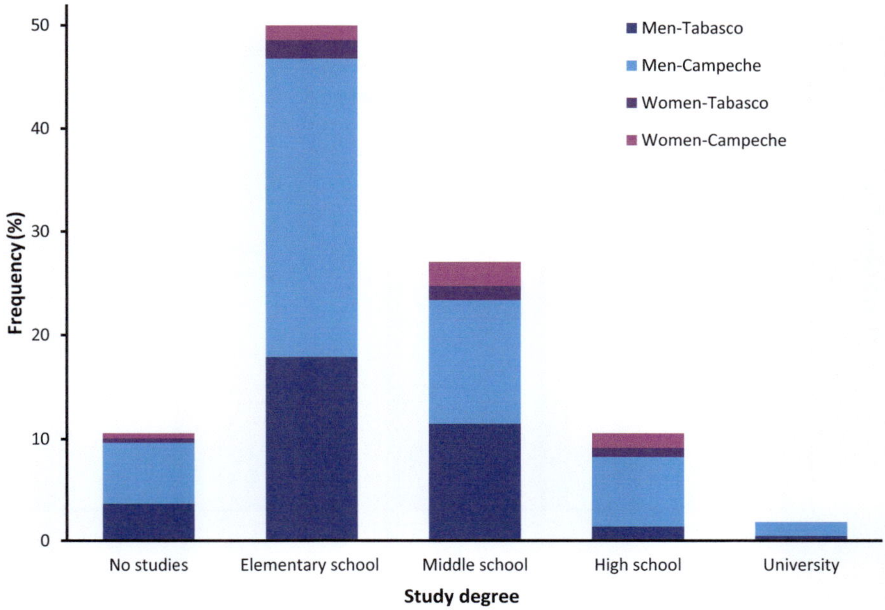

Fig. 7.4 Frequency of educational level by origin, gender, and state

Regarding educational level, half of the fishers (50%) had only completed elementary school. The remaining fishers were evenly distributed among secondary education, high school, and no formal education although a small percentage of fishers in both Campeche and Tabasco had university-level education (Fig. 7.4).

7.3.2 Fishery Situation

A total of 65% and 62% of surveyed fishers in Tabasco and Campeche, respectively, lacked official fishing permits. Permit ownership showed no significant association with fishing type, gender, or age. Over half of the fishers (53%) interviewed lacked health or life insurance and had no retirement plans in place. Only 33% and 8% reported being insured under the former Institute of Health for Well-being (INSABI, *Instituto de Salud para el Bienestar*) or the Mexican Social Security Institute (IMSS, *Instituto Mexicano del Seguro Social*), respectively.

The monthly income distribution from fishing was similar across states. A total of 74% and 75% of fishers earned between USD 49 and 349 (MXN 1000–7000) per month. The highest recorded average monthly income ranged from USD 784–980 (MXN 16,000–20,000), reported by 1.22% and 3.12% of fishers in Tabasco and Campeche, respectively (Fig. 7.5). Conversely, 13% and 7% of fishers in Tabasco

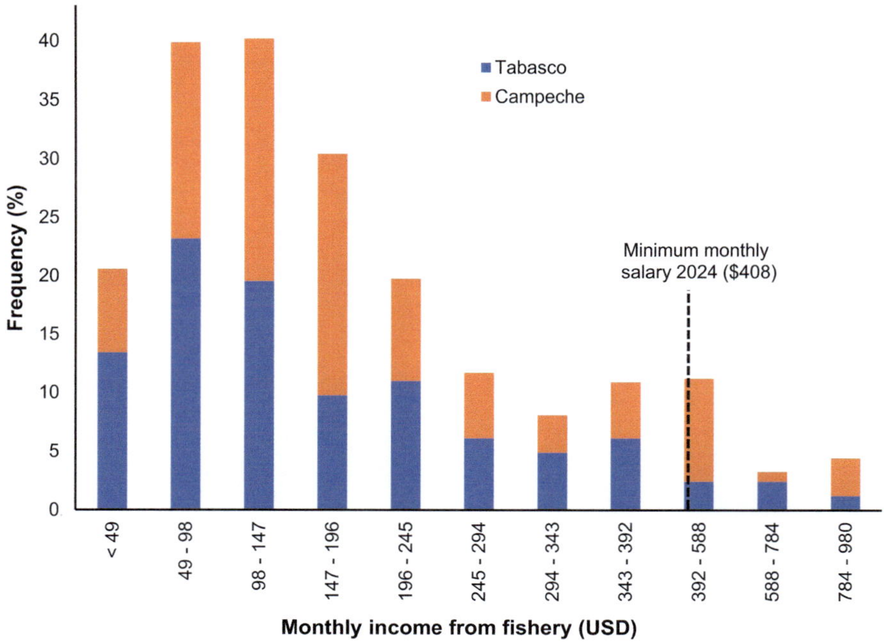

Fig. 7.5 Frequency distribution of monthly salaries (USD) of small-scale marine fishers in Tabasco and Campeche

and Campeche, respectively, earned less than USD 49 (<MXN 1000) per month from fishing. There was no clear relationship between age range and monthly income since fishers of all age groups reported earnings ranging from less than USD 49 to 588 (MXN 1000–12,000). However, only fishers over the age of 50 reported incomes exceeding USD 588 (MXN 11,380). Overall, 93% and 87% of fishers in Tabasco and Campeche earned below the monthly official minimum salary.

Forty percent of respondents specialized in high-seas or coastal-inshore fishing each (Fig. 7.6), while 18% reported engaging in both types of fishing. The remaining practiced coastal-inshore and inland (lagoon) fishing. Notably, women identified themselves as both high-seas and coastal-inshore fishers. In both states, there was no relationship between the most frequently mentioned fishing types (high-seas and coastal-inshore) and the species caught. Similarly, no correlation was observed between age and the kind of fishing.

Fishers reported 44 commercially important species, two species and two groups of elasmobranchs, which include a variety of sharks ("*tiburón,*" sharks >20 kg and

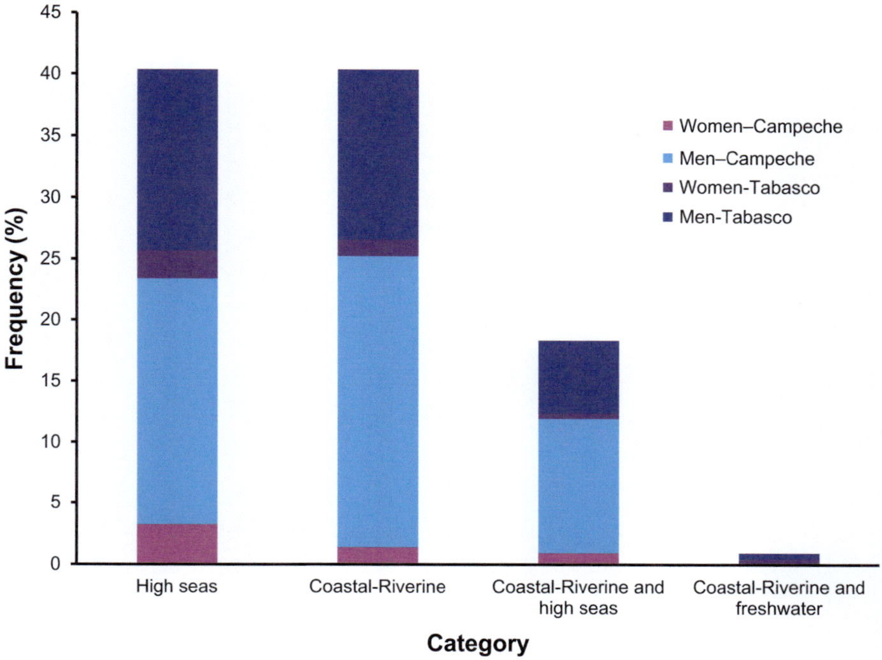

Fig. 7.6 Frequency of fishing types by location and gender in Tabasco and Campeche

"*cazón,*" sharks from 2 to 20 kg) (classification based on Mendoza-Carranza et al., 2018). Additionally, five types of crustaceans and three types of mollusks were recorded (Table 7.2).

Fishers in Tabasco reported 21 commercially important fish species, while those in Campeche reported 33 species (Table 7.2). The species with the highest reported catch volumes were blue runner (*Caranx crysos*), with 122.64 ± 214.99 tons in Campeche, and red grouper (*Epinephelus morio*), with 28.82 ± 62.38 tons in Tabasco. The gafftopsail catfish (*Bagre marinus*) was the third most reported species in terms of catch volume, with 24.51 ± 63.67 tons in Tabasco and 10.68 ± 15.6 tons in Campeche (Table 7.2). Regarding fishing seasons, fishers indicated that 10 species exhibit strong seasonality (Fig. 7.7, top panel), while the remaining species are caught throughout the year without a clear seasonal pattern.

In both states, the highest value species were red snapper (*Lutjanus campechanus*), snook (*Centropomus* spp.), and vermilion snapper (*Rhomboplites aurorubens*) (Fig. 7.8). Red snapper exhibited the broadest price range, with the lowest

Table 7.2 Commercial importance species list and average ± SD of tons by fishery season based on the estimations of fishers

Common Spanish name	Common English name	Scientific name	Tabasco	Campeche	Total
		Pisces			
Cojinuda	Blue runner	*Caranx crysos*	7.73 ± 10.72 (3)	122.64 ± 214.99 (5)	79.55 ± 173.15 (8)
Besugo	Vermilion snapper	*Rhomboplites aurorubens*	28.82 ± 62.38 (19)	6.8 ± 8.44 (4)	24.82 ± 56.88 (23)
Bandera	Gafftopsail catfish	*Bagre marinus*	24.51 ± 63.67 (16)	10.68 ± 15.6 (10)	18.98 ± 50.04 (26)
Jurel	Crevalle jack	*Caranx hippos*	22.67 ± 47.64 (20)	10.73 ± 9.15 (10)	18.56 ± 38.98 (30)
Sierra	Atlantic Spanish mackerel	*Scomberomorus maculatus*	1.98 ± 3.2 (16)	26.25 ± 125.21 (31)	17.99 ± 101.8 (47)
Medregal	Banded rudderfish	*Seriola zonata*	11.67 ± 10.97 (3)	30	16.25 ± 12.82 (4)
Bagre	Hardhead catfish	*Ariopsis felis*	3	15.4 ± 20.65 (2)	11.27 ± 16.26 (3)
Liseta	Silver mullet	*Mugil curema*	2.51 ± 3.92 (7)	20.57 ± 44.41 (6)	10.04 ± 28.5 (13)
Huachinango	Red snapper	*Lutjanus campechanus*	10.75 ± 28.07 (20)	3.98 ± 3.16 (13)	8 ± 21.75 (33)
Ronco	Barred grunt	*Conodon nobilis*	10.8	5	7.9 ± 4.1 (2)
Robalo	Common Snook	*Centropomus undecimalis*	1.26 ± 1.99 (20)	9.16 ± 30.1 (56)	7.05 ± 25.98 (76)
Mero	Red grouper	*Epinephelus morio*	4	6 ± 5.66 (2)	6 ± 5.66 (2)
Extraviado	Greater amberjack	*Seriola dumerili*		7.35 ± 9.4 (2)	4.9 ± 7.89 (3)
Corvina	Weakfish	*Cynoscion* sp.	1.9	4.54 ± 16.56 (36)	4.47 ± 16.33 (37)
Villajaiba	Lane snapper	*Lutjanus synagris*	0.4	8	4.2 ± 5.37 (2)
Bonito	Little tunny	*Euthynnus alletteratus*	4.05 ± 5.6 (15)	0.4	3.83 ± 5.49 (16)
Pargo	Gray snapper	*Lutjanus griseus*	2.19 ± 2.88 (9)	4.04 ± 9.19 (10)	3.16 ± 6.84 (19)
Cherna	Yellowmounth grouper	*Mycteroperca interstitialis*	2.5 ± 1.74 (5)	4	2.75 ± 1.67 (6)
Chopa	Atlantic tripletail	*Lobotes surinamensis*	2	2.76 ± 5.21 (12)	2.7 ± 5 (13)
Negrillo	Lesser amberjack	*Seriola fasciata*	0.2	5	2.6 ± 3.39 (2)
Cintilla	Atlantic cutlassfish	*Trichiurus lepturus*	2.25 ± 3.31 (4)		2.25 ± 3.31 (4)
Macabil	Bonefish	*Albula vulpes*	2		2
Sargo	Sheepshead	*Archosargus probatocephalus*	1.75 ± 3.12 (5)		1.75 ± 3.12 (5)
Robalito	Fat Snook	*C. Parallelus*	1	1.49 ± 3.45 (8)	1.43 ± 3.23 (9)
Rubia	Yellowtail snapper	*Ocyurus chrysurus*		1.33 ± 2.24 (9)	1.33 ± 2.24 (9)
Barracuda	Great barracuda	*Spyraena barracuda*	1.28 ± 1.59 (2)		1.28 ± 1.59 (2)

Common Spanish name	Common English name	Scientific name	Tabasco	Campeche	Total
Carito	Cero	*Scomberomorus regalis*		1.03 ± 0.35 (3)	1.03 ± 0.35 (3)
Chucumite	Mexican Snook	*C. Poeyi*	2.17 ± 3.32 (3)	0.35 ± 0.2 (5)	1.03 ± 2.01 (8)
Peto	King mackerel	*Scomberomorus cavalla*	1.21 ± 1.25 (9)	0.38 ± 0.54 (3)	1 ± 1.16 (12)
Boquinete	Hogfish	*Lachnolaimus maximus*		1 ± 0 (2)	1 ± 0 (2)
Sabalo	Tarpon	*Megalops atlanticus*	1.9	0.024	0.96 ± 1.33 (2)
Pampano	Permit	*Trachinotus falcatus*		0.8 ± 0.28 (2)	0.8 ± 0.28 (2)
Pejelagarto	Tropical gar	*Atractosteus tropicus*		0.7	0.7
Pez Rey	Anchovy	*Anchoa* spp.	0.7 ± 0.5 (4)		0.7 ± 0.5 (4)
Ratón	Southern kingcroaker	*Menticirrhus americanus*		0.54 ± 0.65 (2)	0.54 ± 0.65 (2)
Carpa	Tilapia	*Tilapia* spp.		0.5	0.5
Mojarra	Goatfish	*Eugerridae*	0.25 ± 0.35 (2)	0.4	0.3 ± 0.26 (3)
Trucha blanca	Spotted weakfish	*Cynoscion nebulosus*	0.2		0.2
		Elasmobranchii			
Tiburon	Sharks	*Elasmobranchii*	203.1 ± 397.95 (4)	5	163.48 ± 355.84 (5)
Cazon	Sharks	*Elasmobranchii*	17.86 ± 30.9 (5)	2.35 ± 4.03 (14)	6.43 ± 16.53 (19)
Bala	American stingray	*Hypanus americanus*	10.2 ± 7.52 (3)	6.4 ± 10.05 (3)	8.3 ± 8.21 (6)
Raya	Whitespotted eagle ray	*Aetobatus narinari*	0.4	1	0.7 ± 0.42 (2)
		Crustacea			
Camarón siete barbas	Atlantic seabob	*Xiphopenaeus kroyeri*		33.46 ± 102.98 (12)	33.46 ± 102.98 (12)
Camarón	Atlantic white shrimp	*Litopenaeus setiferus*		1.8 ± 2.71 (28)	1.73 ± 2.68 (29)
Jaiba	Blue crab	*Callinectes* spp.	1.25 ± 1.06 (2)	1.78 ± 2.66 (9)	1.68 ± 2.41 (11)
Langosta	Spiny lobster	*Palinurus* argus	0.3		0.3
Cangrejo	Florida stone crab	*Menippe mercenaria*		0.3	0.3
		Mollusca			
Pulpo	Mexican four-eyed octopus	*Octopus maya*		7.35 ± 14.14 (16)	7.35 ± 14.14 (16)
Pota	Snails	*Ommastrephidae*		0.2	0.2
Caracol	Snails	Various species		0.05	0.05

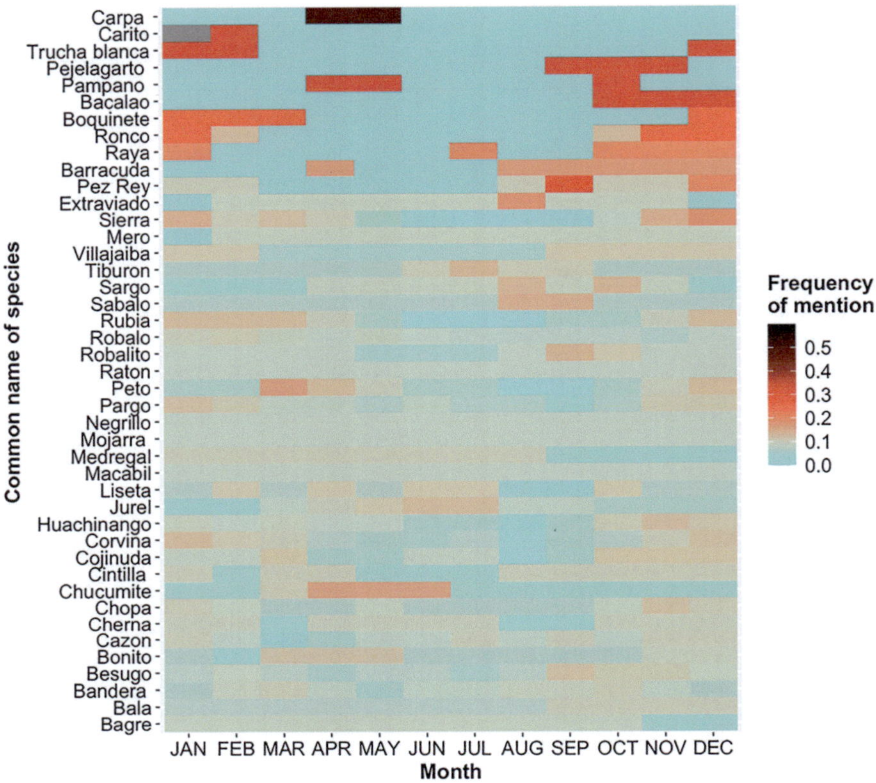

Fig. 7.7 Seasonality in the fishing of small-scale marine fishery resources in Tabasco and Campeche (English common name and scientific names in Table 7.2)

and highest average prices recorded in both states (Tabasco: 4.38 ± 1.55 to 6.82 ± 2.25 USD; Campeche: 3.95 ± 1.89 to 6.53 ± 1.95 USD). The second most valuable species was snook (*Centropomus* spp.), with prices ranging from 3.29 ± 1.32 to 5.52 ± 1.63 USD in Tabasco and from 3.34 ± 1.89 to 5.13 ± 1.95 USD in Campeche. Notably, exceptionally high prices were recorded for red snapper (USD 12.74) and vermilion snapper (USD 10.78) in Tabasco, as well as small shark (*cazón*) at 13.50 USD in Campeche. Other species also reached unusually high prices though they were infrequent and did not represent significant catch volumes. Overall, average fishery product prices were lower in Campeche than in Tabasco.

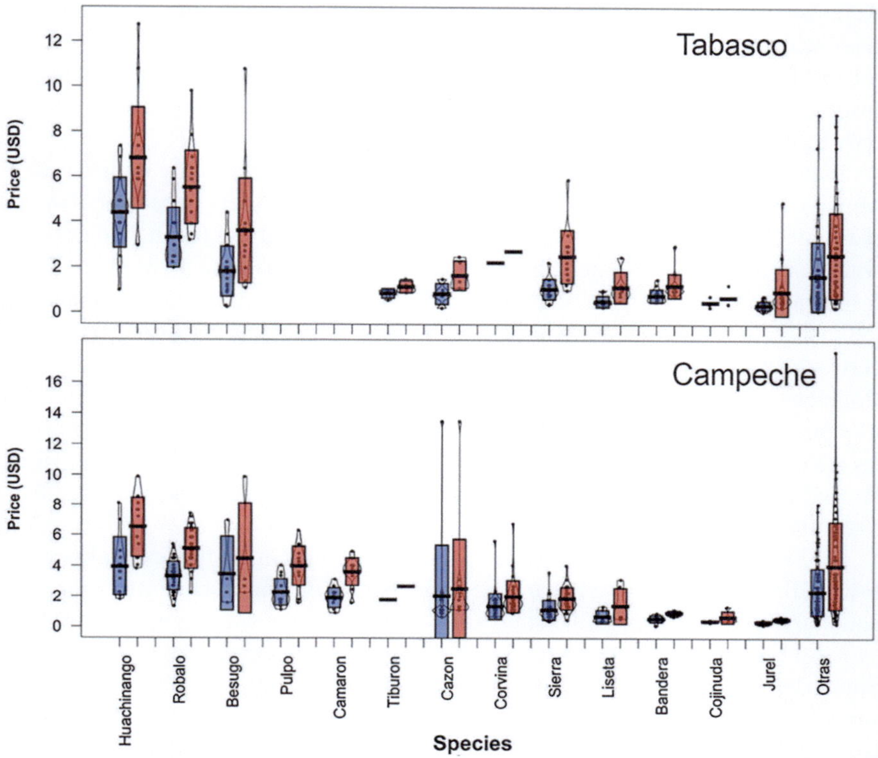

Fig. 7.8 Variation in prices of fishery resources from small-scale marine fishing in Tabasco and Campeche. The horizontal black lines represent the average, the color squares indicate the standard deviation, and the black dots represent individual cases (English common name and scientific names in Table 7.2)

7.4 Discussion

Although the demographic structure of small-scale marine fishers is essential for fisheries management and resource governance, available data on this topic remains scarce. This issue is particularly prevalent in Latin America, where, despite studies on the social aspects of fisheries, few analyses focus on age structure, gender differences, regional variations, and fishing activities. In many cases, access to such data is restricted, as fisheries sector information in Mexico is primarily collected through research projects conducted by the Mexican Institute for Sustainable Fisheries and

Aquaculture Research (IMIPAS, formerly INAPESCA), the country's leading fisheries science authority. Additional data are also gathered by the National Council for the Evaluation of Social Development Policy (CONEVAL) and the National Institute of Statistics and Geography (INEGI); however, these sources often provide only a generalized overview.

Twenty-five years ago, the FAO (2010) conducted a social study on fisheries in Mexico, reporting that 80% of the country's small-scale fishers were over 30 years old, with 54% aged over 40. Our findings indicate that this trend has intensified, with more than 70% of fishers, both men and women, now aged 40 or older. These results align with data from other SSFs in Mexico. For example, in the cannonball jellyfish fishery in Sonora, fishers' ages ranged from 25 to 67 years, with 38% between 36 and 46 years old. Notably, 87% of fishers were over 36 years old (Cruz-Colín et al., 2019). Similarly, in Yucalpetén, Yucatán, the average age of fishers was 41 years (Hernández & Flores, 2022). The aging trends in small-scale marine fisheries are not unique to Mexico. Studies from India, Indonesia, and Greece report similar age distributions, with most fishers aged over 40 (Bhendarkar et al., 2017; Tzanatos et al., 2006; Utami et al., 2023).

For Fisheries Region IV of Mexico (Campeche, Yucatán, and Quintana Roo), the FAO (2010) study reported that the proportion of young fishers was twice the national average. Although the study does not specify the exact age range of these "young" fishers, it can be assumed that they were younger than 29 years old. Similarly, Caballero-Chávez and Morales-Martínez (2021), in a comprehensive study on marine finfish fisheries in Campeche, reported that half of the fishing population in this state was 25 years old or younger. This age distribution is associated with the capture of high-value commercial species, such as spiny lobster (*Panulirus argus*), queen conch (*Lobatus gigas*), red grouper (*Epinephelus morio*), and mayan octopus (*Octopus maya*), which provide high seasonal incomes (FAO, 2010). The type of fishing methods used may also play a role in this demographic pattern, as lobster and conch fisheries require scuba diving and freediving (Diario Oficial de la Federación [DOF], 2016), which demand greater physical strength. However, the findings of the current study, at least for Campeche (part of Region IV), show that most fishers are over 35 years old. This may be linked to a lack of generational replacement in the fishing population or sampling bias, as younger fishers may not yet be heads of households and, therefore, were not included in our survey.

Additionally, the age structure of women engaged in various fishing-related activities in Yucatán has been reported to range between 30 and 49 years, with 72% of the interviewed women falling within this range (Gutiérrez-Pérez, 2024). Similar age structures have been observed in small-scale fisheries in India, where fisher demographics also tend to be skewed toward individuals over 30 years old (e.g., Balasubramaniam et al., 2009; Kumbhar, 2017; Rahaman et al., 2021).

The increasing age of fishers compared to previous decades may be attributed to several factors. One possibility is that young people from fishing communities no longer aspire to become fishers, or that their parents encourage them to pursue alternative careers that are perceived as less risky, offer greater job security, and provide more stable incomes (Oliveto-Andrade et al., 2024). Another potential factor is the public policies and welfare programs introduced by the Mexican government in 2019, which target young individuals aged 18–29 who are neither studying nor employed. These programs provide a monthly stipend and additional benefits to support their training and workforce integration (DOF, 2019). In some cases, such initiatives may have discouraged younger individuals within this age group from entering the fishing sector, given the significant physical effort required. However, the broader social effects and long-term impacts of these programs require further analysis to determine their influence on the declining participation of younger generations in the fisheries sector.

Additionally, entering the formal fishing sector is particularly challenging for young individuals, as obtaining a fishing permit requires proof of ownership of a registered vessel and an engine (CONAPESCA, 2025). In most cases, this is unfeasible due to the high cost of this equipment, which is reflected in the high percentage of fishers who lack official permits but continue fishing, nonetheless. Additionally, fishing permit expeditions are subject to the status of the species mentioned in the National Chart of Fisheries, where most species are in overexploitation or near their maximum sustainable yield, which conditions the permit expedition (CONAPESCA, 2022a).

Moreover, based on previous studies (Ikiara & Odink, 1999) and personal observations (Mendoza-Carranza), it can be inferred that older fishers continue fishing primarily due to economic and cultural reasons. With advanced age, the inherent risks of fishing, and the intense physical demands of the profession, the economic and health vulnerabilities of fishing communities increase. This situation is exacerbated by the lack of health insurance and retirement plans, as observed in this study, where over 50% of respondents lacked medical insurance, and the vast majority had no retirement plans (Mendoza-Carranza et al., 2024).

The educational levels of small-scale marine fishers in Campeche and Tabasco reflect the overall educational conditions of their regions, which are characterized by significant educational gaps and an urgent need for employment to secure a livelihood (Barboza, 2005). This situation is primarily linked to two factors: economic marginalization, which prompts individuals to enter the workforce at an early age, and the lack of alternative opportunities in their communities of origin (Oliveto-Andrade et al., 2024). In Campeche, Caballero-Chávez and Morales-Martínez (2021) emphasize that human capital development through education is a driving force of progress. Consequently, public policies aimed at improving and expanding educational access focus on ensuring that children enroll in elementary school until full coverage is achieved. Like other states in Mexico and Latin America, Campeche and Tabasco face two major challenges: (a) overcoming exclusion caused by both the educational system itself and the state's geographic characteristics, and (b) enhancing the quality of education at all levels and for all demographic groups.

It is crucial to emphasize that the socioeconomic marginalization of fishing communities is not solely due to low-income levels; its causes are far more complex and involve multiple challenges. Among the most significant are limited access to healthcare, education, and financial capital, as well as exclusion from political decision-making and fisheries management processes (Béné & Friend, 2011). These issues are not unique to Mexican fisheries but are common in nearly all countries where small-scale fishing is practiced.

7.4.1 Implications for Fisheries Management

The species reported by fishers as having the highest catch volumes in Campeche were blue runner (*Caranx crysos*), atlantic seabob shrimp (*Xiphopenaeus kroyeri*), and Spanish mackerel (*Scomberomorus maculatus*), while in Tabasco, the most frequently caught species were sharks, vermilion snapper (*Rhomboplites aurorubens*), and gafftopsail catfish (*Bagre marinus*). However, CONAPESCA (2022b) identified the mayan octopus (*Octopus maya*), crevalle jack (*Caranx hippos*), and snails (various species) were identified as the most important species in Campeche, while in Tabasco, gafftopsail catfish (*B. marinus*), snook (*Centropomus* spp.), and crevalle jack (*C. hippos*) were reported as the primary fisheries species. This discrepancy may stem from methodological differences: while the present study focused on specific areas and individuals, CONAPESCA (2022b) encompassed a broader geographic scope. Additionally, the timing of the surveys may have introduced a perception bias, as respondents might report primarily on the species being actively fished during that period (Fulton, 2021; Holland, 2008). Despite these methodological contrasts, the divergence is valuable and informative, as small-scale fishers, due to their extensive experience, often have highly accurate knowledge of species abundance and catch volumes. Studies have demonstrated that small-scale marine fishers can estimate the weight of their catches by species with an error margin of only 0.5 kg and assess fishing depths with an accuracy of 1 m (Mendoza-Carranza et al., 2009). Additionally, landing records from fish reception centers enable precise species identification based on common names, at least for fish species (e.g., Mendoza-Carranza et al., 2018).

However, deficiencies in official data collection are primarily due to the grouping of multiple species into broad categories, which compromises data resolution and limits their usefulness for fisheries modeling (Alvarez-Torres et al., 2002). Thus, it is critical to reform fisheries data collection, particularly for tropical small-scale marine fisheries, where the high species diversity requires standardized and high-resolution data collection frameworks. This reform should involve co-management approaches that include fishers, fish warehouse managers, researchers, and authorities (Mendoza-Carranza et al., 2025; Ramos et al., 2013). In this effort, it is essential to consider both ecological and social objectives to ensure that management policies align with sustainable fisheries strategies (Jiménez-Badillo et al., 2017).

Except for particular species, such as snook (*Centropomus* spp.) and the red snapper (*L. campechanus*), the differentiation between "high-seas" fishing and coastal-inshore fishing is essentially a social construct. The primary distinction between these fishing categories lies in three factors: (1) the number of fishing days,

(2) the distance traveled by fishers, and (3) the target species. While "high-seas" fisheries primarily target species such as red snapper and sharks, coastal-inshore fisheries exhibit no species specialization. In coastal-inshore fishing, fishers remain close to the shore, whereas in high-seas fishing, they may travel 180 to 250 km offshore, particularly reaching the vicinity of Arcas Cay (Pérez-Jiménez et al., 2020). Despite the differences in target species, both fishery types share some similar species (e.g., lane snapper *Lutjanus synagris*, atlantic sharpnose shark *Rhizoprionodon terraenovae*, scalloped hammerhead *Sphyrna lewini*, gafftopsail catfish *Bagre marinus*, southern stingray (*Hypanus americanus*) (Mendoza-Carranza et al., 2018; Pérez-Jiménez et al., 2022). The species overlap likely reflects the extensive continental shelf in this region, which creates ecological continuity. The Campeche Bank, for example, covers approximately 57,000 km^2 and extends 100–300 km offshore, with depths ranging from 200 to 300 m (Logan et al., 1969).

Among the many skills possessed by fishers, one of the most valuable is their ability to precisely assess both buying and selling prices of their catch while tracking price fluctuations over time—skills fundamental to commercial fishing operations. Consequently, the data collected in this study are considered highly reliable. Price variability across species is driven by multiple factors, including seasonality, species abundance, location, product quality, and market demand, among others (Rice et al., 2024).

The species with the highest reported market price is red snapper (*L. campechanus*), which is harvested year-round in both states. Adults are typically associated with hard substrates (e.g., rocks, reefs, and artificial structures), where they form large schools, whereas subadults and juveniles occupy sandy, muddy bottoms (Aguilar-Medrano & Vega-Cendejas, 2019; Froese & Pauly, 2002). Fishing for this species is conducted using 7-m fiberglass boats. It is considered one of the riskiest fisheries due to both the distance from shore (28–204 km offshore) and the trip duration, which can range from 3 to 10 days (Pérez-Jiménez et al., 2022). These conditions increase the operational costs of each fishing trip, particularly for fuel and provisions, thereby reducing potential profits despite the high market value of the catch. Moreover, boat owners and permit holders retain a portion of the profits, further impacting fishers' earnings. According to Mexico's National Fisheries Chart, *red snapper* biomass is currently near Maximum Sustainable Yield (MSY) levels, suggesting a relatively healthy fishery. However, model projections indicate a potential decline in the years to come (CONAPESCA, 2023).

The second most economically important species is the common snook (*C. undecimalis*), which is primarily harvested in the coastal waters of Tabasco and western Campeche (Laguna de Términos) using mainly gillnets. However, this species is also caught in coastal lagoons and rivers, with peak catches coinciding with its spawning migrations, which significantly impact population levels (Caballero-Chávez, 2009; Perera-García et al., 2011). Common snook fishing is primarily coastal, occurring in river deltas, lagoon mouths, and areas near oil facilities, and is predominantly nocturnal. Due to its coastal reproductive habits, which depend on estuarine and freshwater environments[57], this species has been subjected to intense fishing pressure. As a result, the National Fisheries Chart classifies it as overexploited and at risk of decline (CONAPESCA, 2022a). Given this status, implementing regulatory measures is essential to ensure the long-term sustainability of fish stocks and the continued employment in the fishery sector.

The gafftopsail catfish (*Bagre marinus*) is a commercially important species that generates economic value primarily through high catch volumes rather than market price. For decades, it has been a key economic resource for many small-scale marine fishers along the coasts of Tabasco and Campeche (Mendoza-Carranza et al., 2012). However, its conservation faces significant challenges due to its highly seasonal and predictable reproductive cycle. A key vulnerability stems from males being oral incubators, forming surface-swimming schools during their migration from marine waters to rivers, making them highly susceptible to unregulated fishing, as males are often captured in large numbers during this migration (Mendoza-Carranza & Hernández-Franyutti, 2005; Segura-Berttolini & Mendoza-Carranza, 2012). Urgent enforcement and fishing restrictions are necessary to protect the species' reproductive stocks, as current practices directly target this life stage, potentially causing severe long-term population impacts. Economic hardship among fishers, combined with the lure of quick profits, perpetuates this unregulated fishery. Thus, addressing this issue requires a holistic socioeconomic-fisheries approach that considers the persistent economic challenges of small-scale marine fishing communities.

Although red snapper (*L. campechanus*) and common snook (*C. undecimalis*) fisheries—along with those targeting octopus and Spanish mackerel fisheries, function as quasi single-species fisheries, most other species are caught in multispecies fisheries. This complexity demands caution when applying single-species stock assessment models (Hollowed et al., 2000) and further complicates both the implementation of fisheries models and regulatory measures. Moreover, multispecies models require large amounts of high-quality data to function effectively, or at least to minimize uncertainty (e.g., Kinzey & Punt, 2009).

This study reveals suboptimal social, economic, and working conditions of small-scale marine fishers in Campeche and Tabasco. Multiple factors, including low educational levels, an aging fishing population, and a widespread lack of fishing permits, collectively create a complex and challenging reality. Given the precarious economic conditions in which they operate, it is extremely difficult—if not impossible—for fishers to adopt management measures that involve reducing fishing effort or temporarily closing a fishery, as such actions would result in significant income loss, jeopardizing their economic and family stability. Additionally, the high percentage of unlicensed fishers affects the accuracy of fishing effort estimates in stock assessment models. This can lead to an overestimation of catches and an underestimation of fishing effort, causing errors in key parameters, such as the calculation of Maximum Sustainable Yield (MSY). These systemic inaccuracies ultimately compromise the effectiveness of fisheries management and conservation strategies.

7.5 Conclusions

Most of the fishers in the study region are over 40 years old, reflecting a global trend in small-scale fisheries. Factors such as occupational risks, economic instability, misdirected welfare programs, and difficulties in obtaining fishing permits have discouraged younger generations from entering the sector, which may threaten its long-term sustainability. The economic situation of fishers in Tabasco and Campeche is precarious, with 93% and 87%, respectively, earning below the monthly minimum salary.

There are discrepancies between fishers' reports on species diversity and catch volumes and official records, highlighting the need to improve fisheries data collection systems. While fishers possess precise knowledge of the species and volumes they catch, official data often aggregate multiple species into broad categories, reducing their usefulness for fisheries models. This leads to underestimations of fishing effort, ultimately compromising the sustainability of fisheries and the health of marine ecosystems.

Fishers face low educational levels, lack of social security, limited access to healthcare and housing services, and a high prevalence of informal and unlicensed labor. These conditions make it challenging to implement fisheries management measures, such as reducing fishing effort or enforcing temporary closures, as such measures would directly impact the economic stability of fishing families. Furthermore, the large number of unregulated, unauthorized, or independent fishers complicates the accurate determination of the actual fishing effort exerted on the stocks.

To ensure fisheries sustainability, it is crucial to strike a balance between ecological objectives and the social needs of fishing communities. Addressing socioeconomic marginalization, lack of access to healthcare, education, housing, and financial resources, as well as exclusion from political decision-making, is key to improving fishers' livelihoods and enhancing the long-term effectiveness of fisheries management strategies. A co-management approach, where fishers actively participate in establishing regional strategies, is necessary to achieve sustainable and inclusive fisheries governance.

Acknowledgments This study was supported by the project "La desafiante coexistencia de los sistemas socio-ecológicos acoplados; las industrias de la pesca y petróleo en la Sonda de Campeche" (Ciencia de Frontera 2019, Project CF-2019-1564454, CONACyT). Special thanks to Jhony Bryan Cruz Pech for map design and to the fishers who contributed their knowledge and information to this project. Large Language Model, ChatGTP was used to verify some sentences and paragraphs for English grammar.

Competing Interests The authors have no conflicts of interest to declare that are relevant to the content of this chapter.

Ethical Approval Statement of informed consent: "We are conducting research to understand the coexistence between marine fisheries and the oil industry in the Sonda de Campeche. This study includes examining certain aspects of fishers' communities. For this reason, we kindly request your informed consent and cooperation in completing a questionnaire. The collected data will remain confidential under the protection of El Colegio de la Frontera Sur (ECOSUR) and will be used exclusively for scientific purposes." Additionally, according to Article 38 of the Statistical and Geographical Information Law (INEGI), "All information will be kept STRICTLY CONFIDENTIAL and will be used exclusively for statistical purposes."

Appendix A. Questionnaire

Preliminary Procedure

1. **Initial Presentation** (Project title and purpose)
2. **Request for Permission to Record** (Explicit consent must be recorded as "yes" or "no"). Record CURP (Clave Única de Registro de Población - Unique Population Registry Code.):
3. **Reading of the Personal Data Protection Law**:
   - According to Article 38 of the Statistical and Geographical Information Law (INEGI), "All information will be kept STRICTLY CONFIDENTIAL and will be used exclusively for statistical purposes."
4. **Encouragement of concise responses** (Preferably closed-ended "yes" or "no" answers when applicable).
5. **Completion in Writing and, When Applicable, Audio Recording.**

Survey ID:

0. Introduction

0.1. **Date:**

0.2. **Name of Interviewer:**

0.3. **Interview Location:**

0.3.1. **GPS Coordinates:**

0.4. **Home Address:**

0.5. **How many years have you been engaged in fishing?**

0.5.1. How do you categorize yourself according to your fishing activities? (Multiple choice and key question)

1. Permit Holder
2. Cooperative Representative
3. Cooperative Fisher
4. Fisher Employed by Permit Holders and Cooperatives

5. Independent Fisher
6. Subsistence Fisher
7. Unpaid Processing Worker
8. Paid Processing Worker
9. International Middleman
10. Regional Middleman
11. Local Middleman
12. Equipment and Supply Maintenance Worker
13. Administrator
14. Other (Specify): ________

1. Personal Profile
1.1. **Full Name:**
1.2. **Gender:**
1.3. **Date of Birth:**
1.4. **Place of Birth:**
1.5. **Highest Educational Level Attained:**
1.6. **Do you have economic dependents?** (Yes/No)
1.6.1. **If yes, how many?**

2. Housing Characteristics
2.1. **What is the primary material of the walls of this house?** (Single choice)

1. Waste material
2. Cardboard sheets
3. Asbestos or metal sheets
4. Cane, bamboo, or palm
5. Mud or wattle and daub
6. Wood
7. Adobe
8. Brick, block, stone, cement, or concrete

2.2. **What is the primary material of the roof?** (Single choice)

1. Waste and/or recycled materials
2. Cardboard sheets
3. Asbestos or metal sheets
4. Palm, shingles, or wood
5. Tile
6. Reinforced concrete, brick, or slab

2.3. **What is the primary material of the floor?** (Single choice)

1. Dirt
2. Cement
3. Wood, tiles, or other coatings
4. Other (Specify): _______

2.4. Does this household have (Read the options until a response is obtained, multiple choice):

1. Piped water inside the home?
2. Piped water outside the home but within the property?
3. Piped water from a public tap (or hydrant)?
4. Piped water carried from another household?
5. Water from a tanker truck?
6. Water from a well, river, lake, stream, or another source?
7. Other (Specify): _______

2.5. Does this house have electricity? (Yes/No) *If yes, proceed to questions 2.5.1 to 2.5.3.1.*
2.5.1. Do you pay for electricity? (Yes/No)
2.5.2. On average, how much do you pay for electricity every 2 months?
2.5.3. Are you part of any civil resistance regarding electricity payments? (Yes/No)
2.5.3.1. If yes, since when?
2.5.3.2. Why?

2.6. Does this house have a gas stove? (Yes/No)
If yes, proceed to questions 2.6.1 to 2.6.3.1.
2.6.1. Do you use the gas stove? (Yes/No)
2.6.2. On average, how much do you pay for gas per month?
2.6.3. Do you use any other type of fuel or combustion method for cooking or a similar activity? (Yes/No)
2.6.3.1. If yes, what type?
2.6.3.2. What is it used for?

2.7. How many years ago was this house built?
2.7.1. How long have you lived in this house?
2.7.2. Do you own this house? (Yes/No)
If no:

1. Is it currently being paid for?
2. Is it fully paid off?
3. Is it in another situation?
4. Is it rented?
5. Is it loaned or being taken care of?
6. Other (Specify): _______

2.8. Does this house have a drainage or wastewater disposal system? (Single choice)

1. Connected to the public sewage system
2. Connected to a septic tank
3. Connected to a pipe that discharges into a ravine or crevice
4. Connected to a pipe that discharges into a river, lake, or sea
5. No drainage system
6. Other (Specify): _______

2.9. Does this household have (Multiple choice):

1. Radio or radio recorder
2. Television
3. DVD player
4. Blender
5. Refrigerator
6. Washing machine
7. Landline telephone
8. Water heater (boiler)
9. Air conditioner or climate control system
10. Own car or truck
11. Computer
12. Satellite TV (Sky TV) or cable TV
13. Microwave oven
14. Own internet connection

3. Households and Residents in the Home

General Characteristics

Name	Has Own Income (Yes/No)
Head of Household	Relationship to the Head of Household
Male (M)	Female (F)
Age	Less than 1 year, record as 000

Origins and Educational Level

Primary	Secondary	High School	University
Other Education	None	Years Completed	
In which state and municipality were they born?			
Municipality	State		

Medical Service Coverage

Do they have medical service in?	
1. IMSS Social Security	2. IMSS Solidaridad
3. ISSSTE	4. Public or Para-State Institutions (Pemex, SEDENA, Navy)
5. Private Insurance/INSABI	6. Seguro Popular
7. None	8. Other Institution (Specify)

Can they read and write a note? (Yes/No)	**Do they speak an Indigenous language? (Yes/No)**

Marital Status
| 1. Free Union | 2. Married | | 3. Separated | 4. Divorced | | 5. Widowed | 6. Single with Children | | 7. Single without Children | |

Employment Status (Last Week)
| 1. Worked | 2. Had a job but did not work | | 3. Looked for work | 4. Student | | 5. Housework | 6. Retired or Pensioned | | 7. Unable to Work | 8. Does not work |

Occupation and Economic Activity
| What is the name of their occupation, trade, or position? | | | For example: farmer, primary school teacher, street vendor. | | | What does the business, company, or workplace do? | | | For example: growing corn, making furniture, selling clothes. | |

Fisheries-Related Activity
Do they perform any fisheries-related activity?

1. Yes, paid. 2. Yes, unpaid. 3. No.

If involved in fisheries, with whom do they work?

1 With family members.
2 Not with family members.
3 Both with and without family members.

| What fisheries-related activities do they perform? (Specify)

Monthly Income

Income Range	
1. Less than 1,000	2. 1,001 to 2,000
3. 2,001 to 3,000	4. 3,001 to 4,000
5. 4,001 to 5,000	6. 5,001 to 6,000
7. 6,001 to 7,000	8. 7,001 to 8,000
9. 8,001 to 12,000	10. 12,001 to 16,000
11. 16,000 to 20,000	12. Other (Specify)

4. Income

4.1. On average, what was your total income during the year 2020?	%
4.2. What percentage of your income comes from the following sources (if applicable)?	%
4.2.1. Remittances	%
4.2.2. Financial support from family members within the country	%
4.2.3. Government aid	%
4.2.4. Employment outside the fishing industry	%
4.2.5. Land and/or other property rental income	%
4.2.6. Fishing	%
4.2.7. Other (Specify): ____________________	%

4.3. Do you receive any type of financial or in-kind support from a company or institution associated with the oil industry? (Yes/No)

4.3.1. If yes, which one?

4.3.2. How does it work?

4.4. Do you receive any type of financial or in-kind support from a company or institution associated with the tourism industry? (Yes/No)

4.4.1. If yes, which one?

4.4.2. How does it work?

5. Government Support

From the following list, which financial or in-kind support did you receive between 2018 and 2021?

	2018	2019	2020	2021
5.1. Income Support and Commercialization Aid				
5.2. CADENA (Natural Disaster Assistance Component for the Agricultural and Fisheries Sector)				
5.3. Sembrando Vida (Sowing Life) **5.4.** Marine Diesel/Coastal Gasoline Support				
5.5. Program for the Promotion of Agriculture, Livestock, Fisheries, and Aquaculture—Component for the Well-being of Fishers and Aquaculture (Bienpesca)				
5.6. Strategic Food Security Project (PESA)				
5.7. PROCAMPO/PROAGRO Productivo				
5.8. PROGAN/PROGAN Productivo				
5.9. PACMA				
5.10. Sanitation Programs				
5.11. Humid Tropics Project/Productive Development for the Southern-Southeastern Region				
5.12. Oportunidades/PROSPERA **5.13.** 65 and Older (formerly 70 and Older Program)				
5.14. Any type of Well-being Pension (disability, children, etc.)				
State Programs				
5.15. Did you receive any state-level programs? (Yes/No) **5.15.1.** If yes, which ones?				
Municipal Programs				
5.16. Did you receive any municipal-level programs? (Yes/No) **5.16.1.** If yes, which ones?				
Programs Associated with Oil Companies or Institutions				
5.17. Did you receive support from programs related to oil companies or institutions? (Yes/No)				
5.17.1. If yes, which ones?				
Programs Associated with Tourism Companies or Institutions				
5.18. Did you receive support from programs related to tourism companies or institutions? (Yes/No)				
5.18.1. If yes, which ones?				

Other Programs				
5.19. Did you receive any other support? (Yes/No)				
5.19.1. If yes, specify: ________________				
5.20. Any additional programs? (Yes/No)				
5.20.1. If yes, specify: ________________				

6. Food Consumption

6.1. In the last month, which of the following foods did you and your family consume?

Food type	Consumption frequency	Portion consumed per occasion per person(liters, kilograms, fractions)	Food origin (%)		
			Own production	Purchase	Do children consume it? (Yes/No?)
Corn/ Tortilla					
Beans/Other Legumes					
Meats					
Milk and Dairy Products					
Eggs					
Fish and Seafood					
Rice					
Tubers (Potato, Cassava)					
Vegetables					
Fruits					
Sugar					
Other Cereals					
Fats and Oils					
Soft Drinks or Sodas					
Alcoholic Beverages					

7. Fishing as an Activity

7.1. Are you a member of any producer organization? (Yes/No)

7.1.1. If yes, which ones? (List them)

7.2. Do you engage in other activities besides fishing? (Yes/No)

If yes, indicate which ones (multiple choice):

1. Agricultural activities
2. Aquaculture activities
3. Livestock activities
4. Forestry activities
5. Post-production agro-livestock activities
6. Rural non-agricultural activities
7. Other activity (Specify): _______________

7.3. What type of fishing did you practice in 2021? (Multiple choice)

1. Coastal-riverine fishing
2. Offshore fishing
3. Inland waters fishing

7.4. Do you have fishing permits? (Yes/No)
If yes:
7.4.1. Which ones?

7.5. What was the main fish species you caught in 2020?

And what were the next two most important species in order of significance?

Species Name	During which season is it caught? (Month range)	Average Tons	Lowest Price per Kilogram Paid (in pesos)	Highest Price per Kilogram Paid (in pesos)

7.6. Where did you primarily sell your products in 2020? *(Multiple choice)*

1. At home
2. Local market
3. Local or regional cooperative society
4. Regional market
5. National market
6. International market
7. Online sales

7.7. Who did you primarily sell these products to? *(Multiple choice)*

1. Middleman
2. Collector
3. Collection center or warehouse
4. Factory or packaging facility
5. Directly to consumers
6. Directly for export

7.8. Under what criteria did you primarily sell your products? ***(Single choice)***

1. Visual estimation
2. By weight
3. Quality
4. Quality and weight
5. Other (Specify): ___________

7.9. What was the main final destination of your products? ***(Multiple choice)***

1. Unknown
2. Local market
3. Regional market
4. National market
5. International market
6. Restaurant and/or hotel
7. Self-consumption

Question	Options	Organism	Species 1	Species 2	Species 3
7.10. What was the type of selectivity of the fishing gear used per species? *(Single choice)*	a. Low selectivity (trawls, longlines, and gillnets)				
	b. Selective (longline, handline, string)				
7.11. What was the main method of product preservation on your vessel?	a. No preservation				
	b. Ice				
	c. Refrigerated seawater				
	d. Refrigerated compartments				
	e. Plate freezing				
	f. Brine				
7.12. Did you use GPS for navigation? *(Single choice)*	a. Yes				
	b. No				
7.13. Do you keep production and accounting records? *(Single choice)*	a. Yes				
	b. No				
If yes:	7.13.1. Since when? *(year)* and how?				
	7.13.2. What data do you record, and how?				

8. Infrastructure and Equipment

8.1. What infrastructure do you own for fishing?

Concept	Year of construction	Construction cost	Estimated value in 2021
1			
2			
3			
4			
4			
5			

8.2. What equipment do you own for fishing?

Concept	**Acquisition year**	**Acquisition cost**	**Estimated value in 2021**
a. Small vessels under 10.5 m in length			
b. Engines for small vessels under 10.5 m in length			
c. Vessels over 10.5 m in length but smaller than offshore vessels			
d. Engines for vessels over 10.5 m in length but smaller than offshore vessels			
e. Offshore fishing vessels			
f. Engines for offshore fishing vessels			
g. Nets and other fishing gear 1			
h. Nets and other fishing gear 2			
i. Nets and other fishing gear 3			
j. Longlines			
k. Strings			
l. Freezing equipment			
m. Equipment for initial processing			
n. Scales			
o. Trailers			
p. Outboard marine engines			
q. Pickup truck			
r. Other: ____________________			
s. Other: ____________________			

9. Perception of Climate Change and Risks

9.1. Do you know what climate change is? (yes or no)

9.2. What do you understand by climate change?

(If the respondent says they do not know or understand what it is, questions 9.3 and 9.4 do not apply.)

9.3. Do you see effects of climate change in your daily life? (yes or no)

9.4. What are those effects? List them.

9.5. What are the risks that fishers face at sea? List them.

9.6. Have you known or experienced cases of piracy? (yes or no)
9.6.1. If yes, in which areas?
9.6.2. Could you narrate how they were?

9.7. What has been your experience with Nortes (northern storms), hurricanes, cold fronts, swells, turbonadas (sudden storms), and other climatic events?

10. Social Conflict and the Oil Industry
10.1. Have you perceived conflicts in the region where you live related to oil companies or entities?
10.1.1. If yes, which ones?
10.2. Have you participated in social movements related to or against oil companies or entities?
10.2.1. If yes, which ones?
10.3. What is your impression of the future of fishing in the region where you live? Could it be improved?

11. Health
11.1. In your household, do you or any member have a chronic condition such as diabetes, hypertension, or another? (yes or no)
11.1.1. If yes, which ones? List them.
11.2. Do you or any member of the household have a disability?
11.2.1. If yes, which ones? List them.
11.3. In your household, does anyone suffer from an addiction?
11.3.1. If yes, which ones? List them.
11.4. Due to the COVID-19 pandemic, did you or any member of the household get infected? (yes or no)
11.4.1. If yes, how many? (number)

12. Closing
12.1. To conclude, would you agree to be interviewed again to delve deeper into some of the topics discussed? (yes or no)
Thank you very much

Name and signature of the interviewee
Contact information:
(Phone, email, etc.)
Interviewer's comments: *(Cross out what does not apply)*

References

Aguilar-Medrano, R., & Vega-Cendejas, M. E. (2019). Implications of the environmental heterogeneity on the distribution of the fish functional diversity of the Campeche Bank, Gulf of Mexico. *Marine Biodiversity, 1–17*. https://doi.org/10.1007/s12526-019-00954-y

Alvarez-Torres, P., Díaz-de-León-Corral, A., Ramírez-Flores, O., & Bermúdez-Rodríguez, E. (2002). National fisheries chart 2000: A new instrument for fisheries management in

inland waters. *Reviews in Fish Biology and Fisheries, 12*, 317–326. https://doi.org/10.1023/A:1025079900993
Balasubramaniam, S., Vasanthakumar, J., & Geethalakshmi, V. (2009). Technology development efficiency and socio personal characteristics of researchers in marine fisheries. *Fishery Technology, 46*(2).
Barboza, M. N. (2005). El rezago educativo en México: dimensiones de un enemigo silencioso y modelo propuesto para entender las causas de su propagación. *Revista Interamericana de Educación de Adultos, 27*(2), 29–70.
Béné, C., & Friend, R. M. (2011). Poverty in small-scale fisheries: Old issue, new analysis. *Progress in Development Studies, 11*(2), 119–144. https://doi.org/10.1177/146499341001100203
Bhendarkar, M. P., Sarang, N., Bhosale, M., Rathod, R. H., & Laxmi, V. H. K. (2017). A study on profile of socio-economic condition of fishermen in selected village in Kabirdham District, Chhattisgarh State, India. *International Journal of Education and Science Research, 7*, 49–56.
Caballero-Chávez, V. (2009). Pesquería de robalo blanco *Centropomus undecimalis* en Campeche. *Ciencia Pesquera, 17*(2), 77–86.
Caballero-Chávez, V., & Morales-Martínez, R. G. (2021). *Pesquería de escama marina en el estado de Campeche*. INAPESCA.
Comisión Nacional de Acuacultura y Pesca (CONAPESCA). (2022a). *Carta Nacional Pesquera*. https://www.gob.mx/inapesca/documentos/carta-nacional-pesquera-2022
Comisión Nacional de Acuacultura y Pesca (CONAPESCA). (2022b). *Anuario estadístico de acuacultura y pesca*. https://www.gob.mx/conapesca/documentos/anuario-estadistico-de-acuacultura-y-pesca
Comisión Nacional de Acuacultura y Pesca (CONAPESCA). (2023). *Actualización de la Carta Nacional Pesquera*. https://www.gob.mx/cms/uploads/attachment/file/892410/CNP_2023.pdf
Comisión Nacional de Acuacultura y Pesca (CONAPESCA). (2025). *Permiso de Pesca Comercial*. https://www.gob.mx/tramites/ficha/permiso-de-pesca-comercial/CONAPESCA291
Cruz-Colín, M. E., Cisneros-Mata, M. Á., & Montaño-Moctezuma, G. (2019). Análisis de actores de la pesquería de medusa en Guaymas, Sonora. *Región y Sociedad, 31*, e1104. https://doi.org/10.22198/rys2019/31/1104
Cunningham, S. (2004). *Ordenación de la capacidad pesquera: Examen de las cuestiones normativas y técnicas* (FAO Documentos Técnicos de Pesca y Acuicultura No. 409). Food & Agriculture Organization.
Diario Oficial de la Federación (DOF). (2016). NORMA Oficial Mexicana NOM-006-SAG/PESC-2016, Para regular el aprovechamiento de todas las especies de langosta en las aguas de jurisdicción federal del Golfo de México y Mar Caribe, así como del Océano Pacífico incluyendo el Golfo de California.
Diario Oficial de la Federación (DOF). (2019). *Lineamientos para la operación del Programa Jóvenes Construyendo el Futuro. Gobierno de México*. https://catalogonacional.gob.mx/FichaRegulacionid?regulacionId=113578
Finkbeiner, E. M., & Basurto, X. (2015). Re-defining co-management to facilitate small-scale fisheries reform: An illustration from northwest Mexico. *Marine Policy, 51*, 433–441. https://doi.org/10.1016/j.marpol.2014.10.010
Food and Agriculture Organization (FAO). (2010). *Estudio social de la pesca en México: Diagnóstico sobre la problemática social de los pescadores, su entorno y su visión sobre las posibles soluciones a los problemas que afectan la pesca*.
Froese, R., & Pauly, D. (2002). *FishBase: A global information system on fishes* [Database]. www.fishbase.org
Fulton, E. A. (2021). Opportunities to improve ecosystem-based fisheries management by recognizing and overcoming path dependency and cognitive bias. *Fish and Fisheries, 22*(2), 428–448. https://doi.org/10.1111/faf.12537
Gutiérrez-Pérez, C. (2024). Caracterización del trabajo que realizan mujeres al interior de empresas comercializadoras de productos pesqueros en Yucatán. *Ciencia Pesquera, 32*(num. esp.), 65–77.

Hernández, J. L. P., & Flores, A. C. C. (2022). Trastornos musculoesqueléticos y dolor asociados a exigencias laborales en pescadores de mediana escala de Yucalpetén, Yucatán, México. *Salud de los Trabajadores, 30*(1), 7–19.

Holland, D. S. (2008). Are fishermen rational? A fishing expedition. *Marine Resource Economics, 23*(3), 325–344. https://doi.org/10.1086/mre.23.3.42629621

Hollowed, A. B., Bax, N., Beamish, R., Collie, J., Fogarty, M., Livingston, P., Pope, J., & Rice, C. R. (2000). Are multispecies models an improvement on single-species models for measuring fishing impacts on marine ecosystems? *ICES Journal of Marine Science, 57*(3), 707–719. https://doi.org/10.1006/jmsc.2000.0734

Ikiara, M. M., & Odink, J. G. (1999). Fishermen resistance to exit fisheries. *Marine Resource Economics, 14*(3), 199–213. http://www.jstor.org/stable/42629265

Instituto Nacional de Estadística y Geografía (INEGI). (2020). *Censo de Población y Vivienda 2020*. https://www.inegi.org.mx/programas/ccpv/2020

Instituto Nacional de Estadística y Geografía (INEGI). (2021). *Censo de Población y Vivienda*. https://www.inegi.org.mx/default.html

Jiménez-Badillo, L., Galindo-Cortes, G., & Meiners-Mandujano, C. (2017). Fisheries assessment within the Gulf of Mexico Large Marine Ecosystem. *Environmental Development, 22*, 63–70. https://doi.org/10.1016/j.envdev.2017.01.002

Kinzey, D., & Punt, A. E. (2009). Multispecies and single-species models of fish population dynamics: Comparing parameter estimates. *Natural Resource Modeling, 22*(1), 67–104. https://doi.org/10.1111/j.1939-7445.2008.00030.x

Koenker, R. (2005). Quantile regression in R: A vignette. In R. Koenker (Ed.), *Quantile regression* (pp. 295–316). Cambridge University Press. https://doi.org/10.1017/CBO9780511754098.011

Kumbhar, A. C. (2017). Socio-economic status of fisherman around Ekrukh water reservoir of North Solapur Tahsil dist. Solapur (MS). *Global Journal of Research Analysis, 6*(6), 222–223.

Logan, B. W., Bass, M. N., & Cebulski, D. E. (1969). *Carbonate sediments and reefs, Yucatan Shelf, Mexico* (Vol. 11). American Association of Petroleum Geologists.

Markaida, U., Méndez-Loeza, I., & Rosales-Raya, M. L. (2017). Seasonal and spatial trends of Mayan octopus, *Octopus maya*, population dynamics from Campeche, Mexico. *Journal of the Marine Biological Association of the United Kingdom, 97*(8), 1663–1673. https://doi.org/10.1017/S0025315416001132

Mendoza-Carranza, M., & Hernández-Franyutti, A. (2005). Annual reproductive cycle of gafftopsail catfish, *Bagre marinus* (Ariidae) in a tropical coastal environment in the Gulf of Mexico. *Hidrobiologica, 15*(3), 275–282.

Mendoza-Carranza, M., Pérez-Jiménez, J. C., Hernandez-Lazo, C., & Mendez-Loeza, I. (2009). *The San Pedro's Artisanal Fishery: A non-regulated activity in a shark nursery ground in Western Campeche Bank. Joint Meeting of Ichthyologists & Herpetologists.*

Mendoza-Carranza, M., Romero-Rodríguez, A., Arévalo-Frías, W., Segura-Berttolini, E. C., & Ramirez-Mosqueda, E. (2012). El bagre bandera *Bagre marinus* (Mitchill, 1815) como especie clave de la pesca marina de pequeña escala en la costa de Tabasco. In A. J. Sánchez, X. Chiappa-Carrara, & B. Pérez (Eds.), *Recursos acuáticos costeros del sureste: Tendencias actuales en investigación y estado del arte* (Vol. 2, pp. 527–547). RECORECOS.

Mendoza-Carranza, M., Ejarque, E., & Nagelkerke, L. A. J. (2018). Disentangling the complexity of tropical small-scale fisheries dynamics using supervised Self-Organizing Maps. *PLoS ONE, 13*(5), e0196991. https://doi.org/10.1371/journal.pone.0196991

Mendoza-Carranza, M., Paredes-Trujillo, A., & Segura-Berttolini, E. (2024). La importancia de las mujeres en la cadena de valor de la pesca marina de pequeña escala y en la ciencia pesquera: un estudio de caso en la costa de Tabasco, México. *Ciencia Pesquera, 32*(núm. especial), 55.

Mendoza-Carranza, M., Aponte González, L., Perera García, M. A., Domínguez-Petit, R., Contreras Perdomo, A. R., Martín, G., Gaibor, N., Lara-Mendoza, R. E., Macchi, G. J., Mejía Arana, F., Queirolo Palma, D., Godoy, L., Infante Altagracia, J. J., Ortiz, J. R., Bustos Montes, D. M., Castro Hernández, J. J., Lessa, R. P. T., Martínez Táboas, M., Vergara-Chen, C., Molina, J., & Aumesquet, M. L. (2025). In R. Domínguez-Petit, M. Cabaleiro, & H. Fernández (Eds.), *Libro blanco sobre retos y estrategias para garantizar un aprovechamiento sostenible de los recursos pesqueros en los países iberoamericanos*. In press.

Oliveto-Andrade, A., Espinoza-Tenorio, A., Ramos-Muñoz, D., & Pérez-Jiménez, J. C. (2024). Understanding the motivations of young people from marginalized rural communities to participate in small-scale fisheries in oil territories of the Gulf of Mexico. *Ocean & Coastal Management, 248*, 106947. https://doi.org/10.1016/j.ocecoaman.2023.106947

Peña-Puch, A. D. C., Pérez-Jiménez, J. C., Espinoza-Tenorio, A., Méndez-Loeza, I., & Williams-Jara, G. M. (2021). The sustainability of small-scale fisheries in oil-producing sections of the Gulf of Mexico. *The Extractive Industries and Society, 8*(4), 100975. https://doi.org/10.1016/j.exis.2021.100975

Perera-García, M. A., Mendoza-Carranza, M., Contreras-Sánchez, W. M., Huerta-Ortíz, M., & Pérez-Sánchez, E. (2011). Reproductive biology of common snook *Centropomus undecimalis* (Perciformes: Centropomidae) in two tropical habitats. *Revista de Biología Tropical, 59*(2), 669–681.

Pérez-Jiménez, J. C., & Mendez-Loeza, I. (2015). The small-scale shark fisheries in the southern Gulf of Mexico: Understanding their heterogeneity to improve their management. *Fisheries Research, 172*, 96–104. https://doi.org/10.1016/j.fishres.2015.07.004

Pérez-Jiménez, J. C., Méndez-Loeza, I., Mendoza-Carranza, M., & Cuevas-Zimbron, E. (2012). Análisis histórico de las pesquerías de elasmobranquios del sureste del Golfo de México. In A. J. Sánchez, X. Chiappa-Carrara, & B. Peréz (Eds.), *Recursos acuáticos costeros del sureste: Tendencias actuales en investigación y estado del arte* (Vol. 2, pp. 463–481). RECORECOS.

Pérez-Jiménez, J. C., Wakida-Kusunoki, A., Hernández-Lazo, C., & Mendoza-Carranza, M. (2020). Shark-catch composition and seasonality in the data-poor small-scale fisheries of the southern Gulf of Mexico. *Marine and Freshwater Research, 71*(9), 1182–1193. https://doi.org/10.1071/MF19184

Pérez-Jiménez, J. C., Núñez, A., González-Jaramillo, M., Mendoza-Carranza, M., Acosta-Zetina, J., Flores-Guzman, A., & Rocha-Tejeda, L. (2022). Inferring ecosystem impacts of a small-scale snapper fishery through citizen science data, productivity and susceptibility analysis, and ecosystem modelling. *Fisheries Research, 250*, 106269. https://doi.org/10.1016/j.fishres.2022.106269

Pérez-Sánchez, E., Muir, F. J., & Ross, G. L. (2005). Livelihood options for the coastal zone of Tabasco, Mexico. *Aquatic Resources, Culture and Development, 1*(2), 91–108.

Phillips, N. D. (2017). Yarrr! The pirate's guide to R. *APS Observer, 30*(3).

R Core Team. (2019). *R: A language and environment for statistical computing*. R Foundation for Statistical Computing. https://www.r-project.org/

Rahaman, S. N., Samanta, S., Banerjee, J., Sandeep, K. D., Biplap, G., Surajit, D., Kazi, M. A., & Buddhadev, M. (2021). Socio-demographic profile, nutritional and health status of fishermen: A review. *Bangladesh Journal of Medical Science, 20*(4), 707–713.

Ramos, M. J., Flores, D., Sosa, A., Sosa, E., López, J., Arreguín, F., & Ramirez, A. (2013). *Manual de indicadores y metodología para la aplicación del enfoque ecosistémico en las pesquerías de Yucatán, Campeche y Quintana Roo*. RNIIPA, COFUPRO, UAC.

Reyes Grande, F., Espinoza Tenorio, A., & Díaz Perera, M. Á. (2020). Ya no hay pescado. Subsistencia e incertidumbre en una localidad pesquera del Golfo de México. *Intersticios Sociales, 20*, 193–223.

Rice, E., Bennett, A., Smith, M. D., Rice, E. D., Bennett, A. E., Smith, M. D., Liverpool-Tasie, L. S. O., Katengeza, S. P., Infante, D. M., & Tschirley, D. L. (2024). Price volatility in fish food systems: Spatial arbitrage as an adaptive strategy for small-scale fish traders. *Ecology and Society, 29*(2), art13. https://doi.org/10.5751/ES-15076-290213

Salazar-de la Cruz, C. C., Zepeda-Domínguez, J. A., Espinoza-Tenorio, A., & Ramos-Muñoz, D. E. (2020). Governance networks in marine spaces where fisheries and oil coexist: Tabasco, México. *The Extractive Industries and Society, 7*(2), 676–685. https://doi.org/10.1016/j.exis.2020.03.012

Segura-Berttolini, E. C., & Mendoza-Carranza, M. (2012). Importance of male gafftopsail catfish, *Bagre marinus* (Pisces: Ariidae), in the reproductive process. *Ciencias Marinas, 39*(1), 29–39. https://doi.org/10.7773/cm.v39i1.2136

Shann, M. V. (2021). The Sureste Basin of Mexico: Its framework, future oil exploration opportunities and key challenges ahead. *Geological Society, London, Special Publications, 504*(1), 119–146. https://doi.org/10.1144/SP504-2019-214

Tzanatos, E., Dimitriou, E., Papaharisis, L., Roussi, A., Somarakis, S., & Koutsikopoulos, C. (2006). Principal socio-economic characteristics of the Greek small-scale coastal fishermen. *Ocean & Coastal Management, 49*(7–8), 511–527. https://doi.org/10.1016/j.ocecoaman.2006.04.002

Utami, R. Z., Suryana, A. A. H., Apriliani, I. M., & Maulina, I. (2023). The social characteristics of fisherman communities in Cangkol Village, Cirebon, Indonesia. *Asian Journal of Fisheries and Aquatic Research, 24*(2), 16–26.

Venables, W. N., & Ripley, B. D. (2002). *Modern applied statistics with S* (4th ed.). Springer.

Wei, Y., Pere, A., Koenker, R., & He, X. M. (2006). Quantile regression methods for reference growth charts. *Statistics in Medicine, 25*(8), 1369–1382. https://doi.org/10.1002/sim.2271

Wickham, H. (2016). *ggplot2: Elegant graphics for data analysis*. https://ggplot2.tidyverse.org

Zalik, A. (2009). Zones of exclusion: Offshore extraction, the contestation of space and physical displacement in the Nigerian Delta and the Mexican Gulf. *Antipode, 41*(3), 557–582. https://doi.org/10.1111/j.1467-8330.2009.00687.x

Chapter 8
Among Tides and Oil Rigs: Navigating Well-Being in Small-Scale Fishing Communities Coexisting with the Oil Industry in the Southwestern Gulf of Mexico

A. Núñez, D. G. Cupido-Santamaria, A. C. E. Dias, and A. Espinoza-Tenorio

8.1 Introduction

Globally, small-scale fisheries (SSF) play a crucial role in enhancing the well-being of coastal communities, supporting livelihoods, food security, and employment (FAO, 2015; De Oliveira Leis et al., 2019). Beyond these benefits, SSF are deeply embedded in local traditions and values, representing a way of life (Chuenpagdee et al., 2019; Delgado-Ramírez et al., 2023). However, the coexistence of SSF with other maritime sectors introduces complex dynamics for space and resource management, as well as political interests (Dias et al., 2023).

Despite their importance, SSF are often excluded from marine spatial planning, which limits their ability to influence decisions regarding resource use and coastal development (Ayilu et al., 2022). This has led to increased competition and conflict with industrial fishing and extractive sectors, such as oil exploration (Cohen et al., 2019; Andrews et al., 2021).

The oil industry is one of the most politically and economically powerful globally. Despite its adverse socio-environmental effects, there persists an expectation that oil wealth can translate into local development, especially in communities near oil and gas operations (Andrews et al., 2021). However, evidence shows that coastal communities often bear the costs without receiving equitable benefits. These include limited employment opportunities, increased living costs, environmental

A. Núñez (✉) · D. G. Cupido-Santamaria · A. Espinoza-Tenorio
El Colegio de la Frontera Sur, Campeche, México
e-mail: adrian.nunez@ecosur.mx; deysi.cupido@ecosur.mx; aespinoza@ecosur.mx

A. C. E. Dias
University of Waterloo/ EcoVadis, Waterloo, ON, Canada
e-mail: ana.estevesdias@uwaterloo.ca

A. Espinoza-Tenorio et al. (eds.), *The Fishing and Oil Industries in Coexistence*, https://doi.org/10.1007/978-3-032-02621-7_8

degradation, and unresolved governance dilemmas involving multiple stakeholders (Ribeiro et al., 2021; Owusu et al., 2023).

In the Southwestern Gulf of Mexico (SWGoM), the spatial historical competition between the oil and fisheries sectors has had significant political and socio-ecological implications. The 1979 Ixtoc-I spill caused extensive ecological damage and social conflict along the coasts of Tabasco and Campeche, remediated by economic compensation for affected communities (Ramos-Muñoz & Espinoza-Tenorio, 2020). In 2001, the creation of maritime security zones in oil fields further restricted fishing grounds, leading to increased conflict between the two sectors (Villalobos & Rivera-Arriaga, 2008).

The 2013 energy reform opened oil production to private companies, reshaping institutional frameworks and intensifying offshore activity (Díaz-Perera and De los Santos González, 2021). Between 2018 and 2023, the region averaged 1.65 million barrels of oil per day, accounting for approximately 89% of national production (SENER, 2024). Recent projects, such as the Olmeca refinery in Paraíso, Tabasco, have accelerated coastal industrialization, transforming critical ecosystems (e.g., mangrove and seagrass beds) due to oil activities, and increasing socio-environmental pressures on locals (Quist & Nygren, 2015; De la Rosa-Velázquez et al., 2017).

Although several studies have addressed the multiple socio-environmental effects of oil development in the SWGoM (Quist, 2019; Tolentino-Arévalo et al., 2019), there is a lack of knowledge about the trade-offs these impacts pose for the well-being of coastal communities. This chapter addresses that gap by examining fishers' perspectives in Tabasco and Campeche, coexisting in different contexts of oil production. It explores well-being across material, relational, and subjective dimensions, with particular attention to the social aspects of SSF that are at risk from the rapid expansion of the oil industry in the region.

8.2 The Well-Being Approach

The concept of social well-being refers to the conditions that satisfy human needs, allowing individuals to pursue objectives and meaningfully enjoy a satisfactory quality of life (McGregor, 2008). This approach adopts a three-dimensional perspective, incorporating material, subjective, and relational dimensions, which are strongly interrelated and must be comprehensively considered to understand what gives meaning and leads to life satisfaction (Fig. 8.1; Gough and McGregor, 2007; Leite et al., 2023).

The social well-being framework enables an understanding of fishers' connections with ecosystems, emphasizing their relational and subjective importance alongside the material contributions at both the individual and community levels to a well-lived life (Coulthard et al., 2011; Weeratunge et al., 2014). This well-being research aims to examine the socio-environmental impacts of the oil industry on the livelihoods and well-being of fishing communities.

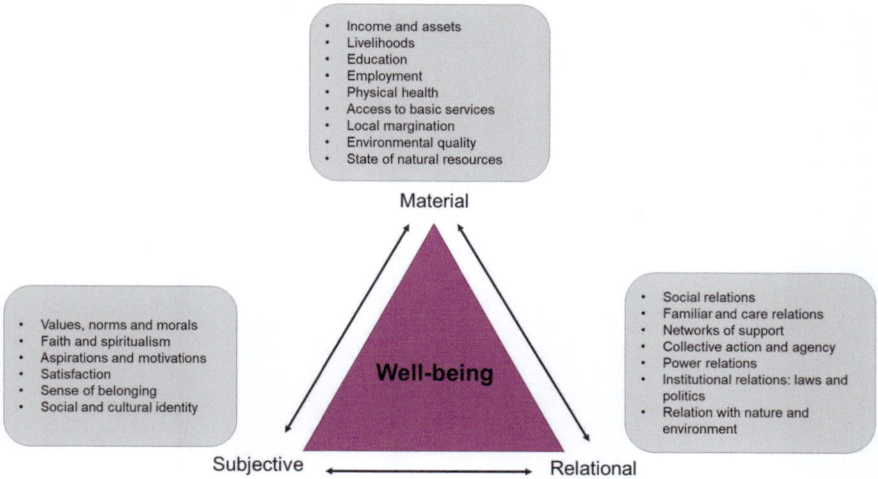

Fig. 8.1 Conceptual framework of well-being (Adapted from Gough and McGregor, 2007)

8.3 Study Area

Campeche and Tabasco (Fig. 8.2) are relevant fishing states in Mexico, ranking sixth and twelfth in landings, respectively (CONAPESCA, 2021). Campeche's key fisheries include octopus (*Octopus maya*), jacks (*Caranx hippos*), snails (*Lobatus gigas; Pleuroploca gigantea*), and shrimp (*Farfantepenaeus duorarum*); while Tabasco's are mojarra (*Calamus* spp.), oyster (*Crassostrea virginica*), catfish (*Bagre marinus*), and jacks (CONAPESCA, 2021). Most regional fisheries are operated by the small-scale fishing fleet, which is composed of permit holders, fishing cooperatives, and independent fishers. Independent fishers typically work under cooperatives or permit holders, often without labor benefits, and receive only provisioning expenses or loans, which are later deducted from their catch earnings (Salas et al., 2011).

Simultaneously, the SWGoM is central to Mexico's energy sector, concentrating most of its oil production and hosting a dense network of offshore oil infrastructure (Fig. 8.2). Campeche and Tabasco contain some of the oldest and largest oil fields, functioning as strategic points for ongoing extraction and distribution (Díaz-Perera & De los Santos González, 2021). This territorial overlap between both industries has led to restricted access, spatial conflicts, and growing socio-environmental pressure on coastal communities (Zalik, 2009; Arroyo & Zalik, 2016).

These tensions manifest through continuous, diverse, and at times catastrophic interactions between the fishing and oil industries. As extractive activities, both compete for seascapes in shallow waters (<200 m), leading to accidents with infrastructure (e.g., loss of fishing nets due to unburied pipelines) and maritime traffic. Additionally, official regulations establish 2500 m security zones around oil infrastructure (e.g., oil wells and platforms), restricting fishing activities (DOF, 2019).

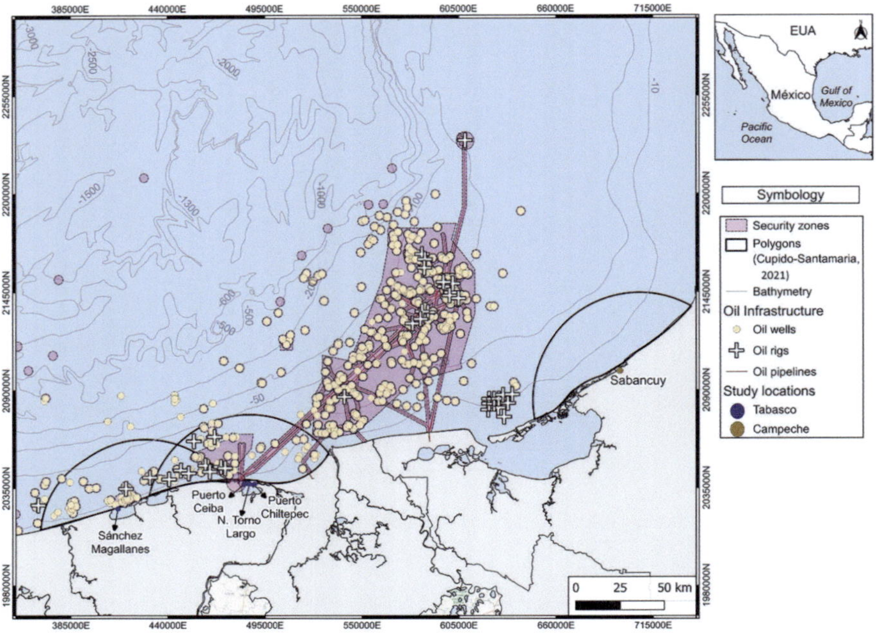

Fig. 8.2 Location of the fishing communities and oil infrastructure in the study area (Source: CNH, 2023). The semicircle polygons (5 km buffer from the coast) represent the nearest fishing zones associated with the communities analyzed in this study. (Modified from Cupido-Santamaría, 2021)

These restrictions are justified by concerns over terrorism and infrastructure theft, reinforcing the exclusion of fishers from their traditional grounds.

Moreover, spills continue to be a persistent source of conflict. Since the Ixtoc-I disaster in 1979, fishers have repeatedly denounced contamination in the area, while authorities and Mexican Petroleum (PEMEX, by its Spanish acronym) have often attributed oil presence to natural emanations. Only recently have remote sensing studies provided systematic evidence of ongoing spills (Uribe-Martínez et al., 2024). When requests for damages are occasionally successful, fishers receive compensation through gasoline vouchers, gears, and motors, often perceived as insufficient given the scale of impact.

8.3.1 Fishing Communities

The study focused on three fishing communities in Campeche and Tabasco, each with different socio-environmental contexts and the shared condition of coexisting with the oil industry (Fig. 8.2; Appendix 8.1).

Fig. 8.3 Studied fishing communities. (**a**) *Sabancuy*, Campeche; (**b**) *Paraíso*, Tabasco; (**c**) *Sánchez Magallanes*, Tabasco

Sabancuy (SAB) in *Ciudad del Carmen*, Campeche (Fig. 8.3a), has 7744 inhabitants (INEGI, 2020a) and is part of the Terminos Lagoon, a natural protected area. Although no oil infrastructure is nearby its coast (Fig. 8.2), PEMEX has historically operated in adjacent maritime zones (García-Cuellar et al., 2004). As part of a municipality with intense offshore activity, many fishing cooperatives are registered there to access oil-related compensation programs. These benefits have often been negotiated through cooperative pressure, citing the environmental impacts of oil operations (Peña-Puch et al., 2021).

Paraíso (PAR), Tabasco (Fig. 8.3b), includes three fishing communities: *Puerto Ceiba, Nuevo Torno Largo,* and *Chiltepec*, with populations ranging from 1696 to 2867 inhabitants (INEGI, 2020a). The area has experienced significant industrial development, following the opening of the Dos Bocas port in 1982 and the recent construction of the Olmeca refinery, which involved the expropriation of land traditionally used for agriculture (Bonfil-Jimenez, 2018). The presence of offshore and onshore oil infrastructure, along with a 5500 m exclusion zone for oil export (DOF, 2019), directly affects local SSF operations.

Coronel Andrés Sánchez Magallanes (SM), in Cardenas, Tabasco (Fig. 8.3c), has 5518 inhabitants (INEGI, 2020a) and is situated between the SWGoM and *El Carmen* Lagoon (Fig. 8.2). Both ecosystems support SSF and oyster farming. SM holds historical relevance as the site of Mexico's first offshore oil well (1959) and the first community to host private oil companies following the 2013 energy reform (CNH, 2019). Currently, PEMEX and European firms like ENI and Wintershall Dea operate offshore platforms in the area. ENI also manages an onshore oil and gas separation station.

8.4 Methodology

8.4.1 Semi-structured Interviews

A random sampling method was applied using a Geographic Information System to select 30% of the geographic blocks within each community, ensuring spatial representativeness. This approach improves data quality in remote areas with limited census information (Oliveto-Andrade et al., 2024). Semi-structured interviews were conducted with the heads of households economically dependent on SSF.

Two rounds of interviews were carried out (see Appendix 8.2). The first, conducted in 2021, was part of a longitudinal study that aimed to characterize both the material and relational dimensions, involving 96 interviews. The second round, referred to as the well-being interviews, took place between 2022 and 2023 and explored relational and subjective dimensions. Nonetheless, due to the high mobility and the aging of fishers, 70 follow-up interviews were completed: SAB (26), PAR (19), and SM (25).

All procedures were conducted in accordance with ethical standards, including obtaining informed consent and maintaining data confidentiality. Responses were analyzed using content analysis and coded into a semi-quantitative database (Tran et al., 2018).

8.4.2 Well-Being Indicators

The well-being indicators, categorized by dimension, employed in this study are described below. The material dimension included household conditions, health insurance, education, fishing assets, and income. Income was categorized into fishing income, alternative activities, and government or oil-related payments. Fishing income was compared to Mexico's 2021 monthly minimum wage (USD 209.7; CONASAMI, 2021). Government payments (2018–2021) included fishing programs, such as *Bienpesca* (USD 370 annually), and Marine Diesel (subsidized fuel), as well as social programs for elderly and low-income households, including 65 and More, Well-being Pension, and PROSPERA. Oil-related payments were considered when fishers had received compensation, including the Community and Environment Support Program (PACMA), which aimed to maintain social license[1] in PEMEX-operated areas (PEMEX, 2018).

---

[1] Social License to Operate refers to the acceptance and approval granted by local communities to a company or industry based on its social and environmental practices. In the case of PEMEX, PACMA was established in 2013 and comprises a set of programs designed to strengthen this acceptance by addressing social needs, promoting human development, and supporting sustainable community initiatives where the company operates.

The relational dimension focused on commercial links and institutional dynamics. This included identifying market actors and the destination of fishing products. It also examined institutional responses to contingencies such as natural disasters, maritime accidents, and oil spills.

The subjective dimension assessed perceptions of community well-being, the impacts, costs, and benefits of the oil industry, as well as nature's contributions from fishing. For this study, the term "nature's contributions" was used as proposed by Díaz et al. (2018), defined as the positive and negative effects of nature on people's quality of life. Following Bryce et al. (2016), these indicators were evaluated using a Likert scale (1 = very low; 5 = very high).

8.5 Results and Discussion

8.5.1 Demography

The profiles of fishers were consistent across the communities (Appendix 8.3). Most interviews were conducted with men aged 50 to 58. The number of economic dependents per household ranged from one to seven, and most fishers were married or in consensual unions. The interviewees were primarily independent fishers and members of fishing cooperatives.

8.5.2 Material Dimension

8.5.2.1 Household

Across communities, household conditions represented a significant foundation for material well-being. Most fishers owned their homes, with SAB and SM showing higher rates than PAR (Table 8.1). Construction materials were generally durable, such as brick and cement for walls and floors, providing physical security.

Access to electricity, drainage, and appliances was similar (Table 8.2), reinforcing basic living standards. However, water access remained uneven, particularly in PAR, where several households still relied on rivers or waterholes, despite being closer to urban areas compared to other communities. This reflects persistent infrastructural gaps, a common issue in Mexican urban settings where SSF operates (Salas et al., 2022).

In addition to development gaps, potential health risks were identified. The use of asbestos roofing materials can lead to long-term diseases, such as asbestosis and cancer (Berry et al., 2022). Moreover, reliance on unregulated water sources is particularly concerning given the proximity to oil infrastructure, where contamination and limited access to healthcare services exacerbate threats to well-being, as explored in the following section.

Table 8.1 Housing characteristics of fishers in *Sabancuy* (SAB), Campeche, *Paraíso* (PAR), and *Sánchez Magallanes* (SM), Tabasco

Housing characteristics	SAB	PAR	SM
n	26	19	25
Home ownership	84.6	68.4	84
Wall materials			
Brick	92.3	100	100
Others	7.7	0	0
Floor materials			
Cement	61.5	84.2	88
Tiles or wood	34.6	15.8	8
Dirt	3.8	0	4
Roof materials			
Asbestos or metallic sheet	88.5	57.9	100
Concrete or brick	11.5	15.8	0
Cardboard	0	26.3	0
Capital assets			
Radio	26.9	42.1	28
TV	92.3	89.5	88
Computer	15.4	15.8	8
Gas stove	92.3	89.5	100
Refrigerator	76.9	94.7	92
Washing machine	80.8	94.7	76
Microwave	11.5	26.3	24
Water heater	3.8	0	0
Air-conditioning	19.2	47.4	32

8.5.2.2 Health

Small-scale fishers are exposed to multiple environmental and occupational risks, often exacerbated by the lack of health insurance resulting from the informal nature of their activity (Turner, 2025; Cochrane et al., 2011). In the studied communities, access to healthcare services was limited across all sites. A significant proportion of fishers reported no health coverage: SAB (53.8%), SM (52%), and PAR (36.8%).

Public health institutions, mainly the Health Institute for Well-being (INSABI, by its Spanish acronym) and the Mexican Social Security Institute (IMSS, by its Spanish acronym), were the primary sources of medical care. Reliance on public services was slightly higher in PAR (57.9%) compared to SAB (42.3%) and SM (44%). However, IMSS coverage is restricted to formally employed individuals, excluding most fishers. Private healthcare use was almost absent.

These patterns reveal structural weaknesses in healthcare provision for SSF communities, limiting their capacity to achieve and sustain well-being. In oil-affected areas, limited medical coverage, combined with environmental health risks, constrains people's ability to respond to illness, undermines physical security, and

Table 8.2 Housing services of families dependent on small-scale fisheries in *Sabancuy* (SAB), Campeche, *Paraíso* (PAR), and *Sánchez Magallanes* (SM), Tabasco

Housing services	SAB	PAR	SM
n	26	19	25
Basic services			
Electricity service (%)	100	100	100
Water service (%)			
Piped water inside the house	42.3	52.6	56
Piped water outside the house (front or back yard)	23.1	5.3	20
Piped water from public tap	19.2	10.5	48
River or waterhole	34.6	63.2	36
Drainage system (%)			
Septic tank	92.3	84.2	92
Public drainage	0	15.8	0
Pipeline to river, lake, or sea	0	0	4
No drainage	7.7	0	4
Telecom services (%)			
Cable TV	76.9	78.9	72
Internet	15.4	15.8	16

erodes long-term quality of life (Adgate et al., 2014; Hemmerling et al., 2020). Strengthening health services is therefore essential to reinforce well-being in coastal communities facing cumulative socio-environmental pressures.

8.5.2.3 Education

Education is a key dimension of well-being and a recognized human right (United Nations, 2019). In Mexico, nearly half of the population over the age of 15 has completed only basic education (INEGI, 2020b), a pattern also mirrored in the fishing sector. In the studied communities, most fishers had completed elementary school, except in PAR, where junior high school was more common (Appendix 8.3). However, PAR also showed the highest proportion of fishers with no formal education and the highest illiteracy rate (10.5%), compared to SAB (3.85%) and SM (0%). This contrast is striking given PAR's greater urban and industrial development, revealing how educational barriers persist even in more urbanized settings.

This contradiction reflects a broader pattern in coastal regions affected by extractive industries: while economic activity expands, improvements in public services (i.e., education and health) remain uneven. Formal education serves as a key strategy for families navigating adverse conditions, yet structural inequalities often constrain access, limiting opportunities to enhance well-being (Idrobo & Johnson, 2020). In oil-dependent communities such as those in Louisiana, USA, and Tabasco, industrial expansion has historically coincided with lower educational attainment, as residents prioritize short-term employment in the oil industry over long-term schooling (Hemmerling et al., 2020; Salazar-De la Cruz et al., 2020).

8.5.2.4 Fishing Assets

Fishing assets are crucial to material well-being in SSF, as they enable access to resources, generate income, and ensure food security (Acott et al., 2018; Salas et al., 2022). In the studied communities, most fishers owned basic tools, including gear, boats, and outboard motors (Fig. 8.4), with higher availability in PAR and SM. However, access to post-harvest equipment and owned infrastructure remained limited, particularly in SAB. This may constrain fishers' autonomy and adaptability to upgrade within the value chain.

Fishing permit ownership was relatively low, with higher rates in PAR (36.84%), followed by SAB (30.8%) and SM (25%). Limited permit access reduces fishers' legal recognition and access to markets and government or oil-related support programs, reinforcing informal dynamics that can hinder economic stability and long-term well-being.

In addition to its productive capacity, fishing plays a crucial role in ensuring food security and nutrition. Most fishers reported reserving part of their catch for family consumption: SM (68%), SAB (66.7%), and PAR (63.2%). In rural coastal contexts, this contribution is vital, as SSF can supply up to half of the animal protein consumed by households (Salas et al., 2019). Subsistence fishing thus strengthens both material well-being and resilience, providing a buffer against socio-environmental pressures.

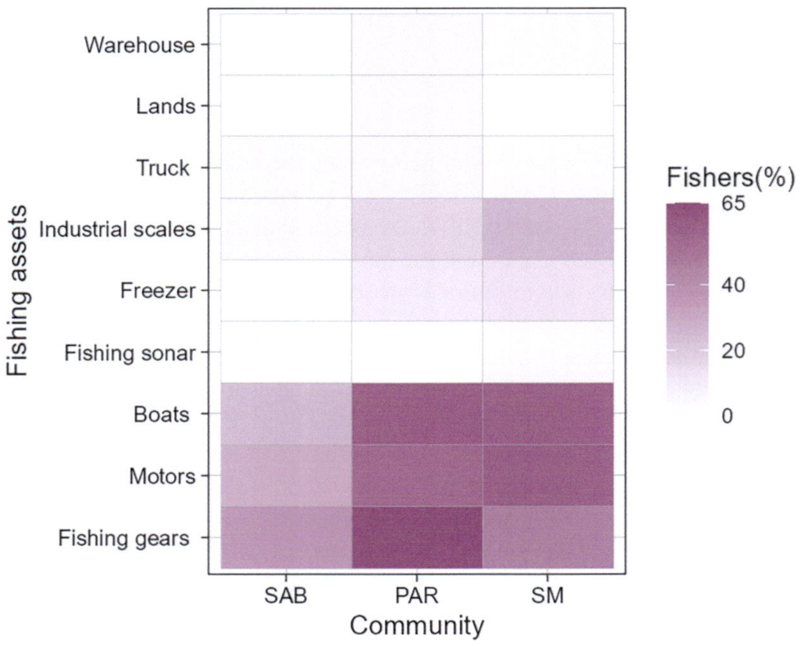

Fig. 8.4 Fishing assets of fishers in *Sabancuy* (SAB), Campeche, *Paraíso* (PAR), and *Sánchez Magallanes* (SM), Tabasco

8.5.2.5 Income

Income is a central component of material well-being, shaping fishers' ability to meet basic needs, respond to shocks, and maintain autonomy. In SSF, incomes are often low, unstable, and complemented by alternative activities or external support mechanisms, dynamics exacerbated by informal working conditions and policy exclusion (Weeratunge et al., 2014; Acott et al., 2018).

In this research, most fishers reported earnings below the minimum wage: SAB (88.4%), PAR (70.6%), and SM (54.2%) (Appendix 8.4). SM stood out with relatively higher income levels, where 33.4% earned between 1.1 and 2 times the minimum wage, compared to SAB (11.5%) and PAR (5.9%). Alternative activities were limited, suggesting fewer opportunities and greater dependence on fishing: SAB (34.6%), SM (20%), and PAR (10.5%). These included jobs in agriculture, construction, aquaculture, and the oil sector.

Government programs such as Bienpesca and Marine Diesel supplemented household incomes, with SM receiving the highest support, followed by PAR and SAB (Table 8.3). Oil-related payments varied considerably: in SM, 28% of fishers received compensation (primarily through ENI), compared to 10.5% in PAR and 7.7% in SAB (mainly from PEMEX). Participation in PACMA remained limited across sites: SM (8%), SAB (3.8%), and PAR (0%).

These differences reflect structural disparities across communities. Higher incomes in SM may be related to organization, agency, fish stock availability, and

Table 8.3 Government payments (2018–2021) received by fishers of *Sabancuy* (SAB), Campeche, *Paraíso* (PAR), and *Sánchez Magallanes* (SM), Tabasco

Community	Payment type	2018	2019	2020	2021
SAB	Bienpesca	19.2	42.3	61.5	30.8
	Marine diesel	3.8	0	0	0
	65 and more program	7.7	7.7	7.7	3.8
	Well-being pension	3.8	3.8	3.8	7.7
	PROSPERA	3.8	0	0	0
	Others	7.7	7.7	7.7	7.7
PAR	Bienpesca	15.8	21.1	42.1	26.3
	Marine diesel	26.3	10.5	10.5	5.3
	65 and more program	5.3	5.3	5.3	10.5
	Well-being pension	5.3	5.3	5.3	10.5
	PROSPERA	5.3	0	0	0
	Others	5.3	5.3	15.8	15.8
SM	Bienpesca	12	32	76	64
	Marine diesel	24	24	16	12
	65 and more program	24	32	32	32
	Well-being pension	0	12	12	16
	PROSPERA	24	16	16	12
	Others	12	8	8	16

market access (Coronado et al., 2020b). Moreover, the presence of international oil companies, which are required to implement social license programs, introduces other challenges not experienced at the other studied sites. Despite these contrasts, low and unstable fishing incomes continue to be a common challenge. Reliance on state and oil sector payments, while providing short-term stability, reinforces informality and dependency, constraining the adaptability and long-term well-being of fishing households (Thanh et al., 2010; Idrobo, 2018).

8.5.3 Relational Dimension

8.5.3.1 Commercial Links

Commercial relationships are fundamental to fishers' relational well-being, shaping income opportunities, autonomy, recognition, and resilience. Across the studied communities, fishers engaged in diverse market arrangements, reflecting different degrees of formality and dependency (Fig. 8.5).

In SAB and PAR, sales were primarily channeled through fishing cooperatives and permit-holder collection centers (Fig. 8.5a), offering purchasing stability but limiting autonomy through control over inputs and prices. In contrast, middlemen dominated in SM, establishing informal but equally asymmetrical dynamics. Recent studies have shown that both forms of intermediation can reduce negotiating power, thereby reinforcing dependency, affecting fishers' decision-making capacity, and ultimately impacting their overall well-being (Coronado et al., 2020a; Biswal and Johnson, 2023).

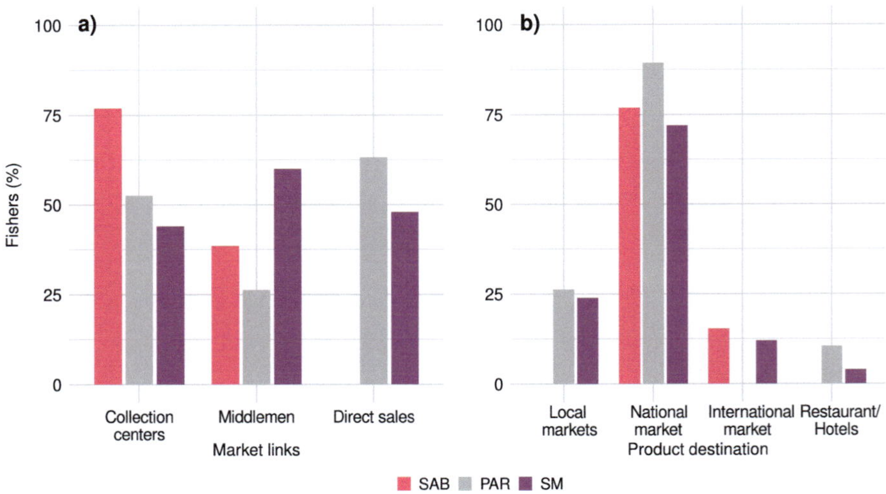

Fig. 8.5 Commercial relations of fishers in *Sabancuy* (SAB), Campeche, *Paraíso* (PAR), and *Sánchez Magallanes* (SM), Tabasco. (**a**) Markets links; (**b**) Product destination

Notably, a significant portion of fishers in PAR and SM engaged in direct sales to consumers. This strategy offers greater autonomy but demands stronger social capital and negotiation skills. These hybrid arrangements reflect "malleable coordination," enabling fishers to navigate uncertainty without structural protection and requiring greater reliance on community ties (Coronado et al., 2020a).

Regarding the destination of fishery products, national markets emerged as the principal endpoint across communities (Fig. 8.5b), with much of the catch distributed to urban centers, particularly Mexico City. Local markets remained important in PAR, while international exports and sales to tourism-linked sectors were limited.

Market dynamics are deeply relational, built on trust, access to infrastructure, and governance structures (Steenbergen et al., 2019). In oil-influenced territories, the exclusion from traditional fishing zones, coupled with the fishers' dependence on compensation programs, may create fragile market relationships, reducing negotiating power and increasing uncertainty and conflict (Zalik, 2009; Idrobo & Johnson, 2020; Biswal & Johnson, 2023). These dynamics not only impact economic outcomes but also erode relational well-being, underscoring the need to restructure SSF value chains toward fairer and more inclusive systems of exchange (Espinoza Tenorio et al., 2023).

8.5.3.2 Institutional Responses During Contingencies

Social relations with institutions are crucial in how fishers experience and manage risks, especially during catastrophes (Westlund, 2007). In communities exposed to socio-environmental pressures, relational well-being becomes critical during crises. For this research, results revealed uneven and often insufficient institutional assistance during contingencies, revealing structural gaps in disaster protocols (Table 8.4).

During natural disasters, most fishers received no assistance. Government and civil protection provided only basic aid, such as temporary shelter or pantry food, across communities (Table 8.4). In contrast, maritime accidents reflected a more consistent institutional response, with the Marine Guard widely recognized for its' rescue operations for vessels. However, disparities emerged: in SAB, fishers accessed state-provided life insurance through the Institute of Fisheries, a benefit unavailable in Tabasco, reflecting persistent regional inequalities.

Institutional weaknesses were even more pronounced during oil spills. Most fishers across sites reported receiving no direct support, with compensation limited to permit holders and fishing cooperatives, systematically excluding independent fishers. Recent major spills in the SWGoM, covering over 600 km^2, further highlighted these gaps, with no coordinated response or communication reaching the affected communities (Uribe-Martínez et al., 2024). Likewise, compensation programs have historically served more to contain conflict than to solve structural problems, fostering dependence and limiting collective agency (Peña-Puch et al., 2021).

Table 8.4 Percentage of assistance received during contingencies in *Sabancuy* (SAB), Campeche, *Paraíso* (PAR), and *Sánchez Magallanes* (SM), Tabasco

Contingency	Assistance Agencies	SAB	PAR	SM	Type of assistance	SAB	PAR	SM
Natural disasters	*n*	26	17	24	*n*	26	17	24
					Previous			
	None	57.7	52.6	44	None	76.9	73.7	80
	Civil protection	19.2	5.3	0	Shelter	15.4	10.5	12
	Municipal government	7.7	21.1	12	Evacuation	3.8	15.8	16
	State government	19.2	15.8	20	Pantry food	7.7	10.5	0
	Federal government	3.8	15.8	20				
	Others	7.7	0	0	*After*			
					None	61.5	63.2	48
					Pantry food	38.5	26.3	40
					Home appliances	0	21.1	24
					Others	0	5.3	20
Maritime accidents	None	23.1	36.8	48	None	23.1	47.4	52
	Marine	34.6	47.4	32	Rescue and search	34.6	52.6	44
	Port captain	3.8	26.3	24	Life insurance	50	0	0
	State government	38.5	0	0	Alerts	0	5.3	4
	INPESCA	31	0	0	Others	16	0	0
Oil spills	None	61.5	26.3	52	None	73	52.6	48
	PEMEX	19.2	57.9	40	Fishing equipment	14.8	31.6	40
	State government	11.5	10.5	0	Money	7.7	5.3	12
	Federal government	3.8	5	4	Fuel	0	10.5	28
	ENI	0	0	8	Others	3.8	31.6	0

Relational well-being depends not only on the presence of institutions but also on how they engage with communities, affecting their sense of security, recognition, and resilience during times of crisis. The dynamics exposed reflect broader deficiencies in risk governance: the lack of transparency, inclusion, and coordination undermines the protection of vulnerable groups, exacerbates disaster impacts, and increases tensions between SSF and the oil industry (Dow, 1999; Ahrens & Rudolph, 2006; Machado et al., 2025). Strengthening this dimension in territories shaped by fisheries and oil development requires transparent, inclusive, and territorially responsive governance networks that recognize fishers as legitimate actors (Arbo & Thủy, 2016; Adjei & Overå, 2019). Without such change, coastal zones will remain spaces of fragmented protection and systemic inequity (de Oliveira Estevo et al., 2021; Owusu et al., 2023).

8.5.4 Subjective Dimension

8.5.4.1 Community Well-Being

Across the studied communities, fishers highlighted health services, public safety, and employment as the most important priorities for improving community well-being (Appendix 8.5). These responses reflect not only structural needs but also shared aspirations, revealing how subjective well-being often intertwines with collective expectations for dignity, stability, and a good life (Armitage et al., 2012).

8.5.4.2 Impacts, Costs, and Benefits of the Oil Industry

Fishers' subjective well-being is shaped by their interactions with the sea and how they interpret and cope with external pressures (Weeratunge et al., 2014). For this research, general perceptions of the oil industry revealed that costs outweighed benefits. However, these opinions varied significantly between sites (Fig. 8.6).

In PAR, fishers expressed the strongest concerns, reporting higher sensitivity to socio-environmental impacts associated with oil activities (Fig. 8.6a, b), particularly following the construction of the Olmeca refinery. Although they recognized more benefits compared to the other sites (Fig. 8.6d), such as community improvements and temporary employment, these were considered insufficient for long-term well-being (Fig. 8.6d).

Concerns regarding SM, centered around ecological damage and maritime exclusion (Fig. 8.6c). However, some positive perceptions emerged regarding community improvements and compensations (Fig. 8.6d), particularly through the private company ENI, suggesting slightly better relational dynamics than in SAB.

In SAB, fishers expressed strong concerns over ecological degradation and maritime insecurity associated with piracy and industrial traffic (Figure 8.6a-c). The benefits perceived were minimal, limited primarily to occasional compensation without broader community improvements (Figure 8.6d).

Across sites, fishers expressed frustration with temporary jobs, unequal compensation, and the lack of dialogue or recognition, which deepened feelings of exclusion and mistrust (Quist, 2019). These experiences reflect broader patterns observed in extractive territories, where oil governance operates through blurred state and corporate practices, systematically prioritizing oil interests over fisheries (Quist & Nygren, 2015).

As institutional agendas remain closely aligned with the oil sector, fishers lose access to marine spaces, visibility in planning processes, and recognition as legitimate stakeholders (Coronado et al., 2024). This erosion of agency, trust, and identity undermines subjective well-being and exacerbates inequalities within coastal territories. Confronting these asymmetries requires inclusive and socially just

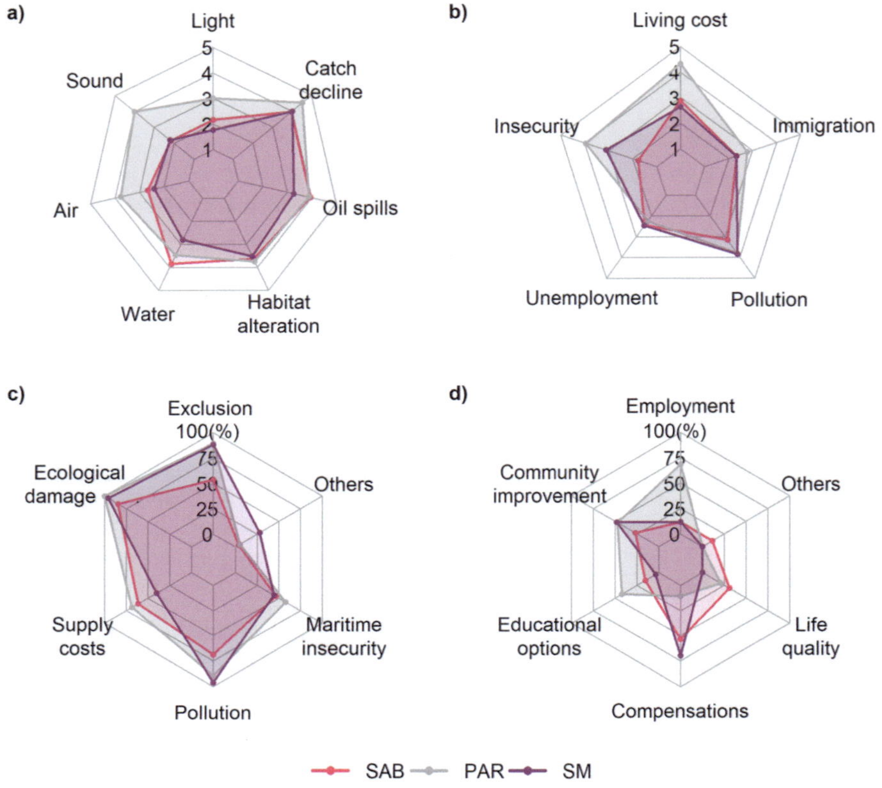

Fig. 8.6 Perceptions of impacts, costs, and benefits of the oil industry of fishers in *Sabancuy* (SAB), Campeche, *Paraíso* (PAR), and *Sánchez Magallanes* (SM), Tabasco. (**a**) Environmental impacts; (**b**) Livelihood impacts—both measured using a Likert scale: 1. Very low; 2. Low; 3. Medium; 4. High; 5. Very high; (**c**) Perceived costs; (**d**) Perceived benefits

governance frameworks (Jentoft & Chuenpagdee, 2022). This challenge is major considering the differentiated local realities; therefore, adaptation and flexibility in governance and public policies will be key.

8.5.4.3 Nature's Contributions from Fishing

Fishing plays a central role in the subjective and relational well-being of coastal communities, serving not only as a livelihood but as a source of identity, belonging, and emotional connection (Acott et al., 2018; Dias & Armitage, 2021). Across the three sites, fishers reported high levels of job satisfaction using a 5-point Likert scale (1: very dissatisfied, 5: very satisfied): PAR (4.25), SM (4.2), and SAB (4). This was tied to the meaning of fishing as a family tradition and a source of pride, freedom, and connection to nature (Fig. 8.7). Beyond income, these non-material contributions to well-being reflect the cultural depth of SSF and its emotional importance (Díaz et al., 2018).

Fig. 8.7 Meanings of fishing in Sabancuy (SAB), Campeche, Paraíso (PAR), and Sánchez Magallanes (SM), Tabasco

However, this attachment coexists with growing concern for the future. Most fishers (68.5–80%) expressed a preference for their children to pursue different livelihoods, citing declining catches, insecurity, and financial instability. Similar patterns have been documented in the region, reflecting limited prospects in SSF (Oliveto-Andrade et al., 2024). This concern becomes greater in places where industrial pressures, such as oil expansion, further undermine its long-term viability (Quist & Nygren, 2019).

Across communities, the meanings attached to fishing were remarkably consistent, rooted in nature, family, and personal realization. However, the expansion of oil activities threatens these contributions by displacing traditional fishing grounds, degrading ecosystems, and eroding relational and subjective well-being (Quist & Nygren, 2019; Leite et al., 2023). Preserving the cultural and emotional foundations of SSF requires recognizing it not merely as an economic activity, but as a way of life increasingly challenged in extractive territories.

8.5.5 Contrasting Experiences Across Communities

Although all communities face similar pressures from oil expansion and structural inequalities, the experiences of fishers differ in intensity, form, and perceived opportunities.

In PAR, industrial development, particularly the construction of the Olmeca refinery, has brought short-term benefits, including community improvements and temporary employment opportunities. However, these "gains" coexist with deep socio-environmental concerns, including pollution, health risks, exclusion zones, and rising living costs. The physical proximity to oil infrastructure intensifies these conflicts, making the effects of industrial expansion more tangible and immediate.

In SM, fishers navigate a mix of ecological risk and partial compensation. Payments from private oil companies, such as ENI, provide some economic support. However, persistent pollution, maritime exclusion, and insecurity continue to undermine well-being. As in PAR, the close spatial overlap between oil operations and SSF deepens tensions, affecting daily livelihoods and community dynamics.

By contrast, SAB presents a different profile. Although geographically closer to a natural protected area and with no oil installations directly offshore, fishers perceived fewer benefits from the oil industry. Feelings of insecurity, linked to piracy and industrial traffic, as well as a sense of invisibility, are more pronounced. However, the connection to *Ciudad del Carmen*, a major oil hub, has enabled fishers to access certain political and institutional benefits, such as life insurance programs, that are unavailable in Tabasco. This more indirect relationship to oil development creates a different, though still complex, dynamic of dependence and exclusion.

These contrasts reveal how territorial context shapes the different dimensions of well-being. They also highlight that pressures from oil expansion are not uniformly experienced, but rather mediated by local histories, access to resources, institutional relations, and governance structures. Strengthening well-being in these coastal zones requires place-based strategies that recognize the specific challenges, capabilities, and aspirations of each community.

8.5.6 Trade-Offs Between SSF and the Oil Industry

The coexistence of SSF and the oil industry in the SWGoM presents a complex landscape of trade-offs affecting well-being and sustainability. These trade-offs encompass economic, environmental, spatial, and social dimensions, all of which are linked to broader governance challenges.

Economic and environmental trade-offs encompass the benefits brought by the oil industry (e.g., temporary jobs and community improvements) at the cost of the health of marine ecosystems and coastal communities, threatening the long-term sustainability of SSF. This dynamic may transform ways of living, offering new opportunities for fishing families to engage in other activities, challenging traditional practices and beliefs.

Spatial trade-offs are evident in the allocation of marine areas to oil infrastructure, where security zones limit fishing access. This reduces available fishing grounds and pushes fishers into more dangerous or less productive areas (Agardy et al., 2011). Although oil infrastructure can create new habitats, it may also disrupt ecological connectivity, affecting marine population dynamics and increasing pressure on SSF (McLean et al., 2022).

Social trade-offs include increased living costs, insecurity, and changes to social dynamics and identity. Despite the deep emotional connection to fishing, reflected in high levels of job satisfaction (Dias & Armitage, 2021; Kpanou et al., 2021), the expansion of the oil sector may intensify health risks, financial instability, and erode the cultural foundations of fishing life.

The situation in Campeche and Tabasco reflects elements of coastal grabbing, where industrial expansion reshapes traditional livelihoods and reconfigures marine spaces (Bavinck et al., 2017). While some local benefits have been recognized, these can influence fishers' perceptions and decisions, sometimes reducing resistance or creating ambivalence (Bennett et al., 2015). Additionally, such "gains" occur alongside ongoing socio-environmental pressures, including exclusion, environmental degradation, and declining fish stocks, which together may push fishers into a socio-ecological trap where poverty and dependence reinforce each other, locking the system into unsustainable trajectories that can eventually collapse (De la Puente et al., 2022; Villasante et al., 2022).

Addressing these challenges requires intersectoral collaboration between the government, oil companies, academics, and the fishing sector. Stringent environmental regulations, effective monitoring programs, bioremediation efforts, and inclusive contingency planning are crucial to protect marine life and coastal communities (White & Molloy, 2003; de Oliveira Estevo et al., 2021). Likewise, promoting occupational pluralism, fair compensation schemes, and integrating fishers' perspectives into marine spatial planning are essential to support sustainable coexistence (Reilly et al., 2016; Tolentino-Arévalo et al., 2019). However, these efforts must also address broader structural inequities, such as the disproportionate allocation of public budgets favoring the oil sector over fisheries, which exacerbates governance asymmetries and limits the resilience of coastal communities (Coronado et al., 2024).

While oil energy remains a strategic resource for many countries' economies, its planning and implementation must prioritize socio-environmental development. In this context, governance interventions are crucial and must be linked to social justice, sustainable livelihoods, and ecosystem health (Jentoft & Chuenpagdee, 2018). A major challenge lies in ensuring that governance responses and policy are sensitive to differentiated local contexts, recognizing the specific complexities, histories, and capabilities of coastal communities (Queffelec et al., 2021). Without such considerations, the oil industry expansion will continue to deepen vulnerabilities and further erode the well-being of coastal territories.

8.6 Conclusions

This study shows how oil expansion and structural inequalities shape the well-being of fishing communities in Campeche and Tabasco. While material conditions, such as housing and basic services, were similar across sites, differences emerged in incomes, fishing assets, and access to social support programs. Relational well-being reflected market dynamics and uneven institutional responses during periods of crisis. Subjective well-being captured a deep emotional connection to fishing alongside growing concerns about oil expansion, highlighting feelings of exclusion, insecurity, and uncertain futures.

Although all communities face pressures from oil expansion, their experiences differ. In PAR and SM, the proximity to oil activities brings short-term benefits but also increases socio-environmental risks. In contrast, in SAB, the absence of nearby infrastructure does not necessarily mean fewer impacts; fishers still face insecurity and marginalization although these are shaped by different political dynamics. These contrasts highlight that vulnerability is not uniform, but shaped by local histories, resource access, and governance structures, ultimately conditioning how well-being is experienced and maintained across communities.

Faced with these differentiated realities, more than generic solutions are required. Strengthening the well-being of coastal communities demands governance approaches that are sensitive and adaptable to local contexts, capable of addressing specific challenges and opportunities. Building on the findings of this study and framed within international principles (FAO, 2015), several strategic priorities are proposed, reflecting recent insights on coastal governance in oil-influenced regions.

A priority is strengthening governance to improve coastal well-being. In oil-influenced territories like Campeche and Tabasco, institutional agendas have often prioritized oil interests (Salazar-De la Cruz et al., 2020; Campelo et al., 2021). Ensuring the inclusion of fishers in marine spatial planning and disaster response strategies is critical. Recognizing their knowledge and rights in decision-making is not only a matter of policy coherence, but also a necessary condition for strengthening relational well-being and rebuilding community trust.

Equally important is improving environmental protection to safeguard marine ecosystems and well-being. Effective policies and systematic monitoring are crucial for preventing and mitigating oil-related disasters (Machado et al., 2025). Beyond enforcing environmental standards, coastal communities should be actively involved in monitoring and contingency planning processes to enhance local capacity for responding to ecological risks.

Additionally, diversifying economic opportunities is crucial to reducing fishing dependency and strengthening community resilience. In contexts where industrial expansion reshapes traditional livelihoods, supporting sustainable alternatives, such as small-scale aquaculture, eco-tourism, or artisanal enterprises, can offer pathways that respect local identities and expand income sources (Foley et al., 2016). Promoting occupational pluralism must be accompanied by investments in education, health, and basic services to ensure that diversification enhances well-being without deepening existing inequalities.

Reframing compensation and development programs is essential to prioritize long-term well-being. In oil-influenced territories, compensation schemes have been employed for short-term mitigation and conflict avoidance rather than addressing the structural needs of communities (Arroyo & Zalik, 2016). New approaches must recognize not only material losses but also the disruption of sociocultural aspects and relational well-being. Ensuring fair compensation and investing in community infrastructure are crucial to protect coastal ways of life.

To conclude, promoting well-being and sustainability in oil-influenced coastal regions requires more than balancing sectoral interests. It demands protecting the socio-ecological foundations that support local livelihoods. Talking about well-being and promoting harmonious coexistence between the oil industry and SSF is impossible if it involves the ecological and social erosion of coastal communities.

Acknowledgments This research is part of the project "The challenging coexistence of coupled socio-ecological systems: the fishing and oil industries in the Sonda de Campeche," supported by the National Council of Humanities, Science and Technology of Mexico (CONAHCyT). The authors thank the fishers and community members who generously shared their time and knowledge for the broader project and this study. Special thanks to Benjamin Hernández García for his support during fieldwork. The authors also acknowledge the use of AI-assisted tools to support the grammatical editing and refinement of the manuscript.

Ethics Approval Informed verbal consent was obtained from all participants after explaining the study's purpose and data use. Although no institutional ethics committee reviewed the protocol, the research adhered to general ethical principles, including confidentiality and voluntary participation. Interview materials included a clear explanation of participants' rights and the use of their data.

Appendix 8.1: Socioeconomic Indicators of Coastal Municipalities in campeche and Tabasco, Mexico (Source: INEGI, 2020a).

Municipality	Carmen	Paraíso	Cardenas
State	Campeche	Tabasco	Tabasco
Population (inhabitants)	248,845	96,741	243,229
State Gross Domestic Product (millions $USD)	24.5	34.7	34.7
Illiterate population[a]	2.9%	2.4%	4.2%
Population without schooling[a]	3.6%	1.8%	4%
Population with elementary school[a]	8.2%	9.4%	9.9%
Access to health services	75.2%	72.9%	66.7%
Access to health services in PEMEX, Defense or Navy	8.1%	11.7%	3%

[a] Older than 15 years

Appendix 8.2: Semi-structured Interview Instruments

Original interview formats used during fieldwork (in Spanish). Translations available upon request.

A) Longitudinal study

ECOSUR ITSCe Instituto Tecnológico Superior de Centla CONACYT

La desafiante coexistencia de los sistemas socio - ecológicos acoplados; las industrias de la pesca y petróleo en la Sonda de Campeche

Procedimiento previo:
1. Presentación previa (título del proyecto y propósito)
2. Explicación del propósito del cuestionario
3. Solicitar permiso para grabar (en la grabación debe quedar explícito sí o no). Si no, apuntar la CURP:

4. Lectura de ley de protección de datos personales:
Con base al artículo 38 de la Ley de Información Estadística y Geográfica "Toda información se mantendrá con carácter estrictamente CONFIDENCIAL" y su uso es sólo para fines estadísticos.
5. Llamado a respuestas breves, cuando sea el caso de preferencia cerradas (sí o no).

LLENADO POR ESCRITO Y CUANDO CORRESPONDA, GRABACIÓN DE AUDIO

Folio:

0. Introducción
0.1. Fecha:
0.2. Nombre del entrevistador
0.3. Lugar de la entrevista 0.3.1. Coordenada GPS:
0.4. Dirección de la vivienda:
0.5. ¿Cuántos años lleva dedicándose a la pesquería?
0.5.1. ¿Cómo se considera usted acorde a sus actividades en la pesquería?[1]
1. Permisionario
2. Representante de cooperativa(s)
3. Pescador de cooperativa(s)
4. Pescador empleados de permisionarios y cooperativas
5. Pescador libre
6. Pescador de autoconsumo
7. Procesamiento no remunerado
8. Procesamiento remunerado
9. Intermediario internacional
10. Intermediario regional
11. Intermediario local
12. Mantenimiento de equipo e insumos
13. Administrador
14. Otro:

1. Perfil:
1.1. Nombre completo:[2]

[1] En caso de que ninguna de las opciones sea acorde con la pesquería, se interrumpirá el cuestionario.
[2] Se pregunta por el jefe o jefa de vivienda, a él o ella se le aplican todas las preguntas.

1.2. Sexo:
1.3. Fecha de nacimiento:
1.4. Lugar de nacimiento:
1.5. Último grado de estudios:
1.6. ¿Tiene dependientes económicos? (sí o no):
1.6.1. ¿Cuántos?

2. Características de la vivienda

2.1. ¿De qué material es la mayor parte de las paredes o muros de esta vivienda? *(opción única)*

1. Material de desecho
2. Lámina de cartón
3. Lámina de asbesto o metálica
4. Carrizo, bambú o palma
5. Embarro o bajareque
6. Madera
7. Adobe
8. Tabique, ladrillo, block, piedra, cantera, cemento o concreto, etc.

2.2. ¿De qué material es la mayor parte del techo de la vivienda? *(opción única)*

1. Material de desecho
2. Lámina de cartón
3. Lámina de asbesto o metálica
4. Palma, tejamanil o madera
5. Teja
6. Losa de concreto, tabique, ladrillo o terrado con viguería

2.3. ¿De qué material es la mayor parte del piso de esta vivienda? *(opción única)*

1. Tierra
2. Cemento firme
3. Madera, mosaico u otro recubrimiento
4. Otro:

2.4. ¿En esta vivienda tienen (lea las opciones hasta obtener una respuesta, marque sólo una opción):

1. ¿Agua entubada dentro de la vivienda?
2. ¿Agua entubada fuera de la vivienda, pero dentro del terreno?
3. ¿Agua entubada de llave pública (o hidrante)?
4. ¿Agua entubada que acarrean de otra vivienda?
5. ¿Agua de pipa?
6. ¿Agua de un pozo, río, lago, arroyo u otra?
7. Otro:

2.5. ¿Hay luz eléctrica en esta vivienda? (sí o no):
2.5.1. ¿Paga la luz? (sí o no):
2.5.2. En promedio, ¿cuánto debe pagar bimestralmente de luz?
2.5.3. ¿Está en algún tipo de resistencia civil? (sí o no):
2.5.3.1. *En caso de sí*, ¿desde cuándo?, ¿y por qué?

2.6. ¿Esta vivienda tiene estufa de gas? (sí o no): *En caso que sí, pase a las siguientes* 2.6.1 a 2.6.3.1.
2.6.1. ¿Usa la estufa de gas? (sí o no):
2.6.2. ¿Cuánto paga en promedio cada mes de gas?
2.6.3 ¿Usa algún otro tipo de combustóleo o tipo de quema para cocinar o alguna actividad similar? (sí o no):
2.6.3.1. En caso de sí, ¿cuál y para qué?

2.7. ¿Hace cuántos años fue construida esta vivienda?
2.7.1. ¿Desde hace cuánto tiempo vive en esta vivienda?
2.7.2. ¿Es de su propiedad? (sí o no):
En caso de no:

1. ¿Está pagándose?
4. ¿Está rentada?

2. ¿Está totalmente pagada?
3. ¿Está en otra situación?
5. ¿Está prestada o la cuidan?
6. Otro:

2.8. ¿En esta vivienda tiene drenaje o desagüe de aguas sucias:
1. A la red pública
2. A una fosa séptica
3. A una tubería que va a dar a una barranca o grieta?
4. A una tubería que va a dar a un río, lago o mar
5. ¿No tiene drenaje?
6. Otro:

2.9. ¿En esta vivienda tienen *(opción múltiple)*:
1. Radio o radiograbadora
2. Televisión
3. Reproductor de DVD
4. Licuadora
5. Refrigerador
6. Lavadora
7. Teléfono fijo
8. Calentador de agua (boiler)
9. Clima o aire acondicionado
10. Automóvil o camioneta propios
11. Computadora
12. Sky TV o cable de TV
13. Horno microondas
14. Internet propio

2.10. Tiene derecho a servicio médico en:
1. IMSS Seg. social
2. IMSS Solidaridad
3. ISSSTE
4. Instit. públicas o paraestatales (Pemex, SEDENA, Marina)
5. Seguro Privado
6. Seguro Popular
7. Ninguno

2.11. ¿Sabe leer y escribir un recado?
(Sí o no)

3. Ingreso

3.1. ¿Cuánto asciende su ingreso mensual?
1. Menos de 1,000
2. De 1001 a 2,000
3. De 2001 a 3,000
4. De 3,001 a 4,000
5. De 4001 a 5000
6. de 5001 a 6000
7. de 6,001 a 7,000
8. de 7,001 a 8,000
9. 8,001 a 12,000
10. 12,001 a 16,000
11. 16,000 a 20,000
12. Otro__________

3.2. ¿Recibe algún tipo de apoyo económico o en especie de alguna empresa o instancia asociada al petróleo?

4.3.1. En caso de sí, ¿cuál y cómo funciona?

3.3. ¿Recibe algún tipo de apoyo económico o en especie de alguna empresa o instancia asociada al turismo?

4.4.1. En caso de sí, ¿cuál y cómo funciona?

5. Apoyos gubernamentales

5.1. De los siguientes, ¿qué apoyos económicos o en especie recibió entre 2018 y 2021?

<table>
<tr><th></th><th>2018</th><th>2019</th><th>2020</th><th>2021</th></tr>
<tr><td>-Apoyo al Ingreso Objetivo y a la Comercialización</td><td>|___|</td><td>|___|</td><td>|___|</td><td>|___|</td></tr>
<tr><td>CADENA (Componente de Atención a Desastres Naturales en el Sector Agropecuario y Pesquero)</td><td>|___|</td><td>|___|</td><td>|___|</td><td>|___|</td></tr>
<tr><td>-Sembrando Vida</td><td>|___|</td><td>|___|</td><td>|___|</td><td>|___|</td></tr>
<tr><td>-Diesel Marino/Gasolina Ribereña</td><td>|___|</td><td>|___|</td><td>|___|</td><td>|___|</td></tr>
<tr><td>- Programa de Fomento a la Agricultura, Ganadería, Pesca y Acuicultura, Del Componente Apoyo para el Bienestar de Pescadores y Acuicultures (Bienpesca)</td><td>|___|</td><td>|___|</td><td>|___|</td><td>|___|</td></tr>
<tr><td>-Proyecto Estratégico de Seguridad Alimentaria (PESA)</td><td>|___|</td><td>|___|</td><td>|___|</td><td>|___|</td></tr>
<tr><td>-PROCAMPO/PROAGRO productivo</td><td>|___|</td><td>|___|</td><td>|___|</td><td>|___|</td></tr>
<tr><td>-PROGAN/PROGAN productivo</td><td>|___|</td><td>|___|</td><td>|___|</td><td>|___|</td></tr>
<tr><td>-Sanidades</td><td>|___|</td><td>|___|</td><td>|___|</td><td>|___|</td></tr>
<tr><td>-Proyecto Trópico Húmedo/Desarrollo Productivo Sur Sureste</td><td>|___|</td><td>|___|</td><td>|___|</td><td>|___|</td></tr>
<tr><td>-Oportunidades/PROSPERA</td><td>|___|</td><td>|___|</td><td>|___|</td><td>|___|</td></tr>
<tr><td>-65 y más (antes70 años y más)</td><td>|___|</td><td>|___|</td><td>|___|</td><td>|___|</td></tr>
<tr><td>-Algún tipo de pensión para el Bienestar (discapacidad, niños, niñas)</td><td>|___|</td><td>|___|</td><td>|___|</td><td>|___|</td></tr>
<tr><td>-Programas Estatales
¿Cuáles?</td><td>|___|</td><td>|___|</td><td>|___|</td><td>|___|</td></tr>
<tr><td>-Programas Municipales
¿Cuáles?</td><td>|___|</td><td>|___|</td><td>|___|</td><td>|___|</td></tr>
<tr><td>Programas asociados a empresas o instancias petroleras: ____
¿Cuáles?</td><td>|___|</td><td>|___|</td><td>|___|</td><td>|___|</td></tr>
<tr><td>Programas asociados a empresas o instancias turísticas: _____
¿Cuáles?</td><td>|___|</td><td>|___|</td><td>|___|</td><td>|___|</td></tr>
<tr><td>Otro. Especificar: ______________________</td><td>|___|</td><td>|___|</td><td>|___|</td><td>|___|</td></tr>
</table>

7. La pesca como actividad

7.1. ¿Cuenta con permiso de pesca?

¿Cuáles?

7.2. ¿Realiza otras actividades además de la pesca? (si o no)
¿Cuáles?

7.3. ¿Dónde vendió principalmente sus productos este 2021?

1. En casa
2. Mercado local
3. Redes sociales/ venta en línea
4. Sociedad cooperativa local o regional
5. Mercado regional
6. Mercado nacional
7. Mercado internacional

7.4. ¿A quién vendió principalmente dichos productos?

1. Intermediario
2. Acopiador
3. Centro o bodega de acopio
4. Fábrica o empacadora
5. Directo al consumidor
6. Directo a la exportación

7.5. ¿Cuál fue el destino final principal de sus productos?

1. No sabe
2. Mercado local
3. Mercado regional
4. Mercado nacional
5. Mercado internacional

8. Infraestructura y equipo

8.1. ¿Qué infraestructura exclusivamente a la pesca tiene y del cual es propietario?

Concepto	Año construcción	Costo de construcción (excluir valor de la tierra)	Valor estimado en 2021 (excluir valor de la tierra)
____	____	$ ____	$ ____
____	____	$ ____	$ ____
____	____	$ ____	$ ____

8.2. ¿Qué equipo exclusivamente para la pesca tiene y del cual es propietario?

Concepto	Año de adquisición	Costo de adquisición ($)	Valor estimado en 2021($)
a. Embarcaciones menores 10.5 m eslora (incluir valor de motor)			
b. Embarcaciones mayores 10.5 m eslora pero menores a barcos de altamar (incluir valor de motor)			
c. Barcos pesqueros de alta mar (incluir valor de motor)			
d. Redes y otras artes de pesca 1			
e. Palangres			
f. Ristras			
g. Equipo de congelación			
h. Equipo para primer procesamiento			
i. Básculas			
j. Remolques			

k. Motores marinos fuera de borda			
l. Camioneta para transporte de productos pesqueros o de los equipos de pesca.			
m. Otro.			

B) Well-being instrument

ECOSUR pescaypetróleo 2.0 CONACYT

La desafiante coexistencia de los sistemas socio-ecológicos acoplados; las industrias de la pesca y petróleo en la Sonda de Campeche
Cuestionario de Bienestar

Procedimiento previo

1. Presentación previa (título del proyecto y propósito)
2. Solicitar permiso para grabar (en la grabación debe quedar explicito sí o no).

Objetivo:
El cuestionario de bienestar tiene como objetivo determinar la condición de bienestar de las familias de pescadores que coexisten con la industria de hidrocarburos, así como identificar los costos y beneficios que ha tenido la industria petrolera en las comunidades pesqueras.

		0.1. Folio	
0.2. Entrevistador		0.3. Fecha	| |
0.4. Nombre		0.5. Localidad	
0.6. Dirección vivienda		0.7. Coordenada GPS	

Perfil:
Nombre completo: Sexo:
Fecha de nacimiento: Lugar de nacimiento:
Grado de estudios: Número de dependientes económicos:
Estado civil:

A) Componente Relacional

- **Estrategias adaptativas: desastres naturales**

1. ¿Recibe apoyo de su patrón, cooperativa u otros pescadores durante desastres naturales (huracanes, inundaciones, etc.)?		Si	No
1.1. ¿Cómo y quién le apoya?	Previo:	Posterior:	
2. ¿Recibe apoyo por parte del gobierno durante desastres naturales (huracanes, inundaciones, etc.)?		Si	No

2.1. ¿Qué agencias brindan apoyo?	

3. ¿Cómo le apoyan?	
Previo:	Posterior:

4. ¿Qué estrategias emplea la comunidad en caso de desastres naturales (huracanes, inundaciones, etc.)?	
Previo:	Posterior:

- **Estrategias adaptativas: Accidentes en altamar**

Nota: Esta pregunta solo aplica a **pescadores/tripulantes**. En caso de ser **procesador(a) o comerciante** pasar a la **pregunta 6.** 5. ¿Recibe apoyo de su patrón, cooperativa u otros pescadores en caso de presentarse un accidente en altamar?	Si	No
5.1. ¿Cómo y quién le apoya?		
Durante:	Posterior:	

6. ¿Recibe apoyo del gobierno en caso de presentarse un accidente en altamar?	Si	No
6.1. ¿Qué agencias brindan apoyo?		
6.2. ¿Cómo le apoyan?		

- **Estrategias adaptativas: derrames de petróleo**

7. ¿Qué estrategias emplean los pescadores de su cooperativa/tripulación en caso de derrames de petróleo?		
Durante:	Posterior:	
8. ¿Recibe apoyo de su patrón, cooperativa o federación pesquera en caso de derrames de petróleo?	Si	No
8.1. ¿Cómo le apoyan?		

Durante:	Posterior:
9. ¿Qué agencias (dependencias) brindan apoyo en caso de derrames de petróleo?	
9.1. ¿Cómo le apoyan?	
Durante:	Posterior:
10. ¿Qué estrategias emplea la comunidad en caso derrames de petróleo?	
Durante:	Posterior:

B) Componente Subjetivo

- **Bienestar comunitario**

11. ¿Cuáles son las principales necesidades para mejorar las condiciones (bienestar) de su comunidad?
Proporcionar importancia: Muy alto (5), Alto (4), Medio (3), Bajo (2) y Muy bajo (1).

Necesidades para bienestar comunitario	Muy alto a muy bajo				
	5	4	3	2	1
Servicios de salud					
Opciones educativas					
Oferta Laboral					
Espacios recreativos					
Seguridad					
Participación comunitaria en toma de decisiones					
Otro:					

Beneficios y costos de la industria petrolera

- **Beneficios**

12. ¿Cuáles son los principales **cambios positivos** que experimentó la **pesca**?	
12.1. Con la industria Petrolera	

13. ¿Cuáles son los principales **cambios positivos** que experimentó la **comunidad**? 13.1. Con la industria Petrolera					
Fuentes de empleo	Ofertas educativas	Calidad de Vida	Mejora condiciones comunidad	Apoyos gubernamentales	Otros:

- **Costos**

14. ¿Cuáles son los principales **cambios negativos** que experimentó la **pesca en su comunidad**? Con la Industria Petrolera (PEMEX)					
Exclusión en zonas pesca	Contaminación	Inseguridad	Incremento costos de insumos	Deterioro hábitat y especies	Otros:
14.1.1. ¿En qué año se percató de estos cambios?					
15. ¿Qué acciones/estrategias ha implementado los **pescadores de su comunidad** ante estos cambios?					
13.1. ¿Alguna dependencia orienta o apoya estas acciones? ¿Cuál(es)?					
16. ¿Cuáles son los principales **cambios negativos** que experimentó la **comunidad**? Con la Industria Petrolera (PEMEX)					
Inseguridad	Incremento costos de vida	Calidad de vida	Desempleo	Contaminación	Otros:
17. ¿Qué acciones/estrategias ha implementado la **comunidad** ante estos cambios?					
15.1. ¿Alguna dependencia orienta o apoya estas acciones? ¿Cuál(es)?					

- **Impactos en medio ambiente**

18. ¿Cuál es el impacto de la industria petrolera en los siguientes aspectos ambientales?
Proporcionar impacto: Muy alto (5), Alto (4), Medio (3), Bajo (2) y Muy bajo (1).

Contaminación por actividad petrolera	Muy alto a muy bajo					¿Cómo le afectan?
	5	4	3	2	1	
Luz						
Sonido						
Aire						
Agua						
Alteración del hábitat (plataformas, gasoductos)						
Derrames						
Disminución de especies marinas						
Otros:						

- **Impacto en medios de vida**

19. ¿Cuál es el impacto de la industria petrolera en los siguientes aspectos de su comunidad?

Proporcionar impacto: Muy alto (5), Alto (4), Medio (3), Bajo (2) y Muy bajo (1).

Impacto en la comunidad	Muy alto a muy bajo					¿Cómo le afectan?
	5	4	3	2	1	
Incremento en costos de vida						
Inseguridad						
Desempleo						
Contaminación						
Inmigración (Foráneos que vienen a trabajar)						
Exclusión de zonas de pesca						
Otros:						

- **La pesca como forma de vida**

20. ¿Qué significa para usted ser pescador?						
21. ¿Quién le enseño a pescar y qué edad tenía?						
22. ¿Qué nivel de satisfacción le produce dedicarse a la pesca? Nivel de satisfacción: Muy alto (5), Alto (4), Medio (3), Bajo (2) y Muy bajo (1).	Muy alto a muy bajo					
	5	4	3	2	1	
23. Además de los beneficios económicos ¿Qué otro aspecto **positivo** le brinda la pesca?						
24. ¿Podría relatar alguna experiencia **positiva** importante que haya vivido como pescador?						
25. ¿Le gustaría que sus hijos continuaran en la pesca?	Si	No				
25.1. ¿Por qué?						
26. ¿Qué acciones considera necesarias para mejorar la relación entre el sector pesquero y la industria petrolera?						
27. ¿Qué acciones propone para mejorar el sector pesquero y su comunidad?						

Appendix 8.3: Profiles of the Interviewees from the Fishing Communities of *Sabancuy* (SAB), Campeche, *Paraíso* (PAR), and *Sánchez Magallanes* (SM), Tabasco

Profile	SAB	PAR	SM
n	26	19	25
Gender (%)			
Man	100	89.5	92
Woman	0	10.5	8
Age (mean ± SD)	50.3 ± 16.7	58.3 ± 13.8	57.4 ± 15.2
Economical dependents (mean [range])	2.6 [1-6]	2.6 [1-7]	2.9 [1-7]
Marital status (%)			
Single	7.7	10.5	0
Married	53.8	52.6	48
Consensual union	34.6	26.3	40
Widowed	3.8	10.5	12
Education (%)			
No studies	3.8	15.8	4
Elementary school	69.2	31.6	52
Junior high school	19.2	47.4	40
High school	3.8	5.3	0
Bachelor's degree (unconcluded)	3.8	0	4
Fisher type (%)			
Independent fisher	50	52.6	40
Cooperative member	23.1	15.8	24
Cooperative representative	7.7	5.3	20
Permit holder	19.2	26.3	16
Processing activities (%)	23.1	5.3	52

Appendix 8.4: Monthly income ($USD) and Minimum Wage (MW) of fishers (%) in *Sabancuy* (SAB), Campeche, *Paraíso* (PAR), and *Sánchez Magallanes* (SM), Tabasco

Monthly income ($USD)	MW ($USD)	SAB (%)	PAR (%)	SM (%)
n	$209.7	26	17	24
< $49.3	<1	30.8	5.9	16.7
$49.4–$98.6	<1	26.9	35.3	16.7
$98.7–$148	<1	19.2	17.6	16.7
$148.1–$197.3	<1	11.5	11.8	4.2
$197.4–$246.6	≈1	0	23.5	12.5
$246.7–$296	>1	0	0	20.8

Monthly income ($USD)	MW ($USD)	SAB (%)	PAR (%)	SM (%)
$296.1–$345.3	>1	0	5.9	4.2
$345.4–$394.7	≈1.5	3.8	0	4.2
$592–$789.3	>2	7.7	0	4.2

Appendix 8.5: Perception of Aspects for Enhancing the Community Well-being of Fishers in *Sabancuy* (SAB), Campeche, *Paraíso* (PAR), and *Sánchez Magallanes* (SM), Tabasco. Likert Scale: 1. Very low; 2. Low; 3. Medium; 4. High; 5. Very high

References

Acott, T. G., Johnson, D. S., Stacey, N., & Urquhart, J. (2018). Reflections on social well-being and the values of small-scale fisheries: Implications for research, policy and management. En D. S. Johnson, T. G. Acott, N. Stacey, & J. Urquhart (Eds.), Social well-being and the values of small-scale fisheries (pp. 317–332). Springer. doi:https://doi.org/10.1007/978-3-319-60750-4_14.

Adgate, J. L., Goldstein, B. D., & McKenzie, L. M. (2014). Potential public health hazards, exposures and health effects from unconventional natural gas development. *Environmental Science & Technology, 48*(15), 8307–8320. https://doi.org/10.1021/es404621d

Adjei, M., & Overå, R. (2019). Opposing discourses on the offshore coexistence of the petroleum industry and small-scale fisheries in Ghana. *The Extractive Industries and Society, 6*(1), 190–197. https://doi.org/10.1016/j.exis.2018.09.006

Agardy, T., Notarbartolo di Sciara, G., & Christie, P. (2011). Mind the gap: Addressing the shortcomings of marine protected areas through large-scale marine spatial planning. *Marine Policy, 35*, 226–232. https://doi.org/10.1016/j.marpol.2010.10.006

Ahrens, J., & Rudolph, P. M. (2006). The importance of governance in risk reduction and disaster management. *Journal of Contingencies and Crisis Management, 14*(4), 207–220. https://doi.org/10.1111/j.1468-5973.2006.00497.x

Andrews, N., Bennett, N. J., Le Billon, P., Green, S. J., Cisneros-Montemayor, A. M., Amongin, S., Gray, N. J., & Sumaila, U. R. (2021). Oil, fisheries and coastal communities: A review of impacts on the environment, livelihoods, space and governance. *Energy Research & Social Science, 75*, 102009. https://doi.org/10.1016/j.erss.2021.102009

Arbo, P., & Thủy, P. T. T. (2016). Use conflicts in marine ecosystem-based management—The case of oil versus fisheries. *Ocean & Coastal Management, 122*, 77–86. https://doi.org/10.1016/j.ocecoaman.2016.01.008

Armitage, D., Béné, C., Charles, A. T., Johnson, D., & Allison, E. H. (2012). The interplay of Well-being and resilience in applying a social-ecological perspective. *Ecology and Society, 17*(4) https://www.jstor.org/stable/26269231

Arroyo, M., & Zalik, A. (2016). Displacement and denationalisation: The Mexican Gulf 75 years after the expropriation. *Area, 48*(2), 134–141.

Ayilu, R. K., Fabinyi, M., & Barclay, K. (2022). Small-scale fisheries in the blue economy: Review of scholarly papers and multilateral documents. *Ocean & Coastal Management, 216*, 105982. https://doi.org/10.1016/j.ocecoaman.2021.105982

Bavinck, M., Berkes, F., Charles, A., Dias, A. C. E., Doubleday, N., Nayak, P., & Sowman, M. (2017). The impact of coastal grabbing on community conservation—A global reconnaissance. *Maritime Studies, 16*(1), 8. https://doi.org/10.1186/s40152-017-0062-8

Bennett, N. J., Govan, H., & Satterfield, T. (2015). Ocean grabbing. *Marine Policy, 57*, 61–68. https://doi.org/10.1016/j.marpol.2015.03.026

Berry, T.-A., Belluso, E., Vigliaturo, R., Gieré, R., Emmett, E. A., Testa, J. R., Steinhorn, G., & Wallis, S. L. (2022). Asbestos and other hazardous fibrous minerals: Potential exposure pathways and associated health risks. *International Journal of Environmental Research and Public Health, 19*(7), Article 7. https://doi.org/10.3390/ijerph19074031

Biswal, R., & Johnson, D. S. (2023). A social wellbeing approach to the gendered impacts of fisheries transition in Gujarat, India. *Maritime Studies, 22*(2), 13. https://doi.org/10.1007/s40152-023-00299-0

Bonfil-Jimenez, U. (2018). *Marcos de negociación en cooperativas pesqueras de Tabasco, las implicaciones territoriales de la reforma energética* [Master's thesis]. Universidad Autónoma de Querétaro.

Bryce, R., Irvine, K. N., Church, A., Fish, R., Ranger, S., & Kenter, J. O. (2016). Subjective well-being indicators for large-scale assessment of cultural ecosystem services. *Ecosystem Services, 21*, 258–269. https://doi.org/10.1016/j.ecoser.2016.07.015

Campelo, R. P. d. S., Lima, C. D. M. d., de Santana, C. S., Jonathan da Silva, A., Neumann-Leitão, S., Ferreira, B. P., Soares, M. d. O., Melo Júnior, M. d., & Melo, P. A. M. d. C. (2021). Oil spills: The invisible impact on the base of tropical marine food webs. *Marine Pollution Bulletin, 167*, 112281. https://doi.org/10.1016/j.marpolbul.2021.112281

Chuenpagdee, R., Salas, S., & Barragán-Paladines, M. J. (2019). Big questions about sustainability and viability in small-scale fisheries. En S. Salas, M. J. Barragán-Paladines, & R. Chuenpagdee (Eds.), Viability and sustainability of small-scale fisheries in Latin America and the Caribbean (pp. 3–13). Springer. doi:https://doi.org/10.1007/978-3-319-76078-0_1.

CNH. (2019). *ENI México se convierte en la primera empresa privada en extraer hidrocarburos en aguas someras de México*. Comisión Nacional del Hidrocarburos. https://hidrocarburos.gob.mx/media/2809/producción-eni-julio-2019_300819_esp.pdf

CNH. (2023). Mapa interactivo de hidrocarburos. Comisión Nacional de Hidrocarburos. https://mapa.hidrocarburos.gob.mx/

Cochrane, K. L., Andrew, N. L., & Parma, A. M. (2011). Primary fisheries management: A minimum requirement for provision of sustainable human benefits in small-scale fisheries. *Fish and Fisheries, 12*(3), 275–288. https://doi.org/10.1111/j.1467-2979.2010.00392.x

Cohen, P. J., Allison, E. H., Andrew, N. L., Cinner, J., Evans, L. S., Fabinyi, M., Garces, L. R., Hall, S. J., Hicks, C. C., Hughes, T. P., Jentoft, S., Mills, D. J., Masu, R., Mbaru, E. K., & Ratner, B. D. (2019). Securing a just space for small-scale fisheries in the blue economy. *Frontiers in Marine Science, 6*. https://doi.org/10.3389/fmars.2019.00171

CONAPESCA. (2021). *Anuario Estadístico de Acuacultura y Pesca* (p. 291). Comisión Nacional de Acuacultura y Pesca.

CONASAMI. (2021). Salarios mínimos 2021 (p. 2). Comisión Nacional de los Salarios Mínimos. https://www.gob.mx/conasami/documentos/tabla-de-salarios-minimos-generales-y-profesionales-por-areas-geograficas

Coronado, E., Salas, S., Cepeda-González, M. F., & Chuenpagdee, R. (2020a). Who's who in the value chain for the Mexican octopus fishery: Mapping the production chain. *Marine Policy, 118*, 104013. https://doi.org/10.1016/j.marpol.2020.104013

Coronado, E., Salas, S., Torres-Irineo, E., & Chuenpagdee, R. (2020b). Disentangling the complexity of small-scale fisheries in coastal communities through a typology approach: The case study of the Yucatan Peninsula, Mexico. *Regional Studies in Marine Science, 36*, 101312. https://doi.org/10.1016/j.rsma.2020.101312

Coronado, E., Zepeda-Domínguez, J. A., Espinoza-Tenorio, A., Santamaria, D. C., Ramos-Muñoz, D., & Monzón-Alvarado, C. (2024). Institutional mapping and implementation of the sustainable development goals across co-existing industries: The case of the fisheries-oil system in Mexico. *The Extractive Industries and Society, 17*, 101390. https://doi.org/10.1016/j.exis.2023.101390

Coulthard, S., Johnson, D., & McGregor, J. A. (2011). Poverty, sustainability and human wellbeing: A social wellbeing approach to the global fisheries crisis. *Global Environmental Change, 21*(2), 453–463. https://doi.org/10.1016/j.gloenvcha.2011.01.003

Cupido-Santamaría, D. G. (2021). *Despojo del paisaje marino a pescadores de pequeña escala en el puerto pesquero Coronel Andrés Sánchez Magallanes, Cárdenas, Tabasco, México* [Master's thesis]. El Colegio de la Frontera Sur.

De la Puente, S., López de la Lama, R., Llerena-Cayo, C., Martínez, B. R., Rey-Cama, G., Christensen, V., Rivera-Ch, M., & Valdés-Velasquez, A. (2022). Adoption of sustainable low-impact fishing practices is not enough to secure sustainable livelihoods and social wellbeing in small-scale fishing communities. *Marine Policy, 146*, 105321. https://doi.org/10.1016/j.marpol.2022.105321

De la Rosa-Velázquez, M. I., Espinoza-Tenorio, A., Díaz-Perera, M. Á., Ortega-Argueta, A., Ramos-Reyes, R., & Espejel, I. (2017). Development stressors are stronger than protected area management: A case of the Pantanos de Centla Biosphere Reserve, Mexico. *Land Use Policy, 67*, 340–351. https://doi.org/10.1016/j.landusepol.2017.06.009

de Oliveira Estevo, M., Lopes, P. F. M., de Oliveira Júnior, J. G. C., Junqueira, A. B., de Oliveira Santos, A. P., da Silva Lima, J. A., Malhado, A. C. M., Ladle, R. J., & Campos-Silva, J. V. (2021). Immediate social and economic impacts of a major oil spill on Brazilian coastal fishing communities. *Marine Pollution Bulletin, 164*, 111984. https://doi.org/10.1016/j.marpolbul.2021.111984

De Oliveira Leis, M., Barragán-Paladines, M. J., Saldaña, A., Bishop, D., Jin, J. H., Kereži, V., Agapito, M., & Chuenpagdee, R. (2019). Overview of small-scale fisheries in Latin America and the Caribbean: Challenges and prospects. En S. Salas, M. J. Barragán-Paladines, & R. Chuenpagdee (Eds.), Viability and sustainability of small-scale fisheries in Latin America and the Caribbean (pp. 15–47). Springer. doi:https://doi.org/10.1007/978-3-319-76078-0_2.

Delgado-Ramírez, C. E., Ota, Y., & Cisneros-Montemayor, A. M. (2023). Fishing as a livelihood, a way of life, or just a job: Considering the complexity of "fishing communities" in research and policy. *Reviews in Fish Biology and Fisheries, 33*(1), 265–280. https://doi.org/10.1007/s11160-022-09721-y

Dias, A. C. E., & Armitage, D. (2021). Ecosystems, communities and canoes: Using Photovoice to understand relationships among coastal environments and social wellbeing. En M. Gustavsson,

C. S. White, J. Phillipson, & K. Ounanian (Eds.), Researching people and the sea: Methodologies and traditions (pp. 159–179). Springer. doi:https://doi.org/10.1007/978-3-030-59601-9_8.

Dias, A. C. E., Armitage, D., Nayak, P. K., Akintola, S. L., Arizi, E. K., Chuenpagdee, R., Kumar Das, B., Diba, S. A., Ghosh, R., Isaacs, M., Islam, G. M. N., Kane, A., Li, Y., Manase, M. M., Mbaye, A. A., Onyango, P., Pattanaik, S., Sall, A., Susilowati, I., et al. (2023). From vulnerability to viability: A situational analysis of small-scale fisheries in Asia and Africa. *Marine Policy, 155*, 105731. https://doi.org/10.1016/j.marpol.2023.105731

Díaz-Perera, M. Á., & De los Santos González, C. C. (2021). *La frontera olvidada: poblamiento costero de Tabasco en cuatro momentos decisivos, 1518–2020*. El Colegio de la Frontera Sur.

Díaz, S., Pascual, U., Stenseke, M., Martín-López, B., Watson, R. T., Molnár, Z., Hill, R., Chan, K. M. A., Baste, I. A., Brauman, K. A., Polasky, S., Church, A., Lonsdale, M., Larigauderie, A., Leadley, P. W., van Oudenhoven, A. P. E., van der Plaat, F., Schröter, M., Lavorel, S., et al. (2018). Assessing nature's contributions to people. *Science, 359*(6373), 270–272. https://doi.org/10.1126/science.aap8826

DOF. (2019). *Medidas para incrementar la seguridad y protección de las instalaciones petroleras marinas de la Sonda de Campeche* (p. 4). Diario Oficial de la Federación.

Dow, K. (1999). The extraordinary and the everyday in explanations of vulnerability to an oil spill. *Geographical Review, 89*(1), 74–93. https://doi.org/10.2307/216141

Espinoza Tenorio, A., Espinosa-Romero, M., Rojas-González, R., Campa-Madrid, A., Salas, S., Lapuente-Landero, J., Maeda-Martinez, A., Poot-Salazar, A., Arce-Ibarra, A., Corral, A., Saldívar-Moreno, A., Miranda-Saucedo, C., Méndez-Medina, C., Torres-Salas, D., Ramos-Muñoz, D., Rivera, E., Albáñez-Varela, E., Arreguín-Sánchez, F., Fernández-Rivera Melo, F., & Castro-Medina, Y. (2023). Decálogo por el Año Internacional de la Pesca y la Acuicultura Artesanales en México.

FAO. (2015). *Voluntary guidelines for securing sustainable small-scale fisheries in the context of food security and poverty eradication*. FAO. https://openknowledge.fao.org/handle/20.500.14283/i4356en

Foley, P., Neis, B., Power, N. G., Knott, C., & Parlee, C. (2016). Opportunities for and challenges of occupational pluralism in seasonal fisheries: Regional cases from Atlantic Canada [report]. *Memorial University of Newfoundland*. http://www.onthemovepartnership.ca/wp-content/uploads/2019/02/OP-final-web-OTM.pdf

García-Cuellar, A., Arreguín-Sánchez, F., Hernández-Vázquez, S., & Lluch-Cota, D. (2004). Impacto ecológico de la industria petrolera en la Sonda de Campeche, México, tras tres décadas de actividad. *Interciencia, 29*, 311–319.

Gough, I., & McGregor, J. A. (2007). *Wellbeing in developing countries: From theory to research*. Cambridge University Press.

Hemmerling, S. A., Carruthers, T. J. B., Hijuelos, A., & Bienn, H. C. (2020, enero 1). Double exposure and dynamic vulnerability: Assessing economic well-being, ecological change and the development of the oil and gas industry in coastal Louisiana. Shore & Beach, 88(1), 11.

Idrobo, C. J. (2018). Adapting to environmental change through the lens of social wellbeing: Improvements and trade-offs associated with a small-scale fishery on the Atlantic Forest coast of Brazil. En D. S. Johnson, T. G. Acott, N. Stacey, & J. Urquhart (Eds.), Social wellbeing and the values of small-scale fisheries (pp. 75–96). Springer. doi:https://doi.org/10.1007/978-3-319-60750-4_4.

Idrobo, C. J., & Johnson, D. S. (2020). Livelihood transitions and social wellbeing on the Atlantic Forest Coast of Brazil. *Maritime Studies, 19*(1), 67–79. https://doi.org/10.1007/s40152-019-00140-7

INEGI. (2020a). *Principales resultados por localidad (ITER) 2020*. Instituto Nacional de Estadística, Geografía e Informática. https://www.inegi.org.mx/app/scitel/Default?ev=9#collapseList-7.

INEGI. (2020b). *Cuéntame de México Población Escolaridad*. Instituto Nacional de Estadística, Geografía e Informática. https://cuentame.inegi.org.mx/poblacion/escolaridad.aspx?tema=P

Jentoft, S., & Chuenpagdee, R. (2022). Interactive learning and governance transformation for securing blue justice for small-scale fisheries. *Administration & Society, 54*(7), 1255–1282. https://doi.org/10.1177/00953997211073947

Jentoft, S., & Chuenpagdee, R. (2018). From poverty to wellbeing in small-scale fisheries: The governability challenge. En D. S. Johnson, T. G. Acott, N. Stacey, & J. Urquhart (Eds.), Social wellbeing and the values of small-scale fisheries (pp. 293–315). Springer. doi:https://doi.org/10.1007/978-3-319-60750-4_13.

Kpanou, S. B.-V. K., Dedehouanou, H., Chogou, S. K., Aoudji, A. K. N., & Dogot, T. (2021). Factors influencing small-scale fishers' individual perceived wellbeing satisfaction in southern Benin. *Sustainability, 13*(11), Article 11. https://doi.org/10.3390/su13116279

Leite, M. C. F., Johnson, D. S., Ross, H., & Seixas, C. S. (2023). Social wellbeing, values, and identity among Caiçara small-scale fishers in southeastern Brazil. *Maritime Studies, 22*(3), 36. https://doi.org/10.1007/s40152-023-00322-4

Machado, L. O. R., de Cássia Franco Rêgo, R., Northcross, A. L., Meyer, A., da Silva Rezende Mota, L., da Silva, C., Gomes de Jesus Lino, P., da Costa, L. C. C., de Kikuchi, R. K. P., & Lima, V. M. C. (2025). Exposure to oil spill among fishers on the Brazilian coast in 2019/2020. *Journal of Agromedicine, 30*(2), 250–262. https://doi.org/10.1080/1059924X.2024.2447908

McGregor, J. (2008). Wellbeing, development and social change in Thailand. *Thammasat Economic Journal, 26*, 1–27.

McLean, D. L., Ferreira, L. C., Benthuysen, J. A., Miller, K. J., Schläppy, M.-L., Ajemian, M. J., Berry, O., Birchenough, S. N. R., Bond, T., Boschetti, F., Bull, A. S., Claisse, J. T., Condie, S. A., Consoli, P., Coolen, J. W. P., Elliott, M., Fortune, I. S., Fowler, A. M., Gillanders, B. M., et al. (2022). Influence of offshore oil and gas structures on seascape ecological connectivity. *Global Change Biology, 28*(11), 3515–3536. https://doi.org/10.1111/gcb.16134

Oliveto-Andrade, A., Espinoza-Tenorio, A., Ramos-Muñoz, D., & Pérez-Jiménez, J. C. (2024). Understanding the motivations of young people from marginalized rural communities to participate in small-scale fisheries in oil territories of the Gulf of Mexico. *Ocean & Coastal Management, 248*, 106947. https://doi.org/10.1016/j.ocecoaman.2023.106947

Owusu, V., Lawer, E. T., Adjei, M., & Ogbe, M. (2023). Impact of offshore petroleum extraction and "ocean grabbing" on small-scale fisheries and coastal livelihoods in Ghana. *Maritime Studies, 22*(2), 17. https://doi.org/10.1007/s40152-023-00307-3

PEMEX. (2018). *Memoria Documental—Programa de Apoyo a la Comunidad y Medio Ambiente* (p. 72). Petróleos de México - Gerencia de responsabilidad y desarrollo social.

Peña-Puch, A. D. C., Pérez-Jiménez, J. C., Espinoza-Tenorio, A., Méndez-Loeza, I., & Williams-Jara, G. M. (2021). The sustainability of small-scale fisheries in oil-producing sections of the Gulf of Mexico. *The Extractive Industries and Society, 8*(4), 100975. https://doi.org/10.1016/j.exis.2021.100975

Queffelec, B., Bonnin, M., Ferreira, B., Bertrand, S., Teles Da Silva, S., Diouf, F., Trouillet, B., Cudennec, A., Brunel, A., Billant, O., & Toonen, H. (2021). Marine spatial planning and the risk of ocean grabbing in the tropical Atlantic. *ICES Journal of Marine Science, 78*(4), 1196–1208. https://doi.org/10.1093/icesjms/fsab006

Quist, L.-M. (2019). Fishers' knowledge and scientific indeterminacy: Contested oil impacts in Mexico's sacrifice zone. *Maritime Studies, 18*(1), 65–76. https://doi.org/10.1007/s40152-018-0123-7

Quist, L.-M., & Nygren, A. (2015). Contested claims over space and identity between fishers and the oil industry in Mexico. *Geoforum, 63*, 44–54. https://doi.org/10.1016/j.geoforum.2015.05.015

Quist, L.-M., & Nygren, A. (2019). Debating the unknowns of marine oil exploration in Mexico. *The Extractive Industries and Society, 6*(3), 855–862. https://doi.org/10.1016/j.exis.2019.06.005

Ramos-Muñoz, D., & Espinoza-Tenorio, A. (2020). ¿Cooperar o no cooperar? Crónica del dilema por el espacio compartido del Golfo de México. *Cadernos de Estudos Sociais, 35*(2). https://doi.org/10.33148/CES25954091V35n2(2020)1955

Reilly, K., O'Hagan, A. M., & Dalton, G. (2016). Moving from consultation to participation: A case study of the involvement of fishermen in decisions relating to marine renewable energy projects on the Island of Ireland. *Ocean & Coastal Management, 134*, 30–40. https://doi.org/10.1016/j.ocecoaman.2016.09.030

Ribeiro, L. C. D. S., de Souza, K. B., Domingues, E. P., & Magalhães, A. S. (2021). Blue water turns black: Economic impact of oil spill on tourism and fishing in Brazilian northeast. *Current Issues in Tourism, 24*(8), 1042–1047. https://doi.org/10.1080/13683500.2020.1760222

Salas, S., Bjørkan, M., Bobadilla, F., & Cabrera, M. A. (2011). Addressing vulnerability: Coping strategies of fishing communities in Yucatan, Mexico. En S. Jentoft & A. Eide (Eds.), Poverty mosaics: Realities and prospects in small-scale fisheries (pp. 195–220). Springer . doi:https://doi.org/10.1007/978-94-007-1582-0_10.

Salas, S., Huchim-Lara, O., Guevara-Cruz, C., & Chin, W. (2019). Cooperation, competition, and attitude toward risk of small-scale fishers as adaptive strategies: The case of Yucatán, Mexico. En S. Salas, M. J. Barragán-Paladines, & R. Chuenpagdee (Eds.), Viability and sustainability of small-scale fisheries in Latin America and the Caribbean (pp. 101–123). Springer. doi:https://doi.org/10.1007/978-3-319-76078-0_5.

Salas, S., Núñez, A., Cepeda, M., Ramos Miranda, J., Cabrera, M., Coronado, E., Rocha, J., & Torres-Irineo, E. (2022). *Pesca Artesanal: Contexto Socio-economico y Bienestar comunitario*. Universidad Autónoma de Campeche.

Salazar-De la Cruz, C. C., Alberto Zepeda-Domínguez, J., Espinoza-Tenorio, A., & E Ramos-Muñoz, D. (2020). Governance networks in marine spaces where fisheries and oil coexist: Tabasco, México. *The Extractive Industries and Society, 7*(2), 676–685. https://doi.org/10.1016/j.exis.2020.03.012

SENER. (2024). *Sistema de Información Energética (SIE)*. Secretaría de Energía. https://sie.energia.gob.mx/inicio/#/

Steenbergen, D. J., Fabinyi, M., Barclay, K., Song, A. M., Cohen, P. J., Eriksson, H., & Mills, D. J. (2019). Governance interactions in small-scale fisheries market chains: Examples from the Asia-Pacific. *Fish and Fisheries, 20*(4), 697–714. https://doi.org/10.1111/faf.12370

Thanh, T. D. P., Hsiao, Y.-J., & Chuang, C.-T. (2010). Fisheries subsidies and their implications, case studies of fuel subsidy programs in Taiwan and Vietnam. *Economics of Fish Resources and Aquatic Ecosystems: Balancing Uses, Balancing Costs, 9*.

Tolentino-Arévalo, O., Markantoni, M., Espinoza-Tenorio, A., & Mesa-Jurado, M. A. (2019). Drivers of adaptive capacity to climate change in coastal fishing communities of Tabasco, Mexico. En S. Salas, M. J. Barragán-Paladines, & R. Chuenpagdee (Eds.), Viability and sustainability of small-scale fisheries in Latin America and the Caribbean (pp. 125–147). Springer. doi:https://doi.org/10.1007/978-3-319-76078-0_6.

Tran, T. H., Dobrovnik, M., & Kummer, S. (2018). Supply chain risk assessment: A content analysis-based literature review. *International Journal of Logistics Systems and Management*. https://doi.org/10.1504/IJLSM.2018.096088

Turner, J. M. M. (2025). Small scale fisheries and the challenges of occupational safety and health. *Journal of Agromedicine, 30*(2), 198–200. https://doi.org/10.1080/1059924X.2024.2434074

United Nations. (2019). *SDG 4. Quality Education Ensure inclusive and equitable quality education and promote lifelong learning opportunities for all*. https://unstats.un.org/sdgs/report/2019/goal-04/.

Uribe-Martínez, A., Espinoza-Tenorio, A., Cruz-Pech, J. B., Cupido-Santamaría, D. G., Trujillo-Córdova, J. A., García-Nava, H., Flores-Vidal, X., Gudiño-Elizondo, N., Herguera, J. C., Appendini, C. M., & Cuevas, E. (2024). An affordable operational oil spill monitoring system in action: A diachronic multiplatform analysis of recent incidents in the southern Gulf of Mexico. *Environmental Monitoring and Assessment, 196*(11), 1069. https://doi.org/10.1007/s10661-024-13161-5

Villalobos, G. J., & Rivera-Arriaga, E. (2008). La actividad petrolera en Campeche: Situación, retos y oportunidades. En J. Fraga, G. J. Villalobos, S. Doyon, & A. García (Eds.), Descentralización y manejo ambiental Gobernanza Costera en México (pp. 161–178). Plaza y Valdés.

Villasante, S., Gianelli, I., Castrejón, M., Nahuelhual, L., Ortega, L., Sumaila, U. R., & Defeo, O. (2022). Social-ecological shifts, traps and collapses in small-scale fisheries: Envisioning a way forward to transformative changes. *Marine Policy, 136*, 104933. https://doi.org/10.1016/j.marpol.2021.104933

Weeratunge, N., Béné, C., Siriwardane, R., Charles, A., Johnson, D., Allison, E. H., Nayak, P. K., & Badjeck, M.-C. (2014). Small-scale fisheries through the wellbeing lens. *Fish and Fisheries, 15*(2), 255–279. https://doi.org/10.1111/faf.12016

Westlund, L., & Nations, F. and A. O. of the U. (2007). *Disaster response and risk Management in the Fisheries Sector*. Food & Agriculture Org.

White, I., & Molloy, F. (2003). Factors that determine the cost of oil spills. *International Oil Spill Conference Proceedings, 2003*. https://doi.org/10.7901/2169-3358-2003-1-1225

Zalik, A. (2009). Zones of exclusion: Offshore extraction, the contestation of space and physical displacement in the Nigerian Delta and the Mexican Gulf. *Antipode*. https://www.academia.edu/1618756/Zones_of_Exclusion_Offshore_Extraction_the_Contestation_of_Space_and_Physical_Displacement_in_the_Nigerian_Delta_and_the_Mexican_Gulf, *41*, 557.

Chapter 9
Voices of the Youths of Puerto Ceiba, Tabasco: Tensions Between Fisheries and the Oil Industry, Navigating Between Tradition and "Progress"

A. A. Ramírez-Pacheco

9.1 Fishing and the Generational Renewal

Southern Mexico, specifically the state of Tabasco, is a territory that since the 1940s has been invaded by extractive economic development projects, a process that has materialized in public works to control water, the introduction of new livestock breeds to catalyze ranching, and communication infrastructure to link the state to the broader region, especially the Southeastern Railroad and the Gulf Circuit (*Ferrocarril del Sureste* and the *Circuito del Golfo*; see Uribe-Iniesta, 2016). In the 1970s, the focus shifted to petroleum extraction (Pinkus-Rendón & Contreras Sánchez, 2012; Rabelo-Avalos et al., 2021). By giving clear priority to certain activities, the Mexican State has intensified tensions in the study area. Concerns over the abandonment of small-scale fishing and the global crisis of generational renewal highlight the need to revitalize this activity in a sustainable way to assure a source of food for future generations (Arias-Schreiber & Gillette, 2021). This chapter gathers the voices of youths in a coastal town on the Gulf of Mexico that has been transformed by this emphasis on petroleum, leading some young people to dismiss fishing as a viable employment option. Diverse instances interested in fishing have documented this process of abandonment and the difficult generational renewal in many countries with coastal areas where fishing has a long trajectory and is a key element of cultural identity (Pollnac & Poggie, 2006; Pollnac et al., 2015). In some European countries, "school boats" have been operationalized to train young people in traditional knowledge by integrating modern technology that may attract them to fishing as a way of life (Loctier, 2020; Arias-Schreiber & Gillette, 2021). The generational renewal of fishing, however, depends mainly on young people, though diverse studies emphasize that uncertainty, the risks of the sea, fluctuating incomes,

A. A. Ramírez-Pacheco (✉)
El Colegio de la Frontera Sur, Unidad Villahermosa, Villahermosa, Tabasco, México
e-mail: anabel.ramirez@posgrado.ecosur.mx

A. Espinoza-Tenorio et al. (eds.), *The Fishing and Oil Industries in Coexistence*,
https://doi.org/10.1007/978-3-032-02621-7_9

long absences, and the loss of lifeways identified with fishing discourage continuity (Arias-Schreiber & Gillette, 2021; Reyes-Grande et al., 2021).

9.1.1 The Future and Anthropology

Pondering the future inevitably leads one to imagine scenarios and make plans to minimize risks. The future is constructed through everyday life, not as a coming event but as a transition that holds aspirations while nurturing certain nuances in, and from, young people, in whose hands lie the hopes for a brighter future as they become responsible for sustainability, resource management, and the possibility of a "better world"—whatever that may mean from subjectivities. But many youths today live in conditions of inequality, poverty, precariousness, and environmental degradation that generate distinct emotions regarding their future, ranging from uncertainty to hopefulness. Anthropology and sociology have explored the future in works by authors such as Marc Augé, Guy Baijot, Edgar Morin, and Arjun Appadurai, among others (Figueroa-Díaz, 2018). Appadurai (2015a) argues that the natural and social sciences have taken leading roles in studies of the future, especially from the perspective of economics, based on calculations and projections of macroeconomic processes, micro-actions, and choices that have led to a systematization of the future in terms of projections in contrast to uncertainty.

For my analysis, I adopt Appadurai's proposal, which examines interaction among three essential aspects of human life that mold the future as a cultural fact: (1) imagination, (2) anticipation, and (3) aspirations. Appadurai views the future not as a neutral space but as one that houses emotions and that is susceptible to systematization (Appadurai, 2015a). This occurs, he argues (2015b), in contexts that he calls "assemblages," a concept that captures how cultural and social elements are articulated—or, possibly, disassembled, lasting or temporary—in global and local settings and then transformed into strategies and actions (Appadurai, 2015b). He proposes emphasizing the ethic of *possibility* to focus on human beings as creators of the future, posing this approach as a way to think and act that broadens hope, imaginaries, and aspirations. The contrasting concept—the ethic of *probability*—is unequal and limiting. It examines the cultural and socioeconomic conditions that affect the likelihood of people achieving their aspirations, but projects and predicts the impossibility of this in settings where structural inequalities predominate (Appadurai, 2015a).

I argue that it is essential to emphasize that we cannot detach the ethic of probability from the ethic of possibility because the context of structural inequalities (the ethic of probability) determines, limits, and reconfigures how people address the future. If the ethic of probability is not considered then there is no level floor for everyone to accede to the ethic of possibility, a path that leads to the construction of hope by pondering the "other," walking with the "other," creating community, trusting in transcendence, and forming horizons of meaning in the face of the uncertainty that the future represents (Hang, 2024). From this theoretical framework, I discuss data that enrich discussions on this topic by elucidating a phenomenon that has emerged in a town of fishers where hope for a generational renewal is vanishing due to the oppressive presence of the oil industry and the imposition of a development model materialized in the *Olmeca* Refinery.

9.1.2 The Setting of Puerto Ceiba

Puerto Ceiba, in the municipality of Paraíso, Tabasco, was one of eight coastal towns registered in the 1900 General Census of the Mexican Republic. The town's main subsistence activities have long been fishing, agriculture, and copra production. The latter enjoyed a boom in the 1940s thanks in part to state government incentives (Díaz-Perera & de los Santos González, 2021). *Puerto Ceiba*'s profound relationship with water—the sea and lagoons—and, obviously, with fishing. Practiced initially in lagoons mainly for self-consumption, fishing gave the local population its identity, exemplified by the enormous figure of a blue crab (*Cardisoma guanhumi*) on the town's pier that represents the zone and highlights its links to fishing, the sea, and local resources.

Díaz-Perera and de los Santos González (2021) write that the coast of Tabasco underwent a process of settlement and territorial change triggered by agrarian reform, government support for copra production, road and electrical infrastructure, the Southeastern Railway, and a program called "The March to the Sea." In Puerto Ceiba, this resulted in fishing achieving economic importance in the mid-1970s (Reyes-Grande et al., 2021; Díaz-Perera & de los Santos González, 2021) though it began to lose ground as the oil industry came to exert a greater impact in the state through four crucial moments: first "the takeoff" (1950–1973) based on exploration and production; second, "the boom" (1974–1985), when oil prices rose and production in Tabasco reached its maximum level; third, "the decline" (1986–1999), when prices and production dropped; and, fourth the so-called "repetrolization" (2000–2008) when prices rose although production remained low (Rabelo-Avalos et al., 2021).

The relation between fishing and petroleum has evolved in a climate of tensions, conflicts, and environmental damage that found a voice in the "River Pact" (*Pacto Ribereño*, 1976) (Velázquez-García, 2016) and was manifested in the restrictions imposed on fishing around oil platforms after the attack on the Twin Towers in 2001. Later, the entrance of foreign investment attracted by the 2013 Energy Reform intensified the problems that fishing faced. This conflict-ridden relationship between fishing and oil has marked the recent history of Tabasco.

Puerto Ceiba has 2806 inhabitants, a captaincy, and a federation that brings together fishing cooperatives in the *Chontalpa* region, but the town's tradition linked to fishing (Fig. 9.1) is now in crisis (Arreguín-Sánchez & Arcos Huitrón, 2011; Reyes-Grande et al., 2021) largely due to the sweeping changes spurred by two large economic development projects: the *Dos Bocas* deep-sea port (in 1979); and

Fig. 9.1 Puerto Ceiba 2023. Photo by Anabel Alejandra Ramírez-Pacheco

the *Olmeca* Refinery (in 2019). Conceived to propel the petroleum industry, these two projects have relegated fishing to a secondary activity, as I argue below.

9.2 Method

My methodological focus is qualitative, based on the tools of the ethnographic method. I conducted eight in-depth interviews with young women and men, and four additional interviews with adults. I also maintained detailed fieldwork diaries during three study periods in March, November, and December 2023. I obtained informed consent from each interviewee using a form approved by ECOSUR's Ethics Committee. I also incorporated a perspective on the coproduction of knowledge using participative cartography[1] with seven of the eight youths interviewed, who provided their knowledge of, and from, the territory on the changes they have experienced since the arrival of the *Olmeca* Refinery. The study began with an intentional sample (Sampieri et al., 2014) designed to gather information for my doctoral research,[2] which proposed a comparative study. I identified a group of young women and men in high school and universities who would soon have to decide whether to continue their studies or enter the workforce, against the backdrop of the enormous territorial transformations spurred by two large-scale development projects. I later widened the sample through the snowball method. These two

[1] I am currently analyzing the results of the cartography as part of my Doctoral project: "Youths and their perceptions of territory in the context of Tabasco" ("*Jóvenes y sus percepciones sobre el territorio en el contexto de Tabasco*"), so they are not addressed here.

[2] Based on an analysis of the impact of the construction of the *Olmeca* Refinery in Puerto Ceiba and the *Tren Maya* through Cerro Norte, San Carlos, and Boca del Cerro in Tenosique.

non-probabilistic techniques are designed to deepen analyses of a specific social reality, rather than making broad generalizations. The cut-off age was determined from previous research conducted in Tabasco by Ramírez-Pacheco et al. (2018) and Sánchez-Matías and Ramirez Pacheco (2023), who recognized that youths in this age group are at a stage of the life cycle where they must make a crucial decision; namely, whether to enter university or seek employment, cognizant of the fact that either choice could involve migration.

Three collective physical maps were developed, and I kept fieldwork diaries, taped the 12 interviews, and then transcribed them. Information was codified with ATLAS.ti software using a five-element scheme: (1) high school education; (2) university education; (3) expectations for future studies; (4) expectations for future work; and (5) interest in fishing or the oil industry. Two age groups were formed for analysis: (a) 15–17-year-olds who imagined their future as a function of the possibility of attending university; and (b) 18–25-year-olds who imagined a future centered on work since some were attending university and others were completing their final degree requirements.

9.3 Results

Table 9.1 summarizes the dimensions of the context of the 469 young people in *Puerto Ceiba*.

The interviewees included high school and university students (Tables 9.2 and 9.3). They expressed interests that centered on engineering programs but also spoke of their close relation to fishing as a productive activity. The latter derives from their parents or grandparents, though—as I elucidate below—I found that they now discourage young people from following in their footsteps. This reflects the growth of the oil industry since the 1970s, which has led many of these youths to shift their views on education, work, and future projects toward the industrial sector.

All interviewees in the younger age group were born and raised in Puerto Ceiba in families that were also native to the town. They attended high school in the *Colegio de Bachilleres de Tabasco* (COBATAB 34). The three older interviewees who were at university were enrolled in one of three public institutions: the *Universidad Politécnica del Golfo de México* (UPGM), in the municipality, the *Universidad Juárez Autónoma de Tabasco* (UJAT)—the most important university in the state—or the *Instituto Tecnológico Superior de Comalcalco*. I found that

Table 9.1 Juvenile population aged 15–25 years in *Puerto Ceiba, Paraíso*, Tabasco

Age	Females	Males	Total
15–17	66	80	146
18–24	167	156	323
			469

Source: elaborated by the author based on the *2020 Censo de Población y Vivienda* (INEGI)

Table 9.2 Profile of the interviewees aged 15–17

Age	Sex	Education	Imagination regarding university education	Level	Imagination regarding work future
17	M	High school	International commerce and Customs, Automotive engineering	Undergraduate degree Technical program	*Dos Bocas* deep-sea port Industry
17	M	High school	Mechatronic engineering	Engineering	*Olmeca* Refinery Industry (northern Mexico)
17	F	High school	Nursing	Technical program	Public hospital (local)
17	F	High school	Beautician	Technical program	Business (local)

Source: elaborated by the author

Table 9.3 Profile of interviewees aged 18–25

Age	Sex	Education	Higher education	Level	Imagination regarding work
24	F	High school	No desire to continue	–	??
25	M	University	Chemical engineering	Engineering	Teaching
22	M	University	Automation Industrial control engineering	Engineering	*Olmeca* Refinery Industry (northern Mexico)
18	F	University	Industrial engineering	Engineering	*Olmeca* Refinery Administration in the oil industry

Source: elaborated by the author

decisions to abandon fishing among these youths reflected a multifactorial situation articulated with the area's historical context, where economic development projects have marked a local territorial trajectory that pits fishing against petroleum production. The importance given to the oil industry permeates diverse domains of people's lives in Puerto Ceiba primarily in four ways: (a) the economic bonanza it brought and the status that accrued to those who worked in the industry; (b) the lifeways and consumption patterns it promotes; (c) environmental degradation; and (d) restrictive measures imposed on fishing. Granting priority to the petroleum industry over the productive activities that once predominated in the area has left deep scars (Uribe-Iniesta, 2016), one of which is the focus of my study: the impact this process has had on young people in Puerto Ceiba and their relation to fishing.

> First, because it's one of the careers with lots of places where you can work. You can work in the petroleum industry, in medicine, in metallurgy… (interview with Ángel, 24, degree candidate in Chemical engineering).

Some of our interviewees create—that is, imagine and anticipate—expectations for their future as a function of their university studies; expectations that have modified their prioritization of economic activities since the oil industry arrived in the state in 1970. One result is the emergence of a new predominant discourse in intergenerational relations (Sánchez-Sánchez, 2020; Aguilar-Freyan & Víquez Calderón, 2008) that now stresses the importance of higher education and the promise of a bright future in industry, particularly in oil production. These expectations rest upon the perspective of obtaining a secure, well-paid job and the higher social status that comes with it, ambitions that are pushing young people away from fishing—characterized by uncertainty—toward personal advancement and an ideal of "making something of oneself," which implies aspirations and recognition. These relations now resonate among local youths, manifesting the impact of the discourse that adults direct toward their children:

> But [my professors and parents] made me see that Industrial engineering is better… it has broader areas for work. You can get a job anywhere, so you think it's better because the career I want to study is Mechatronics and you can't find that around here (interview with Yuliana, 16 December 2023).

Initially, some of these youths had distinct projects for their future, ideas that centered on agriculture or livestock raising, but their perceptions have changed due to the orientation they received through intergenerational relations. Today, some say that those activities are no longer lucrative and lack government support. Others express the view that they never considered fishing, agriculture, or livestock raising, but did imagine earning a university degree, though perhaps in a program distinct from the one they are currently studying. Yuliana is a case in point as a young woman who originally thought she would obtain a B.A. in Education, but after consulting with her teachers, learned that it was very hard to find a job in that field, so she opted to switch to Industrial Engineering after comparing and evaluating the program in Mechatronics.

> My first idea was to be a schoolteacher, but it was impossible to get a job there. Today schoolteachers are not highly-valued, the job isn't worth much and it's badly paid. Moreover, getting work as a teacher is impossible, it would cost me a lot to get [a permanent position] (interview with Yuliana, 16 December 2023).

Intergenerational relations deeply affect these youths as they make their decisions, as well as a function of anticipation and imagination. They are unsure about the possibility of obtaining jobs if they opt for a career that the industry does not require, but at the same time, they are driven by aspirations such as achieving the status that comes with a secure, well-paid job. Coupled with this is the reality that fishing generates greater insecurity among adults, and they transfer their concerns to their children, causing fear and shrinking any hope in fishing and the "ethic of possibility."

> I was a fisher, but the fishery sector has been forgotten, the fishing banks are gone, and the [] refinery continues to displace people and generate poverty in the community. Fishing is really hard work, so many days away from the family, dangers, pirates who steal everything and strand [you] out in the ocean.[3] And you never know if you're going to make a profit or [go into debt]… and if you work for someone who has a permit, if you don't have your own boat and motor, well, it's much more difficult. That's why I told my children to study and make something of themselves, be professionals or engineers, so they don't have to suffer in fishing. I don't want them to be like me (interview with Pablo, a former fisher, 56 years, 25 March 2023).

[3] Fishing is in crisis, in part, due to raids by pirates along the coasts of Tabasco and Campeche where I have collected statements from witnesses. We cannot be sure if the pirates are linked to organized crime because this issue must be approached with great discretion.

In response, youths anticipate and aspire to open a path toward the "ethic of probability."

> On the one hand, [interest in fishing is okay], but not as your work, just a hobby, a pastime. I think fishing is really nice, but not settling down to it as work, not as a job, because I don't think it's profitable, even less now because there are no fish… so fishers say, 'since there's no more fish, let's go to the refinery' (interview with Jesús, 2 March 2023).

I found that the impact of the oil industry in Puerto Ceiba with the opening of the Olmeca Refinery and the *Dos Bocas* deep-sea port, coupled with intergenerational relations—youths/parents, professors, educational authorities, institutions/students—have fed the crisis that affects fishing today, manifested in economic uncertainty, scarce resources, and the risks of open-sea operations. Hence, young people are anticipating and imagining careers that will distance them from fishing; for example, some of the high school students expressed interest in careers in International commerce and Customs administration or in Mechatronic Engineering, where they think they will find jobs quickly. They emphasized that they would prefer to work in town, but if not, they have the option of migrating to northern Mexico, considered a region with prolific industrialization.

The youths who opt for technical careers like nursing, or for a trade—beautician—think they will be able to work near their families in Puerto Ceiba. These options generally reflect the views of young women who live under greater parental control, a phenomenon that was accentuated when so many strangers—mainly men—came to work in constructing the refinery. Young women do not see the refinery as a suitable workplace because the oil industry is an overwhelmingly masculine one. For this reason, their options are restricted to the service sector; for example, selling food to construction workers. We verified that local work options for women are, indeed, limited to services and self-employment, though some expressed the view that a university degree could open new possibilities for them.

Teenagers still in high school imagine and anticipate educational projects that include university and a future in formal employment. They showed no interest in fishing. Two youths said they were attracted to livestock raising or agriculture, but one emphasized that only if those activities were more extensive and highly technified:

> The truth is fishing isn't something I'm interested in. I'd like cattle-ranching, but only if it's extensive, large-scale, with machinery; like milking [where] you can use machines. Yeah, I'd like that, strikes me as being a bit better. And agriculture today, with technology [like] better tractors, fumigators, and [irrigation] systems than the traditional ones (interview with Carlos, 2 March 2023).

The youths who are enrolled in university are in a process of formation that foments work expectations linked directly to their imaginings of a future that, as my data show, are oriented by intergenerational relations that raise hopes for a "better future" in terms of job security, income, and social status, and freedom from uncertainty and danger. Adults are elements of the assemblage in Appadurai's terms (2015b), not only parents in the family nucleus, but also a State that prioritizes extractivist development models based on petroleum and professors who foment an approach to education that attends to the needs of industry. This set of imaginings, anticipation, and aspirations for the future centers on industry, whether local oil operations or advanced industries in northern Mexico, especially Monterrey, the capital of the state of Nuevo León and one of the most heavily industrialized areas in Mexico. There are, of course, exceptions; for example, one young man is considering a career as a high school teacher and one young woman is not planning to continue her studies or obtain employment in the short term because she is a single mother who considers—with her family—that her priority is raising her child.

High school and university play a primordial role in the imaginings, anticipation, and aspirations for the future of these youths as institutions with programs that channel them toward an educational project designed by the State in accordance with the demand for skilled personnel for the industrial sector. This is exemplified by the study plans offered at the UPGM: Petroleum engineering, Automation and Industrial control engineering, Agrotechnology engineering, Industrial safety and automation engineering, and Financial engineering, as well as undergraduate programs in International commerce and Customs administration or Physical therapy. Universities have an even more fundamental role in the aspirations of these youths because they have forged links with the petroleum industry and foreign companies that came to work in constructing the *Olmeca* Refinery; for example, Samsung Engineering and Associates. Through these ties students can obtain training periods in companies, an activity that intensifies their expectations and the importance of their study programs. More importantly, it increases the possibility of obtaining employment in those workspaces after graduating. The entry of foreign oil companies also impacted the town by introducing a series of norms for performing activities, including a Social Outreach Plan that, among other benefits, offers university scholarships for fishers' children that further nourish their imaginings, anticipation, and aspirations of one day forming part of this industry.

> Yeah, that's right, many of us connected with the Samsung company; in fact, those who study this program have everything paid for; I mean it's a great help the companies give you. I had a scholarship from the Grupo Chevron through my Dad [a fisher] [that covered his university studies]. That opportunity was only for children of fishers (interview with Ricardo, 14 March 2023).

The relationship between public universities and the oil industry that began during the "boom" has intensified since the 2013 Reform due to the State's insistence on mitigating the social and environmental impact of the activities of that industry's typical activities. This opened a window of opportunity for these educational institutions to forge linkages that foment hopes for a better future among local youths. The presence of universities like the UPGM has a specific weight in the imaginings and aspirations of the educational and employment future that young people in *Puerto Ceiba* have created, reinforced through ties with foreign companies involved in the oil industry. These youths affirmed that this opens broader possibilities of insertion into the workforce than in the past when their expectations for employment centered on one sole business: PEMEX, the national oil company. But that option depended on inheritance, as positions were reserved for the children of parents who had worked in the State company or one of its subcontractors.

> Universities came to my high school [COBATAB] and talked a bit about the study program I was really interested in [International commerce and Customs administration], and I saw that the Polytechnical College [UPGM] offered it, so I'm going to take advantage of the opportunity (interview with Jesús, 2 March 2023).

Just as important as the role that some educational institutions have taken in training young people in accordance with the demands of the petroleum industry is the fact that other schools that originally offered training programs related to fishing—and played a significant role in forming local youths—are in processes of transformation. This is the case of Technical Secondary School 26 (*Escuela Secundaria Técnica 26*) in *Chiltepec*. Founded in 1976 only a few kilometers from *Puerto Ceiba*, this institution of basic education was oriented toward training youths in fishing. In fact, it had a teaching boat—now obsolete—and teachers skilled in fishing. Today, however, no teachers fit this profile and technological training in fishing is no longer offered, as those courses have been replaced by subjects like office automation, though programs in Aquaculture and Processing of fishery products continue.[4]

In this vein, while the division of Academic and Agricultural Sciences (*Ciencias Académicas y Agropecuarias*, DACA) at the most important university in Tabasco—UJAT—still offers a program in Aquacultural engineering, the study plan has not been updated since 2016[5] and the aquaculture school farm there was closed in February 2024 due to problems with organized crime that threatened teaching in engineering.[6]

The lack of interest in fishing among youths is not a recent process, for we found that this activity began to lose ground with changes in the educational institutions, in the discourses that emerged from intergenerational relations, the State's prioritization of the oil industry, the growing importance of this industry in the municipality, the precarious nature of fishing, and the shrinking fishery resources. All these factors show why the possibility of achieving a generational renewal of fishing has become so difficult.

> Yeah, when I was little, around 12, I'd go out fishing with my Dad, but actually [he] took me so I could see how [fishers] suffer [and] that it's better to study than go out there and suffer. I mean it takes a lot of effort, in economic terms too, because catches aren't always good, you don't always have a secure income. When the fishing's good you can make money, maybe $1,000 or $1,300 [MXN] pesos a day, or more, but then there's nothing for five or six days, so what do you do? (interview with Ricardo, 14 March 2023).
>
> Well, I know almost nothing about fishing, and I can't even help my Dad out in the fields because I've always preferred studying (interview with Ángel, 14 March 2023).

[4] Interview held via *WhatsApp* with Marcela Rodríguez, 15 June 2024.

[5] According to the study plan presented on its official page: https://www.ujat.mx/97 [consulted 10 June 2024].

[6] https://www.elheraldodetabasco.com.mx/policiaca/por-que-cerro-la-granja-acuicola-de-la-ujat-11515984.html [consulted 3 June 2024].

As mentioned above, parents are also discouraging a generational renewal of fishing, as my findings shows that their imaginary now stresses the importance of education so their children can obtain employment in the petroleum industry. They perceive this as an ideal lifestyle because of the economic stability and status it provides. As the oil industry has gathered strength in Puerto Ceiba with the arrival of the *Olmeca* Refinery, and youths have found new possibilities to enter the workforce as skilled personnel, not just as construction workers, they are increasingly attracted to work in this sector though this means gradually distancing themselves from fishing.

> I've been thinking about that because a friend had the same choice as me between going full out for the refinery or, if not, for the *Dos Bocas* port. [He works] for a company there unloading tubes from ships, so there's an opportunity over there. He says: 'whenever you want, just let me know. The truth is I'm interested, but I'm also scared to death' (interview with Ángel, 14 March 2023).

In contrast to these views, we found young women who foresee a future without higher education due to personal circumstances. This group includes the aforementioned single mother and others who stated that they "don't like school." They do not envision a future in the oil industry except, perhaps, in a certain stage of construction as unspecialized laborers who clean or direct traffic around the sites (*banderera*s); two jobs sometimes assigned to women. These youths perceive that their work options are restricted to the service sector.

> I think the [Refinery] is more for young men since so many are studying careers that will benefit them, but that's more beneficial for them, right? because they have a chance to work at the refinery. I think it's more for the men, it's just too complicated for us [women] (interview with Leslie, 2 December 2024).

These data reflect the imaginings, anticipation, and aspirations of a group of young people whose assemblage—context—leads them to imagine and aspire to a future that distances them from fishing and guides them toward the petroleum industry.

9.4 Discussion

9.4.1 The Ethic of Possibility Among Youths in Puerto Ceiba

To approach the causes that lead this group of young women and men, residents of a small town traditionally populated by fishers and farmers, away from maintaining the tradition of fishing in their projects for the future, I examined key elements of their past, explored how they live their present, and pondered how this configures their imagining of the future. Based on these findings, I discuss my results regarding these youths from the perspective of the ethic of possibility.

Figure 9.2 summarizes the principal components identified in the analysis of the empirical data. The young people interviewed in *Puerto Ceiba* imagine, anticipate, and aspire to (A) a future based on an assemblage (B) (Appadurai, 2015b) that

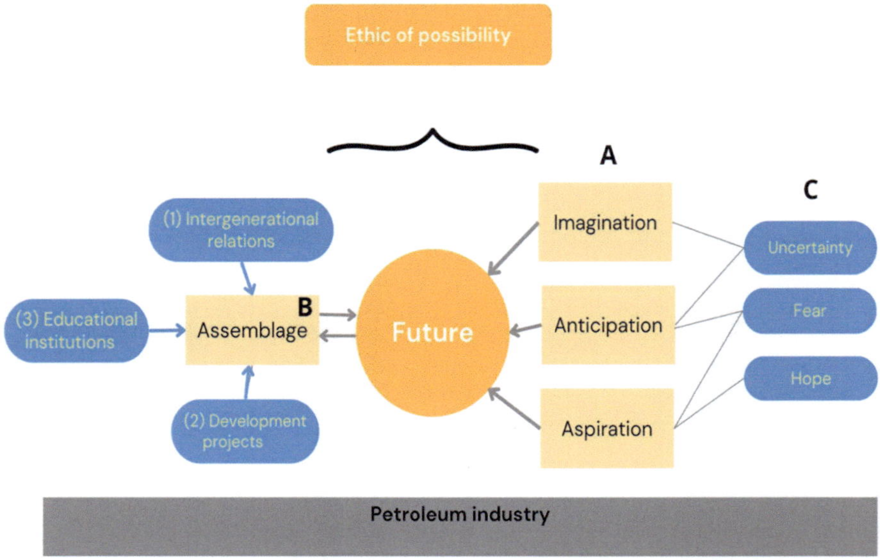

Fig. 9.2 The future as a social fact among young people in *Puerto Ceiba*. Elaborated by the author based on the proposal for the future as a social fact in Appadurai (2015a, b)

consists of: (1) intergenerational relations with parents, local authorities, and teachers who guide and persuade them; (2) a State that sponsors development projects to boost petroleum production, materialized in infrastructure like the *Olmeca* Refinery and the *Dos Bocas* deep-sea port, and prioritizes extractive activities over others; and (3) institutions of higher education that respond to, and align with, the needs of the State's extractivist projects and the demands of the oil industry. It is this assemblage that leads youths to imagine, anticipate, and aspire to (A) a future centered on that industry. This assemblage is adult-centered and limits the ethic of possibility for these youths (Appadurai, 2015b). As a result, we found that the young people interviewed imagine their future from the intergenerational relations that form the assemblage, anticipate as a function of the emotions this generates—fear, uncertainty, hope (C)—that lead them to imagine a future which includes university studies in programs related to engineering and industry due to their perceptions—real or not—of the high likelihood of obtaining employment in a place they aspire to, like the *Olmeca* Refinery or *Dos Bocas*. That aspiration is founded upon the *hope* that they will achieve higher status and "a better life" and, in this way, fulfill the expectation that they will "become more" than their parents.

Anticipation derives from the difficult scenarios—economic uncertainty, danger—that lead them to choose programs related to industry. A career in industry, of course, also entails risks but these are hidden by adults and other youths. The result is that these youths discard fishing and other productive activities, like agriculture and cattle-ranching, from their imaginings of their future.

Education, intergenerational relations, the problems that plague fishing, the growing importance of the oil industry, the emphasis that the federal and state governments place on extractivist development projects, and the prioritization of certain productive activities in the region combine to form a complex mosaic that generates doubts among youths concerning the viability of a life that depends on an uncertain economic activity like fishing. While they show attachment to their territory, this complex mosaic is leading them to imagine a future there but one distanced from the activity that for many years nourished Puerto Ceiba and its identity.

In spite of this process of imagination, anticipation, and aspiration (A) permeated by the assemblage (B) that adults have created, as Appadurai (2015a) suggests, space for the ethic of possibility still exists since some young people expressed interest in careers in teaching, where they can influence the formation of children and youths, while others manifested preferences for agriculture or livestock raising as long as it is highly technified. Still others mentioned nursing and labors in the domestic sphere, projecting the idea of working in the community, perhaps as beauticians given the impossibility of—or lack of interest in—entering the petroleum industry. Although these activities also distance these youths from fishing, they show that it is possible to place human beings in the center of the future, where, perhaps only as whispers, we can hear the voices of these youths.

9.5 Final Comments: Is There a Generational Renewal in *Puerto Ceiba*?

The potential for a generational renewal of fishing is clearly facing a crisis, for as we see in this analysis an assemblage has formed that, limits, and guides the imaginings, anticipation, and aspirations of this group of young people in *Puerto Ceiba*, pushing them away from fishing toward the oil industry. Thus, it is important to ask just how significant my sample is. I would argue that it is significant because high school education is obligatory in Mexico, so a high percentage of young people reach at least this level of schooling, encountering a space where they are encouraged to imagine, anticipate, and aspire to the future, and because they contemplate studying at the next level, where training related to fishing is losing importance in Tabasco. It is true that not all these youths will achieve this goal and that those who do not may well take up small-scale fishing, but under what conditions? Youths who do not opt for higher education find that their job options center on the service sector as store clerks, workers in the formal or informal economy, or in trades like construction workers, ironworkers, or mechanics, among others.

How could a generational renewal be produced when adults—parents, educational institutions, and a State that foments certain kinds of development and educational projects—shout to youths that they should imagine a future in the petroleum industry, not in small-scale fishing, which could ensure its sustainability. It is essential to recover the ethic of possibility so that more youths have the option to imagine a future in which fair conditions reign, not one based solely on anticipation, which generates fear and uncertainty. Hope must be allowed in by attacking the context of structural inequalities, for only in this way can small-scale fishing—perhaps—recover its value through a process of reestablishing its identitary value and dignity as a profitable economic activity and support local youths to allow them to recover aspirations distinct from those imposed by the State.

It is also important to document the loss of a generational renewal in fishing across Latin America in order to devise actions that reverse this situation. One suggestion is to focus attention on the processes and actions that some European countries have implemented to promote such a renewal from the perspective of sustainability. Here, I refer to places where studies have documented actions that drive a renewal of this kind, not by considering the sea as a model to be followed by all but, rather, by recognizing the need to rescue knowledge and create horizons of possible actions from the State, public policies, and organizations of civil society. Some European countries are successfully encouraging the teaching of small-scale fishing among young people by granting permits, financing inputs, opening local markets, and forging a complex that goes beyond simply catching fish in an effort to achieve a profitable, sustainable activity that guarantees alimentary sovereignty. The state of Tabasco in general, and *Puerto Ceiba* specifically, however, are moving in the opposite direction as the State and public policies abandon their role in fomenting a generational renewal, leaving this task to individuals. If the goal is to achieve a generational renewal of small-scale fishing from the perspective of sustainability, then these policies must be revisited and retracted.

References

Aguilar-Freyan, W., & Víquez Calderón, D. (2008). Construcción Social de La Juventud y El Papel Percibido de Los Medios Desde La Perspectiva de Los Jóvenes. *Revista Actualidades En Psicología, 22*(109), 43–66.

Appadurai, A. (2015a). *El futuro como hecho cultural. Ensayos sobre la condición global*. Fondo de Cultura Económica.

Appadurai, A. (2015b). Mediants, materiality, normativity. *Public Culture, 27*(2), 221–237. https://doi.org/10.1215/08992363-2841832

Arias-Schreiber, M., & Gillette, M. B. (2021). Neither fish nor fowl: Navigating motivations for fisheries participation and exit in Sweden. *Society & Natural Resources, 34*(8), 1019–1037. https://doi.org/10.1080/08941920.2021.1925381

Arreguín-Sánchez, F., & Arcos Huitrón, E. (2011). La pesca en México: Estado de la explotación y uso de los ecosistemas. *Hidrobiológica, 21*(3), 431–462.

Díaz-Perera, M. Á., & de los Santos González, C. C. (2021). *La frontera olvidada*. El Colegio de la Frontera Sur.

Figueroa-Díaz, M. E. (2018). El futuro como dispositivo: la mirada de algunos estudiantes universitarios. *Política y Cultura, 50*(2), 177–201.

Hang, B. C. (2024). *El espíritu de la esperanza*. Herder Editorial.

Loctier, D. (2020). *Nuevos jóvenes para dar vida a la pesca tradicional en Europa*. EuroNews. https://es.euronews.com/green/2020/01/21/nuevos-jovenes-para-dar-vida-a-la-pesca-tradicional-en-europa

Pinkus-Rendón, M. J., & Contreras Sánchez, A. (2012). Impacto Socioambiental de La Industria Petrolera En Tabasco: El Caso de La Chontalpa. *LiminaR Estudios Sociales y Humanísticos, 10*(2), 122–144. https://doi.org/10.29043/liminar.v10i2.99

Pollnac, R. B., & Poggie, J. J. (2006). Job satisfaction in the fishery in two southeast Alaskan Towns. *Human Organization, 65*(3), 329–339.

Pollnac, R. B., Seara, T., & Colburn, L. L. (2015). Aspects of fishery management, job satisfaction, and Well-being among commercial fishermen in the Northeast Region of the United State. *Society & Natural Resources, 28*(1), 75–92. https://doi.org/10.1080/08941920.2014.933924

Rabelo-Avalos, V., Ramos Muñoz, D. E., Díaz Perera, M. Á., & Mesa Jurado, M. A. (2021). El Petróleo En Tabasco: Propuesta Para Una Periodización. *Revista Mexicana de Sociología, 1*(1), 129–155.

Ramírez-Pacheco, A. A., Ramos Muñoz, D. E., Mena Farrera, R. A., & Analco Martínez, A. (2018). ¡Ahora Que Ya Tenemos Internet! Usos de La Internet Entre Jóvenes Rurales de Quintín Arauz, México. *EntreDiversidades. Revista de Ciencias Sociales y. Humanidades, 1*(11), 199–208. https://doi.org/10.31644/ED.11.2018.a07

Reyes-Grande, F., Espinoza Tenorio, A., & Díaz Perera, M. Á. (2021). Ya no hay pescado. Subsistencia e incertidumbre en una localidad pesquera del Golfo de México. *Intersticios Sociales, 10*(20), 193–224.

Sampieri, R., Fernández Collado, C., & Baptista Lucio, M. P. (2014). *Metodología de la investigación* (6th ed.). Mc Graw Hill.

Sánchez-Matías, D., y Ramirez Pacheco, A. A. (2023). Educación En La Reserva de La Biósfera Pantanos de Centla, Tabasco. El Caso de Los Jóvenes de Quintín Arauz. Maya America: Journal of Essays, Commentary, and Analysis, 5(1). https://doi.org/10.32727/26.2023.12.

Sánchez-Sánchez, D. (2020). *Palos Altos Entre La Muchachada y La Juventud: La Condición Juvenil Rural En Una Comunidad Ranchera de Jalisco [Tesis Doctoral en Desarrollo Rural].* Universidad Autónoma Metropolitana Xochimilco.

Uribe-Iniesta, R. (2016). *Tiempos y Procesos En La Constitución de Un Espacio Regional: El Caso Tabasco*. Centro Reg. Ciudad de México, Universidad Autónoma de México, Centro Regional de Investigaciones Multidisciplinarias.

Velázquez-García, M. A. (2016). Recursos del Estado mexicano contra los movimientos sociales: las distintas manos de leviatán. *Estudios sociológicos, 34*(101), 247–272. https://doi.org/10.24201/es.2016v34n101.1448

Chapter 10
Digital Alphabetization in Small-Scale Fishing in Sánchez Magallanes, Tabasco

D. E. Ramos-Muñoz, D. Matías-Sánchez, J. R. Ruiz-Blanco, and O. J. Ruiz-Blanco

10.1 Introduction

This chapter examines the case of *Sánchez Magallanes,* Tabasco, as part of a broader study that explores how fishing communities in Mexico engage with information and communication technologies (ICTs) (Ramos-Muñoz et al., 2022). The study focuses on three key dimensions: (1) how fishers utilize, evaluate, and perceive ICTs in their economic activities; (2) the challenges they face in implementing such technologies; and (3) the identification of pathways for digital literacy that can promote their integration into the fishing value chain.

The central research question addressed here is: How do fishing communities in *Sánchez Magallanes* incorporate digital skills and socio-digital networks into their livelihoods? To address this, the study examines two key aspects: the level of internet access and local infrastructure, as well as the socio-digital practices that either facilitate or hinder the adoption of ICTs in fishing activities.

D. E. Ramos-Muñoz
El Colegio de la Frontera Sur, Villahermosa, Villahermosa, Tabasco, México
e-mail: dramos@ecosur.mx

D. Matías-Sánchez (✉)
Centro de Investigaciones Interdisciplinarias en Ciencias Sociales, Universidad Veracruzana, Minatitlán, Veracruz, México
e-mail: damatias@uv.mx

J. R. Ruiz-Blanco
El Colegio de la Frontera Sur, San Cristóbal, San Cristóbal de Las Casas, Chiapas, México
e-mail: jaime.ruiz@ecosur.mx

O. J. Ruiz-Blanco
El Colegio de la Frontera Sur, Chetumal, Chetumal, Quintana Roo, México
e-mail: oscar.ruiz@ecosur.mx

A. Espinoza-Tenorio et al. (eds.), *The Fishing and Oil Industries in Coexistence*,
https://doi.org/10.1007/978-3-032-02621-7_10

A qualitative methodology was employed to identify the digital competencies acquired and developed through fishing practices in *Sánchez Magallanes*. We also analyze the use of ICTs and the obstacles encountered by fishers in applying them to their work.

The findings are organized around the social relations, exchanges, and networks within the community that enable real-time interactions, content creation, profile sharing, and participation in virtual communities that support fishing activities (Morduchowicz, 2008; Winocur & Sánchez, 2015). These dynamics are examined through the lens of van Dijk and van Deursen's (2014) framework on digital skills, which conceptualizes digital literacy as a key component of social, economic, and political inclusion in contemporary digital societies.

We argue that the motivations for and conditions of digital adoption vary among social groups, shaped by differentiated access to infrastructure and opportunities (Heeks, 2022). Moreover, the development and use of digital skills depend significantly on who uses the technologies and for what purposes.

A distinctive characteristic of *Sánchez Magallanes* is the significant presence of the petroleum industry. This industrial context emerges in the analysis as a factor that both complicates and enhances digital engagement, particularly by increasing fishers' safety and making visible the socio-environmental impacts of industrial activity.

10.1.1 Background

Numerous studies have emphasized the relation between information technologies and small-scale fishing (Mazuki et al., 2020). In general, they present the relation between the utilization of specific technologies (GPS, video probes, location sensors) and their effects on captures or the logistics of fishing (Lopez-Ercilla et al., 2021; Moran et al., 2017). Our study stands out because it does not emphasize the relationship between ICT and its impact, but rather identifies the processes of digital alphabetization that promote greater utilization of ICT in fishing.

Our theoretical approach is structuralist, as this enables us to understand societies as multidimensional in terms of their digital capabilities, a reality that can lead to inequalities or functional diversity within a given socioeconomic context, with effects that can yield differentiated social outcomes (Heeks, 2022; Warschauer, 2004). In this view, digital alphabetization is a social ability that individuals employ to find, evaluate, utilize, and share technological information and the Internet, understand digital security, and develop the capacity to critically analyze digital information (OpenAI, 2024). This phenomenon has been addressed from various perspectives, initially to refute the dichotomic idea of the existence of "digitally literate" and "digitally illiterate" people (Gunkel, 2003; UNESCO, 2023). Our study contrasts with this by analyzing a continuum comprising distinct levels of alphabetization and the physical and social structures in which inequalities emerged within the study group (Tinmaz et al., 2022; Vogels, 2021).

In this analytical context, our interest centers on the three principal areas that fishers in *Sánchez* exploit: lagoons (LF), the coast (CF), and the open sea (OSF). To analyze alphabetization, we follow van Dijk and van Deursen (2014) and UNESCO (2023) as a function of four axes:

1. Motivation: the attitudes, intentions, and social support that exist to utilize digital tools in each context.
2. Physical conditions of access to the digital world.
3. People's level of formal education—primary, middle, high school, university—and the abilities required to: (1) operate a particular digital device; (2) understand the information obtained; (3) communicate the information a device provides; (4) create content; and (5) generate strategies based on that information (van Dijk, 2005).
4. De facto access to use the devices studied.

These axes structure our methodological proposal to define the continuum of digital alphabetization in the small-scale fishing sector in *Sánchez*.

10.2 Methodology

Our qualitative methodological strategy consisted of a survey, a focus group, and participant observation. The survey consisted of four sections: identifying the ICTs that fishers utilize, their experience with telephone or Internet signals in the locality, how ICTs are utilized when fishing, and recording the information transmitted from three cellphone applications: WhatsApp, Facebook, and *Llamada*. In the focal group,[1] we documented the principal problems in fishing identified by the fishers themselves and the potential for integrating digital devices to resolve those challenges with the same three applications. This group consisted of 11 participants and users of digital social networks, specifically individuals trained in using and managing digital technology. Finally, we applied five surveys to complement this information.

Participant observation was conducted during fieldwork from April 18 to 23, 2022, following a guide that focused on observing access to the Internet, existing infrastructure, and the difficulties that fishers face as they strive to incorporate ICT into their activities. This information was recorded in fieldwork diaries kept by four individuals.

We analyzed the utilization of ICT in three moments of fishing activity: pre-capture, during capture, and post-capture. This division was deemed adequate to clarify the distinct uses of devices by type of fishing (fishing arts) and to distinguish digital conditions along the value chain.

---

[1] Informed consent was obtained from the focus group participants.

Table 10.1 Social and digital ethnographic matrix (elaborated by the authors)

Analytical categories	Descriptive categories	
Types of fishers	Lagoon fishers (LF)	
	Coastal fishers (CF)	
	Open sea fishers (OSF)	
Digital capacities	Position	Coordinates of fishing areas Security
	Communication:	Telephone (CT) • Security/Family • Supply chain/Work Social networks (SN) • Supply chain • Training • Prestige • Security
	Recording	Safety/Risks Fishing resources/fishing arts
Social and digital alphabetization	Digitally quasi-illiterate	
	Digitally literate	
	Advanced digitally literate	

With the data from the focus group, the researchers' diaries (participant observation), and surveys, we constructed an analytical matrix (Table 10.1) that allowed us to: (1) classify the information obtained, (2) identify the fishers' digital abilities, and (3) propose levels of digital alphabetization for these users-fishers. A broader description of this matrix is presented in the conclusions. Before the fieldwork, the research team introduced themselves to local authorities and fisher organizations.

10.2.1 The Setting

Since *Sánchez* has access to the sea and coastal lagoons, three types of fishers have emerged: lagoon, coastal, and open sea. This has led to the coexistence of distinct fishery production systems and the involvement of people of various ages, both sexes, and, on occasion, different ethnic groups (Alcala-Moya, 2003). Production is small-scale, characterized by limited assets and low productivity, conditions that create economic and social barriers, impeding fishers' access to markets and economic opportunities, while also affecting their incomes and overall well-being (FAO, 2024).

Sánchez is the fishing locality closest to Tabasco's border with Veracruz. The landscape includes three lagoons: Del Carmen, Pajonal, and Machona (Hernández et al., 2008). Its importance in fishing is reflected in the 74 economic units registered in the National Statistical Directory of Economic Units (DENUE, by its

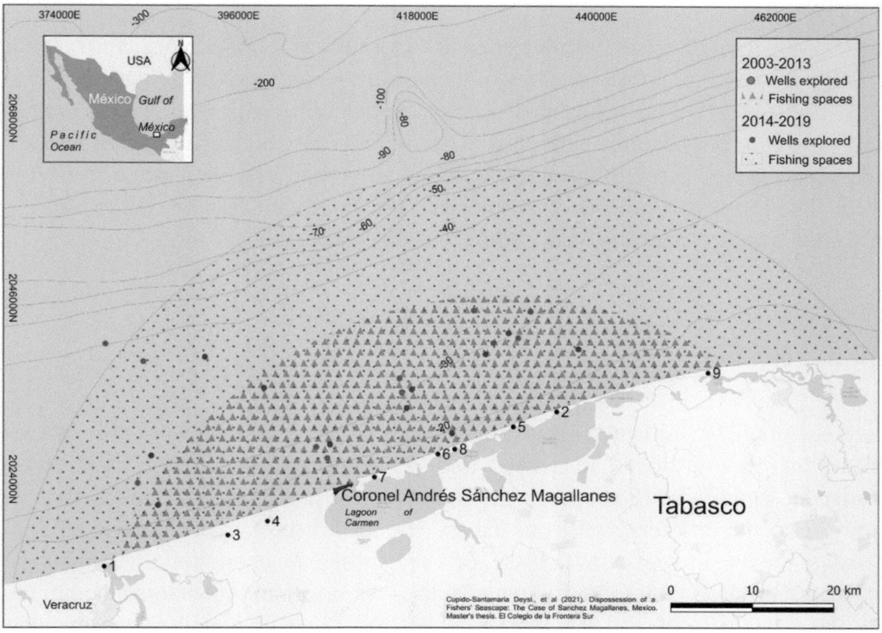

Fig. 10.1 Small-scale fishing zones and areas with petroleum activity on the coast of *Sánchez Magallanes*, Cárdenas, Tabasco (Modified from Cupido Santamaria, 2021)

Spanish acronym),[2] which are dedicated mainly to oyster farming and capturing scaly fish.

Sánchez lies near the first shallow water oil platform erected in Mexico, which was later dismantled. Today, following the 2013 energy reform, it has the first platform that extracts oil under private international initiatives (Comisión Nacional de Hidrocarburos, 2019), reflecting the close relationship that now exists between two key industries—fishing and petroleum—as they share the same workspace: the sea. According to studies by Salazar de la Cruz et al. (2020), De los Santos González et al. (2023), and Cupido Santamaria, (2021), this relation is characterized by fragility and disequilibrium that disproportionately affect fishing.

The reduction of the contiguous fishing zone evidences this fragility. A consequence of efforts by the Mexican State to provide security for the petroleum platforms (Cupido Santamaria, 2021; Fig. 10.1), this restriction boosted open sea small-scale fisheries (OSF) and promoted the integration of digital devices—echosounders, video probes, GPS devices—to register the coordinates of prime fishing sites and increase safety for fishers as they exploited new, more distant, marine horizons.

[2] *Instituto Nacional de Estadística y Geografía* (INEGI) 2024, "Mapa de la Clasificación Nacional de Actividades Económicas por Entidades Federativas," INEGI, consulted 17/04/2024 at https://www.inegi.org.mx/app/mapa/denue/default.aspx.

The LF and CF groups have distinct working conditions. Since the former focus their production on oyster farming and catching mollusks, crustaceans, and fish like mullet, they require little in the way of technification and work an average of 3–5 h/day. The coastal fishers center their activity on capturing scaly species like sawfish and snook, so they require fiberglass boats with motors powerful enough to resist the ocean waves. Regarding their workdays, we found that they range from 6 to 12 h. The longest work times correspond to the OSF, who capture sharks, sailfish, tuna, red snapper, and wahoo, spending 2–7 days on each trip.

10.3 Results

In this section, we initiate our analysis of fishers' digital social network motivations, examining their attitudes, intentions, and social support for utilizing digital tools under the conditions of access to the digital world provided by *Sánchez*, without comparing the different groups of fishers. We then discuss the respective abilities we registered in these groups (LF, CF, OSF) regarding the operation of several apparatuses, as well as their capacity to understand the information obtained, communicate it, create content, and generate strategies utilizing it. To conclude, we elucidate their de facto utilization of digital instruments.

10.3.1 Motivations

Although the lagoon fishers have the best telecommunications and electrical conditions during their work (Fig. 10.2), they perform their tasks with minimal digital support. Oyster farming, for example, begins in laboratories—handling larvae—before the fishers head out in small motorboats to seed the larvae in the lagoons using metallic or plastic traps, tend to them, and, finally, harvest their products. Their activity is manual, mainly, whether in the laboratory or the lagoon. All the women and men who participate are older adults. Their activities require knowledge of local ecological dynamics, but we found no records of digital education among them. Moreover, the incentives for utilizing digital technologies are limited in this regard. A leader of a cooperative has a computer used for accounting operations, creating presentations and projects, and communicating with the state government. This equipment, along with the skills he has developed in its use, has enabled him to maintain his position of leadership. The use of cellphones in the pre-capture and capture stages is described in the LF row of Table 10.2.

The CF operators capture fish in waters up to 20 m deep off the coast. Their work organization requires three to four individuals, often family members. Ensuring good communication is especially important for them during the pre- and

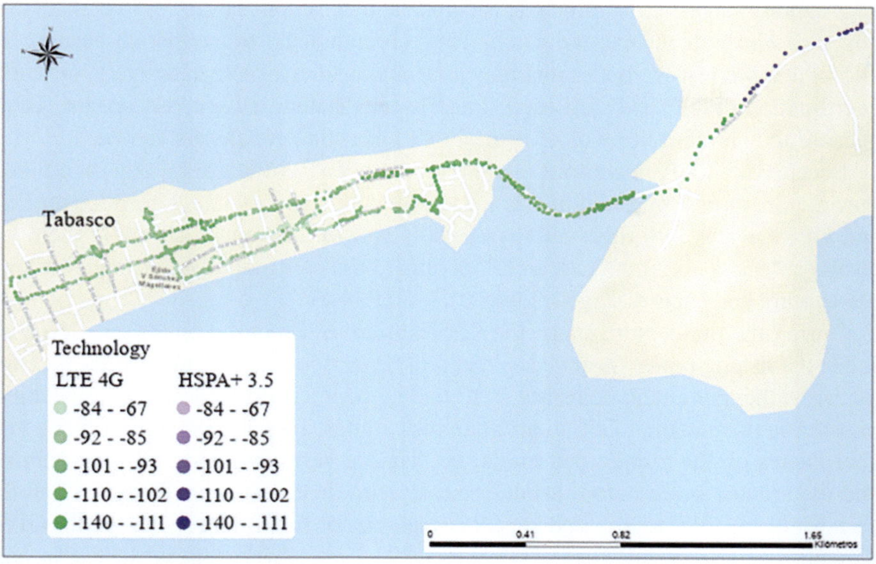

Fig. 10.2 Transects of mobile Internet service by technology type and the intensity of the Telcel signal (color tonality)

Table 10.2 Digital abilities in the three stages of small-scale fishing based on the ethnographic matrix (elaborated by the authors)

Stage		Pre-capture	Capture	Post-capture
Type of fisher	Device	Digital abilities		
LF	Cellphone	NC[a]: supply chain/ security TC: supply chain	Register: resources of fishing/ fishing arts	
CF	Cellphone	NC: supply chain/ security	Register: resources of fishing/ fishing arts TC: security NC: entertainment	TC: security/ supply chain
OSF	Cellphone	NC: supply chain/ security	Register: resources of fishing/ security NC: security	TC: security/ supply chain NC: security/ supply chain
	GPS	Position: register the coordinates of fishing points	Position: register, identify, and verify coordinates/ security	
	Video probe		Position: verify fishing points	

[a]Note: *NC* network communication, *TC* telephone communication

post-capture stages; for example, to secure adequate supplies of ice or to contact the buyer(s) who will receive the catch. They communicate by cellphone because it allows them to resolve issues that may arise about safety, product delivery, or work coordination (Table 10.2, CF row). Interviewees stated that they do not use GPS because they prefer to rely on terrestrial references that are always visible.

Because the OSF, in contrast, are out on the open sea for several days at a time, they require more ample equipment and coordination, so they integrate video probes and GPS devices into their activity as tools and technologies that allow them to ensure at least a minimum capture. Both apparatuses are employed primarily in the pre-capture and capture stages (Table 10.2, OSF row).

During the pre-capture stage, the GPS is used to detect the boat's position and select the fishing points (*pesqueros*) that will be exploited. Here, the fishers use two methods: the first entails selecting 5–10 fishing points from their fishing logs[3] after registering them in their GPS to program their routes. This allows them to verify the coordinates on the plotter, determine the distance between distinct fishing points, and then plan a route before heading out to sea. Their routes are designed before each foray. No OSF uses cellphones or computers for support, even though the GPS they utilize (e.g., Furuno) can connect to a PC. The main reasons why they do not integrate this information are that they lack knowledge of how to utilize this function and the investment required to acquire the necessary digital skills. Other uses of the GPS include selecting a specific fishing point from their log, programming the site into the device, and, after probing it, determining whether it would be advisable to visit and text another point to obtain the best possible catch.

The GPS is then employed exclusively to explore and register new fishing points. The video probe, in contrast, is used when the fishers are at the capture point to verify the presence of products. This is the most complex ability, so it is performed by the boat's captain. The probe has two cables: one is connected to the energy source (a battery), and the other is connected to the scanner. This apparatus is turned on to perform "probing," which entails placing the scanner at the fishing point and sailing around that point while monitoring the screen to detect the shapes of fish banks or refuge zones.

It is essential to add that these fishers do not store the coordinates of new fishing banks in their GPS devices; indeed, they erase the information after each trip. In their view, this allows them to protect their data from the prying eyes of third parties. This increases the value—economic and emotional—of their fishing logs, as evidenced by one fisher who declared: "[…] when they find a point they spend gasoline, spend fuel, sometimes they'll be there all day […], sometimes they don't find the point until the afternoon, and [maybe] have some success, so they program it, [and] each one protects his points…" (PMS, *Sánchez Magallanes*, 2022).

[3] The fishing log is a ledger or register where fishers write down important information on their fishing trips, including details like the species caught, climatic conditions, the location of fishing banks, the techniques utilized, and any other observation that is significant for the individual fisher.

10.3.2 Conditions of Access

Table 10.3 outlines the observation guide used to assess the level of Internet access and infrastructure in *Sánchez*.

During our field trips through various areas around *Sánchez*, we observed that some Internet cable installations were on the verge of collapse, and that the electricity supply, which originates from the town of Benito Juárez, suffers intermittent blackouts that disrupt internet service (both cable and wireless). This supports the testimonies of residents who characterize local connectivity[4] as "deficient."

Table 10.3 Type and quality of telecommunications infrastructure (elaborated by the authors based on the observation guide)

Telecommunications technology	Infrastructure (tangible)	Description
Internet cable (Fixed)	Posts	Greenish-brown wooden pole with a grayish-black concentrator (hub; Connection points where telecommunications signals like Internet and telephone service are grouped and distributed.) in the upper part with numerous cables. This concentrator is usually installed by the Telmex (Mexico's most important company for landline telephone service, Internet (trademark: *Infinitum*), and television.) company Gray concrete pole (for electricity) with a grayish-black concentrator fastened to one of the cables. This concentrator is usually installed by the Megacable (Mexican company that offers cable TV, Internet, and telephone service.) company
	Traditional Internet cable	Black cables with three copper wires that lead from the poles set up by Telmex Gray coaxial-type cables provided by Megacable
	Fiberoptics	Special thin cables labeled as "Fiberoptics," also identified by boxes with orange/yellow markings mounted on electricity or telephone poles
Internet without cables (Mobile/Wireless)	Towers	Pyramidal metallic structures equipped with telecommunications devices on the roofs of homes, schools, and government offices
	Antennas/Receivers	Apparatuses placed in the upper section of a tower, usually placed on the roofs of houses (WiFi)
	Telecommunication tower	White and red pyramidal metallic structures equipped with dishes, antennae, or cables that run from the lower part toward the upper part

[4] Refers to a device's capacity (computer or telephone) to maintain a constant Internet or telephone connection.

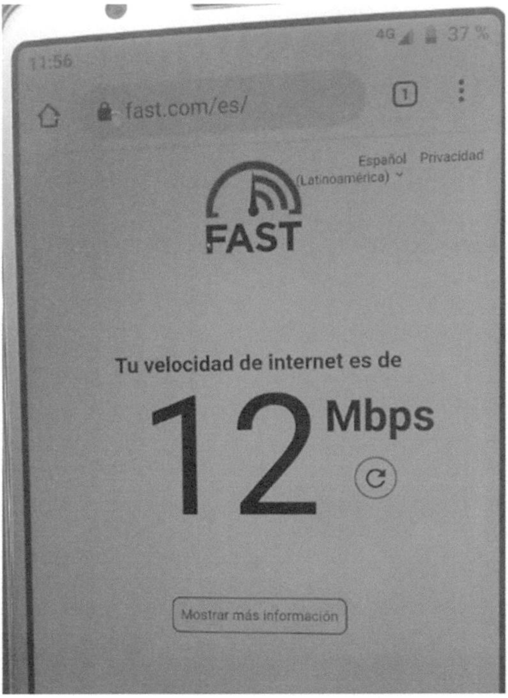

Fig. 10.3 Highest test result for mobile Internet velocity

To validate the connection velocity of each type (mobile, fixed), we consulted the website *fast.com* in the Chrome browser installed on the base apparatus.[5] We registered the mobile velocity using a cellphone with an Android operating system on the Telcel network. Two speed tests were conducted, one in downtown Sánchez and the other on the beach (playón), where the LF disembarks. The maximum velocity was recorded on the beach at 12 Mbps (Fig. 10.3), while the maximum reading in the town center was 1.7 Mbps.

These results show that while people on the beach can experience fluid operations using resources and platforms for educational, entertainment, and communications purposes (YouTube, Facebook, WhatsApp, Google), the area lacks the physical conditions required to utilize them efficaciously; that is, electrical infrastructure and adequate furnishings (benches, roofed areas). Users in the city center, in contrast, can access more comfortable spaces, such as the town square; however, the slow, intermittent service lessens their motivation and provokes frustration in using technology.

[5]The base device used was a Motorola smartphone, model G6 with 4G LTE technology (maximum theoretical download velocity = 300 Mbps) and WiFi Dual 2.4 + 5 GHz (maximum theoretical download velocity = 450 Mbps).

(a) (b)

Fig. 10.4 Digital Infrastructure in *Sánchez*: (**a**) The ENI company's telecommunications antenna (orange arrow); (**b**) point-to-point antenna (yellow arrow)

For the fixed velocity, we also chose two test sites: a hotel and a cellphone repair business. The results were velocities of 1.5 and 10 Mbps, respectively, for the companies Telmex[6] and JG Networks. We found that the latter service provides stability through fiber optics and point-to-point access via an antenna installed outside the town.

Figure 10.2 shows the sites we visited and the transect of mobile Internet service we tested. Coverage by Telcel 4G (LTE) is almost complete there, but as we move further from downtown, the intensity decreases, and the technology changes to HSPA+ (H+). We also observed four telecommunications antennae (Figure 10.4b), which were attributed to Telmex, and one antenna in the Gulf of Mexico that was associated with the *Ente Nazionale Idrocarburi* (ENI) petroleum company (Figure 10.4a).

Table 10.4 presents the characteristics of the digital apparatuses integrated into fishing activity in *Sánchez*. To respect people's privacy, we did not request specific data from their cellphones (e.g., brand and capacity). We verified that fishers tend to purchase cellphones with only basic functions ("dumbphones"): making calls, sending text messages, low-resolution cameras for photos, and, in some cases, direct

[6]Tenants stated that *Telmex* service comes from a "satellite" in Agua Dulce, Veracruz, but we could not verify this.

Table 10.4 Technological devices used by fishers in *Sánchez* (elaborated by the authors, based on survey data)

Devices	Characteristics
Cellphone	Types of mobile phones: basic (classic) and intelligent (smartphones) Telephone company: *Telcel* Investment required: MXN 100 every 2 weeks
GPS	Brand: *Furuno* Model: 32/33
Video probe	Brand: *Furuno* Model: 4100/6100

access to social network services, such as WhatsApp and Facebook. We did, however, detect an emerging trend to replace those devices with smartphones that feature touchscreens and some computer functions, including connectivity, storage capacity for information, and multitasking capabilities.

In summary, the conditions of access to the digital world in Sánchez in terms of infrastructure and ICT equipment can best be described as precarious.

10.3.3 People's Abilities

Table 3.2 outlines the integration of devices, people's digital abilities, and the purposes for which fishers utilize digital technology in their fishing activity, according to our three-stage division: pre-capture, capture, and post-capture. The cellphone is the only apparatus incorporated into the activities of all three types of fishers (LF, CF, OSF), and the CF and OSF utilize it in all three stages, thanks to three digital abilities: recording, social networking (SN), and telephone communications (TC). The differences we found lie in the objectives that fishers pursue in each stage of their work.

Despite differences in the specific aims of each type of fisher, we identified patterns in their intentions regarding cellphone use. In the pre-capture stage, all fishers utilize SN because it aids in two aspects. First, security personnel verify climatic conditions to determine if they are optimal before heading out to sea to begin their daily capture. These consultations are conducted on the Facebook profiles of local radio stations or in the WhatsApp group that the captain of the port set up for the fishers. Second, for supply chain support by organizing fishing forays through personal or group messages—using the app in WhatsApp—between the captain and crew members, and by receiving or confirming orders with buyers of fishery products.

In contrast, in the capture stage the digital ability of recording varies between the CF and OSF, since in addition to taking photos and videos of fishery resources or fishing arts to share through SN to achieve recognition—or to "show off"—the CF and OSF use this resource to collect evidence of irregularities committed by the petroleum companies that operate[7] along their coasts, such as crude oil spills, sonorous submarine contamination, and damage to their fishing arts. As one fisher told us: "… because we've always had the problem that we hold meetings with businessmen [and tell them] hey! There's a ship out there [but they answer] Really! Who knows who it belongs to? […] But when you have a photo […] logically the owner has to show his face…" (PMS, Sánchez Magallanes, 2022).

The importance of "showing off" through their recordings and the SN was evidenced by one fisher in our focal group who used the phrase: "… cause you take the photo [of fishing], upload it, and those who have family on the other side [United States], well, they see it right away, that's how it's done, through WhatsApp,… we call it pride [laughter]" (Focal Group, Sánchez Magallanes, 2022).

We observed a similar phenomenon in CF and OSF during the post-capture stage, as they utilize cellphones, with either SN or TC for two purposes: (1) security since they can inform family members of their return, so if no message is sent, they would know that some incident occurred during the capture. These calls go to wives, partners, or mothers, and (2) the supply chain, as fishers communicate with ice suppliers or the warehouse supervisor to report the amount of ice they require, and the space needed to conserve their catch.

The abilities described to this point—as displayed in the ethnographic matrix (Table 10.1)–cover one of van Dijk and van Deursen's (2014) references for alphabetization; namely, operating the digital device. Regarding the fishers' ability to understand the information that these devices collect, and the capacity required to communicate that data (van Dijk, 2005), we learned that this remains a significant challenge. This gap was confirmed in the Focal Group when one OSF spoke of the practice and importance of registering GPS data obtained in their fishing logs: "… I have eight logs, each one with a hundred pages, all of them full; I carry them in my bag, where I also keep my schedule … This programming [notebook]… well just imagine… there are a hundred pages [he showed the log], how many [fishing points] can you put on a page… I have a world of programs… a whole life history, my family's sustenance, my struggle" (Focal Group, *Sánchez Magallanes*, 2022).

The log, moreover, exemplifies fishers' fears about integrating their activity into the digital world: "… You know, in a moment [someone] could steal all that information. Imagine how many liters of gasoline, how many days under the hot sun… If there's resistance, it's precisely because of that, [some fishers] say what'll happen if they rob me or I lose [the GPS]. I just feel more secure with my log, you know?" (Focal Group, *Sánchez Magallanes*, 2022). Clearly, the fear of losing valuable information impedes these fishers from transferring information from the GPS to other apparatuses.

Turning to the fishers' ability to create content and generate fishing strategies based on digital information (van Dijk, 2005), we detected only an incipient willingness among the OSF. One fisher expressed his desire to accede to the market through his social networks: "See, that would be really good, right when you are

[7] Example, ENI.

uploading it [to the social networks]… a lot of guys have this and they say: hey, I need some of that, or I have three tons of bonito [a fish species] here in Campeche. Okay, bring it along or send me a bit… I mean, that would be a real benefit…" (Focal Group, *Sánchez Magallanes*, 2022). Creating content from his social network would allow that fisher to promote sales of his product before he returns to port. We further identified that these fishers have ideas for generating value in their fishing arts through digital images, as these comments by one fisher reveal: "… our fishing needs promotion. Why?... because in reality the product is sold very cheap… You know, people [enjoy] seafood, but don't know all the [activities] behind it. A huge effort; it's something that never ends…." (Focal Group, *Sánchez Magallanes*, 2022).

While in reality these digital conditions—stable Internet access, capacity to communicate the information from GPS to other electronic devices, technical knowhow to create and share content—do not exist in Sánchez, some fishers imagine the possibilities that could emerge if they were able to achieve a more complete integration; that is, greater digital alphabetization.

10.3.4 De Facto Integration

Table 10.4 presents a preliminary summary of the digital artifacts currently integrated into small-scale fishing activities in Sánchez.

The investment of USD 5 (MXN 100) for approximately 2 weeks evidences the fishers' interest in obtaining communication through cellphones and social networks. We must recall that the cellphone is the only digital device used by all three types of fishers, in contrast to the GPS and video probe, whose acquisition and use are directly related to the activities of the OSF, as they help identify fishing sites more effectively and increase personal security.

For the GPS and video probe, we observed a tendency among these fishers to purchase the Japanese brand Furuno, which specializes in navigation and maritime communication. Such is the fishers' interest in these devices that they travel as far as Ciudad del Carmen in the neighboring state of Campeche to make their purchases in the only store in Mexico that handles the Furuno brand: *Proesa Electrónica, S.A. de C.V.*

10.4 Discussion

In the area around *Sánchez*, we identified three digital abilities related to fishing activities that arise from the use of mobile telephones, GPS, and video probes.

- Location ability: fishers use digital devices to identify (e.g., GPS), verify (video probe), or record (e.g., GPS/video probe) the coordinates of key fishing points, but they also enhance their security, as they use the coordinates to chart routes that allow them to leave or access the pier with the intention of referencing points where there is a risk of accidents.

- Communication ability: a capacity acquired by using the cellphone for two types of communication: via telephone and social networks. Telephone communication is limited to conventional calls for two primary purposes: security and supply chain management. These calls are directed to buyers, cooperative members, permit holders, shippers, clients, or even accountants. Communication through social networks, in contrast, involves two apps: WhatsApp and Facebook. WhatsApp supports the supply chain by facilitating the exchange of information on climatic conditions and safety incidents. At the same time, Facebook is used to search for weather-related information (for safety), for entertainment, and to generate content (images) showcasing the resources obtained during fishing.
- Recording ability: based on capturing images and videos with a cellphone. It provides documentation—evidence—of their fishing resources and arts, as well as the risks and accidents that fishers experience during their daily coexistence with oil platforms.

If we compare the three abilities found in these fishers (LF, CF, OSF) to those described by van Dijk and van Deursen (2014), we must conclude that the level of digital alphabetization among them is low, as they only demonstrate one ability: operating the digital device. This finding underscores the importance of considering people's motivations, the existing physical conditions—access and infrastructure—and the value of using each type of apparatus in our analyses.

This case highlights the challenges faced by fishing communities in accessing digital materials, particularly those that overlap with petroleum infrastructure, as this restricts their fishing zones and forces them to venture farther into the open sea. Despite the precarious quality of internet access and infrastructure, as well as the low motivation to integrate technology, these fishers have achieved a certain level of understanding in using digital apparatuses to counteract insecurity in accessing their coasts, which is due to the presence of the oil industry.

In reality, as the ethnographic matrix suggests, the levels of understanding in the use of these devices vary with the distinct working conditions of the three types of fishers who operate in *Sánchez*. As a result, each type of fisher (LF, CF, OSF) has attained a distinct level of digital alphabetization, which we have classified as:

- Quasi-alphabetization: the capacity to distinguish some functions of a digital application or device and then act based on this knowledge. This action converts fishers into users and equips them to perform searches to obtain information though this is limited to the social networks to which they have access: WhatsApp and Facebook.
- Medium alphabetization: the capacity to distinguish some functions of one or more digital applications (climate, alarm, WhatsApp, Facebook) or devices (GPS, video probe, cellphone) and then act on the basis of that knowledge. This includes conducting searches for information at levels beyond social networks, such as WhatsApp and Facebook, and discerning the information obtained through or available from apps or devices.

- Advanced alphabetization: this level combines abilities and performance that utilize and diversify the functions of apps or devices. In this sense, it represents the ideal type of "digitalized fisher" who can:
  - Distinguish the functions of the devices and apps used in fishing.
  - Conduct searches to obtain information that supports production.
  - Discern the information and diversify the use of available devices, such as cell phones, GPS devices, and video probes.
  - Generate content to solve production problems in fishing.

In Fig. 10.5, we propose a classification where we consider that the LF and CF barely achieve the ability to navigate at the medium level, while, at least in some cases, the OSF approach the advanced level since they use more devices and recognize the importance of integrating them to promote their fishing products and arts.

It is not our intention, in proposing this classification, to describe these fishers as homogeneous, unmovable, or lineal subjects; instead, but only to highlight aspects that permit identifying the digital gaps that they confront and the areas that require attention for them to achieve a greater understanding and integration of digital technology in their fishing activity.

The fishers' logs, where they register key information on fishing, exemplify their fear of technology. While they have integrated the cellphone into almost all their activities, the idea of replacing their logs with a digital tool is not even discussed. Fishers in *Sánchez* operate under conditions of deep insecurity (De los Santos González et al., 2023; Cupido Santamaria, 2021), a condition that increases exponentially in the digital context. Despite their fears, all these fishers use technology to some extent. Our findings reflect how different working conditions and exposure to technology affect the distinct types of fishers' motivation to attain digital alphabetization, as they seek greater efficiency in their captures and increased security in their activities.

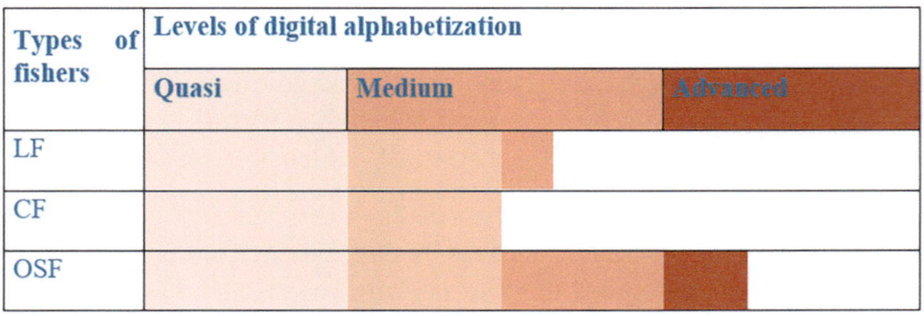

Fig. 10.5 Typology of social and digital users in *Sánchez* (elaborated by the authors)

10.5 Conclusions

We consider that the motivations generated through each type of fishing open digital possibilities. Integrating new devices does not mean going against tradition; for instance, their logs can indeed coexist with GPS service. However, there are elements that these fishers have not considered; for example, using technology to search for lower fuel prices, utilizing GIS systems to chart fishing routes, analyzing statistics to better understand the ecology of their catches, or incorporating remote sensors to redesign their fishing methods. This leads us to propose that to expand their horizons, these fishers should incorporate new digital practices to improve (1) the safety of their boats; (2) the efficacy and traceability of their captures; and (3) the ecological conditions of their fishing arts.

On the topic of climate and risk management associated with it, additional research is required to address two key questions: What systems for climate alerts do fishers envision? Moreover, is it necessary to develop climate models or early warning systems for small-scale fishers?

The process of digital alphabetization requires access to digital artifacts, so there is an urgent need not only to maintain but also to upgrade the infrastructure of ICTs along the coasts of *Sánchez*. The deficiencies we recorded require the attention of all levels of government.

Acknowledgments The national-level project was proposed and funded by EDF México, to whom we extend our gratitude for their financial support, methodological guidance, and suggestions regarding fieldwork locations, as well as for their efforts in facilitating various dissemination activities. Special thanks go to Berenice García Villareal. We are also deeply grateful for the enthusiastic support provided by Alejandro Espinoza Tenorio and Deysi Guadalupe Cupido de los Santos. The authors, however, bear full responsibility for the data presented herein. ChatGPT was used for translation (abstract, keywords, and acknowledgments); when used for definitions, they have been quoted.

References

Alcala-Moya, G. (2003). La pesca: ¿actividad de hombres y de mujeres? *Estudios del hombre, 17*, 163–183.

Comisión Nacional de Hidrocarburos. (2019). Reporte de producción ENI. https://hidrocarburos.gob.mx/media/2809/producci%C3%B3n-eni-julio-2019_300819_esp.pdf

Cupido Santamaria, D. G. (2021). *Despojo del paisaje marino a pescadores de pequeña escala en el puerto pesquero Coronel Andrés Sánchez Magallanes, Cárdenas, Tabasco, México*. [Master Thesis]. El Colegio de la Frontera Sur. https://ecosur.repositorioinstitucional.mx/jspui/handle/1017/2691

FAO. (2024). The state of world fisheries and aquaculture 2024. In *Blue transformation in action*. FAO. https://doi.org/10.4060/cd0690es

Gunkel, D. (2003). Second thoughts: Toward a critique of the digital divide. *New Media & Society, 5*(4), 499–522.

Heeks, R. (2022). Digital inequality beyond the digital divide: Conceptualizing adverse digital incorporation in the global south. *Information Technology for Development, 28*(4), 688–704. https://doi.org/10.1080/02681102.2022.2068492

Hernández, G., García, M., & Pérez, J. (2008). *Tecnologías de la información en la educación: Análisis y perspectivas*. Editorial Académica Española.

Instituto Nacional de Estadística y Geografía. (2024). *Censo de Población y Vivienda*. https://www.inegi.org.mx/app/mapa/denue/default.aspx.

Lopez-Ercilla, I., Espinosa-Romero, M. J., Rivera-Melo, F. J. F., Fulton, S., Fernández, R., Torre, J., Acevedo-Rosas, A. H., Velasco, I., & Amador, I. (2021). The voice of Mexican small-scale fishers in times of COVID-19: Impacts, responses, and digital divide. *Marine Policy, 131*, 104606.

los Santos González, D., Díaz Perera, C. C., Ángel, M., Ramos Muñoz, D. E., & Gama Campillo, L. M. (2023). Vulnerabilidad en la costa de Tabasco: Coronel Andrés Sánchez Magallanes. *Publicaciones E Investigación, 17*, 1. https://doi.org/10.22490/25394088.6806

Mazuki, R., Osman, M. N., Bolong, J., & Omar, S. Z. (2020). Systematic literature review: Benefits of fisheries technology on small scale fishermen. *International Journal of Academic Research in Business and Social Sciences*.

Moran, M., Bolivar, M. M., Verónica, G.-P., & Faruk, R. (2017). Influencia del uso de las TICs en los ingresos económicos de la pesca artesanal en Santa Rosa. *Salinas., 4*, 63–69.

Morduchowicz, S. (2008). *Los jóvenes y las pantallas: De la educación a la cibercultura*. Fundación Santillana.

OpenAI. (2024). *ChatGPT, prompt "what is the difference between digital divide and digital literacy?"*. https://chat.openai.com

Ramos-Muñoz, D. E., Matías-Sánchez D., Espinoza-Tenorio, A., Ruiz-Blanco J. R., Cupido-Santamaria, D. G., Ruiz-Blanco O. J., Mendoza-Martínez, M. P., Rabelo-Ávalos A. V., Oliveto-Andrade A. J. G., & Castillo-Domínguez S. (2022). *Diagnóstico de capacidades digitales en las comunidades pesqueras de pequeña escala en México, Reporte Técnico, Environmental Defense Fund, México*. https://mexico.edf.org/sites/default/files/DiagnosticoCapacidadesDigitales_0.pdf

Salazar de la Cruz, C. C., Zepeda-Domínguez, J. A., Espinoza-Tenorio, A., & Ramos-Muñoz, D. E. (2020). Governance networks in marine spaces where fisheries and oil coexist. *The Extractive Industries and Society, 7*(2), 676–685. https://www.sciencedirect.com/science/article/pii/S2214790X19301108

Tinmaz, H., Lee, Y. T., Fanea-Ivanovici, M., & Baber, H. (2022). A systematic review on digital literacy. *Smart Learning Environments, 9*(1), 21.

UNESCO. (2023). Organización de las Naciones Unidas para la Educación, la Ciencia y la Cultur. *Alfabetización: Promover el poder de la alfabetización para todos*. https://www.unesco.org/es/literacy

van Dijk, J. A. (2005). *The deepening divide: Inequality in the information society*. SAGE Publications. https://doi.org/10.4135/9781452229812

van Dijk, J. A. G. M., & van Deursen, A. J. A. M. (2014). *Digital skills: Unlocking the information society*. Palgrave Macmillan. https://doi.org/10.1057/9781137437037

Vogels, E. (2021). *Digital divide persists even as Americans with lower incomes make gains in tech adoption*. Pew Research Center. https://www.pewresearch.org/fact-tank/2021/06/22/digital-divide-persists-even-as-americans-with-lower-incomes-make-gains-in-tech-adoption/

Warschauer, M. (2004). *Technology and social inclusion: Rethinking the digital divide*. MIT Press.

Winocur, R., & Sánchez, J. (2015). Introducción. In R. Winocur & J. Sánchez (Eds.), *Redes sociodigitales en México* (pp. 9–15). Fondo de Cultura Económica.

Chapter 11
Technological Syncretism in Small-Scale Fisheries

J. Martínez-Amezcua, C. Monzón-Alvarado, and R. Armada-Ramírez

11.1 Introduction

Small-scale fisheries in Mexico constitute 70% of the fishing population, followed by aquaculture (20%) and deep-sea fishing (5%). We refer to small-scale fisheries practiced on board vessels smaller than 10 feet, whether in coastal waters, inland waters, or continental waters (SEPESCA, 2022). Technological tools can transform small-scale fisheries practices by influencing the interaction between people and nature. However, these tools are designed by specialists who live in different contexts than the potential users, each with their own "technological culture" languages, tools, and values (Quintanilla, 2005). For these innovations to be adopted, technology developers must understand both modern technology and traditional knowledge. This implies integrating different knowledge systems[1] in the design, transfer, and use of technological tools. In other words, the social appropriation of technology requires an intercultural dialogue among all the stakeholders concerned, which contributes to combining local knowledge with technical and scientific knowledge (Olivé, 2007).

To study the degree of intertwining between knowledge systems, in this chapter, we adopt the concept of technological syncretism (Berger & Azurdia, 2008), which

[1] Knowledge systems comprise the set of actors, practices, and institutions that organize the production, transfer, and use of knowledge (Cornell et al., 2013).

J. Martínez-Amezcua · R. Armada-Ramírez
Universidad Nacional Autónoma de México, UNAM, México, México
e-mail: rodrigoarmada@filos.unam.mx

C. Monzón-Alvarado (✉)
Secretaría de Ciencia, Humanidades, Tecnología e Innovación-El Colegio de la Frontera Sur, Campeche, México
e-mail: cmonzon@ecosur.mx

A. Espinoza-Tenorio et al. (eds.), *The Fishing and Oil Industries in Coexistence*,
https://doi.org/10.1007/978-3-032-02621-7_11

arises from the encounter between techniques and practices based on traditional knowledge and those produced by contemporary Western technologies. Technological syncretism is understood as "the intercultural, multilingual and multiethnic co-production of technological alternatives (machines) and techniques (from the body) for the transformation of the living environment" (Azurdia Bravo, 2013, p. 35). Knowledge co-production implies the joint and equitable creation of knowledge, where it is assumed that one type of knowledge is not superior to another. Technological syncretism yields a new, functional, and highly adaptable technological system, instrumental in building community resilience (Azurdia Bravo, 2013).

This chapter analyzes the social appropriation of "Pepe Bots," a chatbot developed through the "Fishing and Oil" project,[2] as well as the degree of knowledge articulation between small-scale fishers from Tabasco and Campeche and the team developing the chatbot (Fig. 11.1). The interdisciplinary team that developed Pepe Bots includes researchers, technicians, and students from marine sciences, systems engineering, ecology, and rural development. The first author in this chapter joined the project and the development team to conduct her undergraduate thesis in intercultural management. The second author was part of the development team and served as an advisor, as well as the third author. We adopt a critical stance on the concept of technological syncretism to identify its practical limitations. Although syncretism offers an attractive theoretical framework for studying knowledge co-production, it is necessary to question the extent to which knowledge articulation occurs horizontally and equitably. Epistemic hierarchies and technical decisions do

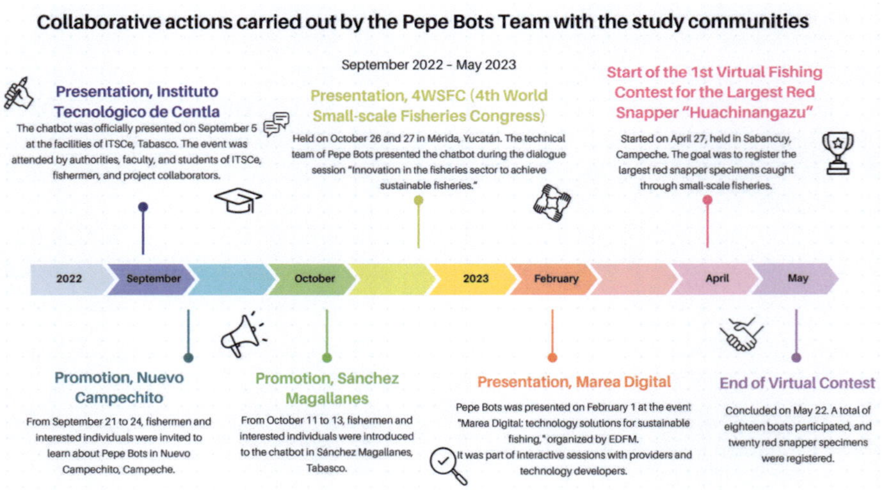

Fig. 11.1 Collaborative actions were carried out by the Pepe Bots Team with the study communities. Own elaboration

---

[2]The Challenging Coexistence of Coupled Social-Ecological Systems: The Fishing and Oil Industries in Sonda de Campeche. National Council of Humanities, Sciences, and Technologies (CONAHCYT).

not always reflect equitable participation. Thus, this chapter seeks to explore how fishers' knowledge is intertwined, strained, or subordinated in the production of technologies, and what tensions emerge between the ideal of syncretism and the realities of external technological innovations.

11.2 Methodological Considerations

We adopt a qualitative approach based on the thematic analysis of interviews and participant observation. A literature review was conducted on the social appropriation of technologies, technological syncretism, the co-production of knowledge, and the challenges faced by small-scale fisheries. We applied 21 surveys to chatbot users (local university students, small-scale fishers, cooperative leaders) to learn their opinions about Pepe Bots, initial impression and the usefulness they perceived in it, sixteen in-depth interviews to fishers to assess their knowledge and identify their fishing practices, and five in-depth interviews to the members of Pepe Bots development team and other developers of related technologies.

During the chatbot co-production process, between August 2022 and May 2023, several collaborative actions were carried out by the Pepe Bots Team with the study communities. These included workshops, pilot testing, and public activities, aiming to foster dialogue among potential users, developers, and other relevant stakeholders in small-scale fisheries (Fig. 11.1). Participant observation was conducted through these activities, and more specifically during four situations: (a) during the visits to the fishing communities ($n = 3$), (b) weekly virtual meetings with development team ($n = 31$), (c) forums and events where Pepe Bots was presented, and (d) the "Huachinangazu" contest, where participants from *Sabancuy*, Campeche contended to register the largest specimen of red snapper (*Lutjanus* sp.). Participants uploaded a photo through Pepe Bots, registering the dimensions and location of the fish. A total of 18 boats participated and 20 red snapper specimens were registered. This activity involved a 2-week stay in the community to support and promote this contest.

Participatory evaluation was not carried out directly by the chapter authors but was conducted on an ongoing basis by the chatbot development team. This evaluation included monitoring the functioning of Pepe Bots during weekly sessions, as well as the analysis of field inputs (interviews, user feedback, and usage logs), direct and indirect exchanges with users and other developers, and the experience accumulated during the "Huachinangazu" contest through observations and suggestions from participants and organizers. Although this chapter enables the documentation and analysis of the technological co-production process of Pepe Bots, this initial analysis must be complemented by future assessments to evaluate its impacts on small-scale fisheries within oil seascapes. The thematic analysis was organized considering the knowledge system, the stage of the process, and the different actions associated with the co-production process (Table 11.1).

Table 11.1 Operationalization criteria to study technological syncretism

Criteria	Description
Knowledge system	Academic, governmental, civil society, small-scale fisheries
Moment of the process	User presentations, workshops, forums, contest
Actions involved in technological syncretism/process of co-production of technological innovations	– Plan logistics for events – Review presentations at conferences and forums – Implement Pepe Bots dissemination strategy – Recognize possible areas of opportunity emerging from the use of Pepe Bots – Integrate feedback from fishers, users, and field work – Confirm technical issues (improvements and results) – Articulate collaborations with government, fisheries, and academia

Source: authors

11.3 Technology in Small-Scale Fisheries

Small-scale fishers use various technological tools that facilitate each stage of the activity, making it more efficient and accessible. These tools include mobile applications and digital devices with diverse functionalities and price ranges, adapting to the specific needs of fishers. The digital capabilities assessment in small-scale fishing communities in Mexico (ECOSUR-EDFM, 2022) analyzes the degree of digitalization in fishing zones by evaluating their access to the internet and the ability of their communities to acquire and use digital devices. The diagnosis reveals that the cell phone is the only digital device used by all types of fishers (lagoon fishers, coastal fishers, and open sea fishers) surveyed within the conducted ethnographic study. The acquisition and use of GPS and video sonar are aimed at identifying fishing sites and improving safety (ECOSUR-EDFM, 2022). The assessment also reveals that these devices are integrated into pre-capture, capture, and post-capture fishing activities; however, their adoption varies according to the type of fishing and the specific needs of each fishing group. The mobile phone is the most widely used device among lagoon and coastal fishers; it is used for registration, communication through social networks, and making telephone calls. On the other hand, GPS is used exclusively by offshore fishers, mainly to record fishing spots. While lagoon and coastal fishers can benefit from functions such as communication and data logging via mobile phone, offshore fishers rely more on GPS for specific activities related to navigation and the location of fishing spots (ECOSUR-EDFM, 2022).

The research conducted for this chapter revealed that, in addition to cell phones, GPS, and video sounders, there are other tools for fishing, such as echo sounders, radios with higher frequencies, or oceanographic charts. However, most of these have very high costs that make them inaccessible to most small-scale fishers, especially those not affiliated with a fishing cooperative (Table 11.2). We also identified

Table 11.2 Technological tools used in small-scale fisheries

Tool	Operation	Advantages	Disadvantages
WhatsApp	Means of communication with fishing team and family, to receive weather information	Rapid communication, with reliable and accurate information	The service network is exclusive for inland and coastal areas, not offshore. Payment of a fee
Phone compass	Consultation when there is a GPS failure	Alternatives in case of emergency	
Phone camera	Take pictures of the catch or anything relevant during the process	Recording of evidence	
GPS	Locate themselves at sea from the moment they leave and until they arrive, mark "fishing spots"	Facilitates fishers' work by assisting with location and navigation	Dependence on these tools can be dangerous in case of failure during the activity
Videosonde	With the help of a video camera they observe the seabed	Provides a detailed view of marine life and characteristics of the underwater environment	
Echosounder	Emitting and receiving sound to obtain detailed information	Mapping the relief of the seabed and for the identification and estimation of fish size	Expensive technological tools, accessible to few
VHF radio	Means of communication at sea	Improves safety at sea	
Oceanographic chart Garmin Navionics	Fishers can visualize the position of their vessel in real time in relation to the cartographic information provided by the chart	Informs on meteorology, water, seabed, harbors, buoys, points of interest, and other relevant info for fishing in oceanic waters	
Online logbooks	Record of catches, fishing spots	Information systematization	Some fishers do not trust the tool, fearing misuse of the information provided
Apps for traceability	The end user is not the fishers, yet this type of tool allows to know the origin of the products consumed	Tracing production, processing, and distribution of a product	

Source: authors

digital applications developed by civil society organizations, companies, and academic groups that aim to facilitate and automate specific processes in small-scale fisheries. For example, simplifying and streamlining the process of data collection in the field, eliminating the need to pass information from paper to a database. Information automation offers economic benefits by improving efficiency, reducing operating costs, and increasing productivity. Politically, it facilitates transparency by providing access to accurate information and streamlining administrative processes, strengthening fishers' governance. It also contributes to sustainability issues by helping to reduce environmental impact through optimized resource use and improved production system management, thereby promoting more sustainable practices.

From the interviews conducted with the developers of some of these tools, we document the most important lessons to consider in the process of designing technologies and strategies to promote their appropriation:

1. Integrate fishers in the tool design process. Listen and understand the needs and suggestions of the end users to adapt the technology to their specific requirements.
2. Adopt an integral and intergenerational approach that incorporates the traditional knowledge of fishers, as well as the knowledge of younger generations, in the design and continuous improvement of the tool, considering digital gaps and capabilities, as well as resistance to change.
3. Recognize diversity among fishing cooperatives in terms of objectives, strategies, governance, and capacities, as well as individual fishers.
4. Facilitate collaboration among fishers, businesses, and organizations in the implementation of responsible fishing practices and fisheries certification.
5. Promote spaces for face-to-face learning. Strengthen users' capacities for the appropriate use of technology, especially among those of advanced age.
6. Ensure data security. Ensure that fishers feel safe sharing their data and make transparent how that data will be used and protected.
7. Maintain a flexible and responsive approach to ensure the relevance and usefulness of the application, utilizing a continuous, adaptive, and feedback-driven process through regular meetings with fishers, which include direct input from users.

According to the interviewed developers, considering the previous elements contributes to the creation of more effective technologies for fishing activities. Olivé (2007) points out that excluding groups, such as indigenous peoples, from technological innovation processes must be avoided, and that guaranteeing their participation in the use of knowledge is a matter of justice. He argues that guaranteeing environmental justice would imply ensuring that technology is available to all in the fishing sector and contributes to sustainability. During the presentation of the Digital Capacities diagnosis, Rafael Ortiz, director of EDFM, referred to the mixed impacts that technology can have: it can create gaps due to heterogeneous digital capacities, but, with its disruptive potential, it can help the fisheries sector to achieve sustainability. This highlights the need to design technological tools integrating the diversity of perspectives, especially those of potential users, to promote the relevance and appropriation of technology (Ortiz, 2022).

11.4 Pepe Bots

11.4.1 Chatbot Origin

Pepe Bots is a non-profit technological infrastructure that seeks to merge citizen science and artificial intelligence (AI) to benefit small-scale fisheries by providing reliable information to users and facilitating their participation in ecological monitoring (Espinoza-Tenorio & Cupido Santamaria, 2021). Under this premise, the chatbot was conceived in 2021, based on workshops where fishers emphasized the importance of climate and market data for their management decisions (Table 11.3). The development team identified the usefulness of web mining to collect relevant public data to build a database of market prices of fishery resources and climatic conditions. This would allow gathering the information and autonomously responding to fishers' requests through the Pepe Bots chatbot.

The development team chose to implement messaging through the WhatsApp Business API (Application Programming Interface), which facilitates secure and reliable communication and data exchange, including the user's location (Espinoza-Tenorio & Cupido Santamaria, 2021). WhatsApp is recognized as one of the most important messaging platforms nationwide. Some benefits of using WhatsApp as a communication channel are the security of personal data, network optimization, multimedia file compression, free access, and low messaging costs. This platform was chosen for its accessibility and economy, as many fishers have phone plans that include this application for free.

According to the development team, in addition to its contribution to fishing, Pepe Bots can also serve as a tool for social organization, improve safety in fishing, and strengthen trust between citizens, the oil industry, and academia. They recognize that the application of AI to assess the impact of the hydrocarbon industry on small-scale fisheries is innovative in the field of ecology and small-scale fisheries (Espinoza-Tenorio & Cupido Santamaria, 2021). From its outset, Pepe Bots was designed to work with fishing communities, starting with its development as a chat

Table 11.3 Small-scale fishers' concerns and Pepe Bots potential contribution

Concern	Potential contribution
Need for accurate weather information	The chatbot works by compiling information from different public sources on weather forecasts at the regional, municipal, and local level and combines them into a single text
Need for accurate information on market prices	The chatbot provides information updated by the Secretary of Economy on market prices of the main catch species in the area

Source: authors

on WhatsApp. The advantage of providing information through this channel lies in the reliability of the sources and the consolidation of dispersed data in one place. This allows Pepe Bots to be a secure and centralized platform for accessing reliable fishing information. It also facilitates the feedback of data collected from fishers in the form of infographics, thus fostering citizen science and promoting co-production between the project and the fishing community.

A crucial aspect of the co-production process is the involvement of all stakeholders, including those who may be directly impacted by the technology in question. To achieve this, Olivé (2007) stresses that it is crucial for the public to trust science and technology as sources of information and problem-solving, while recognizing the limitations and value influences that technology can bring. The scientific and technological community has a responsibility to generate trust through transparency in its procedures, methodologies, and acknowledgment of limitations. The following events and forums constitute the major spaces for exchange and co-production of Pepe Bots.

11.4.2 Actions and Events in Pepe Bots Co-Production

The core of the co-production analysis presented in this chapter lies in the examination of the weekly virtual meetings with the development team ($n = 31$). It is crucial to analyze what happened in the virtual meetings before and after significant events (Fig. 11.2). The exchanges of ideas and decisions are often left in the shadows, but these meetings are the stage where agreements are reached and decisions are made

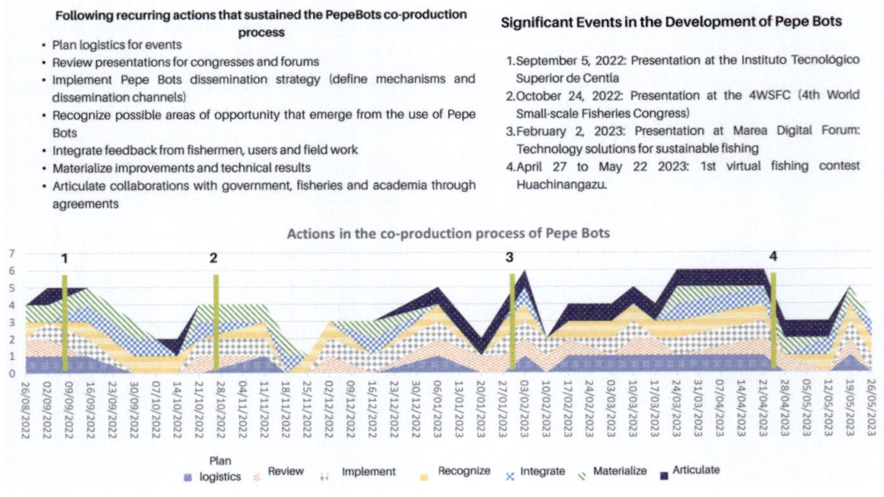

Fig. 11.2 Chronology of the recurring actions addressed in the weekly Pepe Bots meetings. Own elaboration

about improvements and next steps for the tool. They also provide a space for continuous learning within the Pepe Bots team. Through a careful listening of meeting records and their thematic analysis, we elicited the recurring actions that sustained the Pepe Bots co-production process. The following are recurring actions and significant events in the development of Pepe Bots:

1. Plan logistics for events: organization of events such as the official presentation of Pepe Bots and the "Huachinangazu" virtual contest.
2. Review presentations for congresses and forums: review of narratives and visual material for these participations by the development team and other project collaborators.
3. Implement the Pepe Bots dissemination strategy (define mechanisms and dissemination channels): how to present the information provided by Pepe Bots, and how to disseminate news of the tool updates to fishers?
4. Recognize the possibilities that emerge from the use of Pepe Bots: reflections on the range of opportunities that the tool has according to its updates.
5. Integrate feedback from fishers, users, and field work: how could Pepe Bots be improved according to the opinion of those who use it?
6. Materialize improvements and technical results: proposals and updates presented by the engineers in charge of the tool.
7. Articulate collaborations: agreements with government, fisheries, and academia that benefit the diversification of information.

Significant events were:

1. September 5, 2022: Presentation at the Instituto Tecnológico Superior de Centla (ITSC).

   Official launch of Pepe Bots at ITSC, in *Centla*, Tabasco, in an auditorium with students, local authorities, and fishers (*dissemination strategy*). After the presentation, adjustments were made based on the integration of user feedback, e.g., in the infographics to improve their attractiveness (*integrate*). One section was renamed "ConCiencia Ciudadana" to highlight the role of social participation (*concretize*). The development team was able to recognize that the most consulted sections were weather and prices (*recognize*). Reflections were made on how to increase the use of Pepe Bots, generating trust among fishers and exploring strategies such as agreements with other institutions (*articulate*). The project was disseminated in fishing communities (*implement*) and opinions were integrated to improve the project in virtual meetings (*acknowledge*).
2. October 24, 2022: fourth World Small-Scale Fisheries Congress.

   Pepe Bots was presented at the fourth World Small-scale Fisheries Congress (4WSFC) as a technological tool for small-scale fishers (*implement*). Presented as a communication platform designed to facilitate participatory decision-making, it seeks to diversify the market, conserve resources, and improve fishers' safety. After the congress, the infographics on market prices were the most popular. Pepe Bots registered 217 users after the event.

3. February 2, 2022: Marea Digital Forum: Technology solutions for sustainable fishing.

   Pepe Bots gained visibility at events such as Marea Digital (*implement*), receiving positive feedback and potential funding opportunities (*articulate*). Alongside, improvements to Pepe Bots were explored, including species price graphs and increased information on fishing heritage (*concretize*). Collaboration with CONAGUA and SEMAR was articulated to diversify climate information (*articulate*). Although the information was regional, it was seen as an opportunity to offer options to fishers. To this end, clear summaries were designed, and it was agreed to present climate information segregated by region, municipality, and locality (*implement*).
4. April 27 to May 22 2023: first virtual fishing contest Huachinangazu.

   The team planned a virtual fishing contest with Pepe Bots as host in *Sabancuy*, Campeche, taking advantage of the local fishing activity (*implement*). An ichthyometer was designed to measure fish, and image processing tests were conducted to evaluate its performance. Contest participants uploaded photos of their Huachinango catches (red snapper) with the official reference ichthyometer using Pepe Bots (*implement*). The records were shared on a web page without personal identification to avoid conflicts. Prior to the contest, the development team received suggestions from local fishing tournament experts to ensure fairness (*plan*). Verification of specimen photographs was done with fisheries experts (*articulate*). The contest ended with a total of 20 registrations and 18 participating vessels, and the winning specimen of red snapper measured 85 cm and weighed 8 kg. The contest provided lessons on and for the systematization of information about the fishing areas and gear.

Figure 11.2 shows the recurrence of topics addressed in the virtual meetings, which correspond to different actions within the Pepe Bots co-production process. In general terms, adjustments to the technological tool were made every week by the development team, based on what was observed in the field, the direct recommendations of the fishers, and the reflections that arose in dialogues through forums or workshops. This continuous process reflects an attempt to adapt the technology to the local context while also incorporating different forms of knowledge. Norström et al. (2020) propose four key principles to ensure quality in knowledge co-production processes in sustainability contexts: (1) that they are based on the particular context (social, economic, and ecological), (2) that they adopt a pluralistic approach, explicitly recognizing the diverse ways of knowing and doing, (3) that they are oriented towards collectively defined objectives, with shared and evaluable milestones, and (4) that they promote constant interaction among participants throughout the process, from the formulation of the agenda to the application and dissemination of the knowledge generated.

Under this idea, technological syncretism should not be understood simply as a combination of traditional knowledge and modern technology, but rather as a dynamic and collaborative process of intercultural co-production. According to

Azurdia Bravo (2013), it implies the joint creation (from mutual recognition and epistemic equity) of technological alternatives that are socially relevant and culturally appropriate. In the case of Pepe Bots, this ideal of syncretism began to manifest itself in the constant interaction between fishers and developers, where field observations and local knowledge influenced, albeit unevenly, the evolution of the tool.

11.5 Final Reflections

Technological tools generally respond to the social, institutional, economic, and cultural contexts in which their developers operate. However, it is essential to develop what Quintanilla (2005) refers to as an adequate "technological culture," which includes the ability to design, develop, use, and evaluate technical systems appropriate to the specific purposes of the users. Quintanilla argues that success in the adoption and application of technologies depends to a large extent on the embodied and unembodied technological culture of the social groups involved. The former refers to the cultural traits necessary for the effective functioning of a technical system, such as the level of training in the use of specific technologies. In the case of Pepe Bots, this is observed in the fishers' prior knowledge of cell phone use, especially WhatsApp. On the other hand, the unembodied technological culture encompasses beliefs and attitudes related to technology, such as distrust towards chatbots or the perception that external technologies fail to address their real needs.

Although technological syncretism is proposed as an ideal form of articulation between technical knowledge and traditional knowledge, in practice, this integration is often unequal. In the process of creating and improving Pepe Bots, although there were moments of consultation and feedback with the fishers, most of the technical decisions rested with the development team. These dynamic challenges the ideal of syncretism as a horizontal and reflexive co-production, and raise the need to recognize the asymmetries that persist even in well-intendend projects. Technological appropriation does not always imply an equitable and horizontal participation of all actors. Although the discourse of syncretism emphasizes complementarity and convergence between knowledge systems, the fact is that, in many cases, technological developments continue to reproduce epistemic hierarchies in which technical knowledge has greater legitimacy than local knowledge. This translates into limited or superficial participation by fishing communities in key decisions on the design, implementation, and evaluation of technological tools intended for their use. For technological syncretism to be genuinely transformative, it is essential to overcome the mere incorporation of opinions as superficial validation and instead generate sustained mechanisms for effective and binding participation.

Olivé mentions that "empowering technological development, then, does not only mean using more technology; it is also, and above all, developing technological culture" (Olivé, 2007, p. 62), for which it is necessary to recognize the different knowledge systems involved. This also implies recognizing the limitations of the syncretism approach, which overlooks the critique of power structures that subordinate certain forms of knowledge. If fishers do not participate on equal terms in technological creation, it is more likely that their knowledge will be aestheticized or simplified, without really being given a decisive place. Despite these tensions, Pepe Bots has demonstrated potential as a space for mutual learning. Its process of continuous improvement, driven by user feedback and team commitment, represents a step towards a more collaborative model. However, it is precisely from this experience that it becomes evident that co-production cannot be understood as a one-time event, but as a permanent practice that must question and redistribute epistemic power. Technology, then, must not only adapt to local contexts but also be built from the ground up, recognizing the fishers not as beneficiaries but as co-creators.

Acknowledgments We thank the fishers and developers who shared their time and experiences related to technological developments for small-scale fisheries. We also thank the members of the Pepe project (Ciencia de Frontera 2019, Project CF-2019-1564454, CONACyT) team for the logistical and material support provided during the months of work to develop this research.

References

Azurdia Bravo, I. (2013). Resiliencia Adaptativa y Sincretismo Tecnológico. *Revista Mexicana de Física, 59*(2), 31–35.

Berger, W. M., & Azurdia, B. I., (2008). Resiliencia Adaptativa y Sincretismo Tecnológico, adaptación al cambio climático. Observatorio de la Sostenibilidad, Red Latinoamérica. San José, Costa Rica.

Cornell, S., Berkhout, F., Tuinstra, W., Tabara, J. D., Jager, J., Chabay, I., de Wit, B., et al. (2013). Opening up knowledge systems for better responses to global environmental change. *Environmental Science & Policy, 28*, 60–70.

ECOSUR-EDFM. (2022). *Diagnóstico de capacidades digitales en las comunidades pesqueras de pequeña escala en México. Reporte*. Environmental Defense Fund México.

Espinoza-Tenorio, A., & Cupido Santamaria, D. G. (2021). *Primer Reporte Anual del proyecto "La desafiante coexistencia de los sistemas socio-ecológicos acoplados; las industrias de la pesca y petróleo en la Sonda de Campeche"* (Reporte 1). Departamento de Ciencias de la Sustentabilidad, ECOSUR.

Norström, A. V., Cvitanovic, C., Löf, M. F., West, S., Wyborn, C., Balvanera, P., et al. (2020). Principles for knowledge co-production in sustainability research. *Nature Sustainability, 3*(3), 182–190. 38.

Olivé, L. (2007). *La ciencia y la tecnología en la sociedad del conocimiento: Ética, política y epistemología*. Fondo de cultura económica.

Ortiz, R. (2022). *Diagnóstico de Capacidades Digitales en Comunidades Pesqueras de Pequeña Escala [Presentation]*. Environmental Defense Fund.

Quintanilla, M. A. (2005). *Tecnología: un enfoque filosófico y otros ensayos de filosofía de la tecnología* (p. 45). FCE.

SEPESCA. (2022). *Programa de Ordenamiento de la Pesca Ribereña*. Consultado en Datos Abiertos Tamaulipas.

Part III
Reconfiguring Governance

Chapter 12
The Coastal Zone of Campeche: A Spatial Conflict Among Fisheries, Oil Extraction, and Conservation

A. C. Peña-Puch, E. Rivera-Arriaga, J. W. Day, and R. G. Hunter

12.1 Introduction

The Southwestern Gulf of Mexico (SWGoM) encompasses a rich and varied ecosystem, characterized by lagoons, river basins, deltas, and shallow continental shelves. This region serves as a vital hub for numerous economic activities, contributing significantly to local and regional development. The two most important economic drivers are the fishing industry and the oil industry. In the early twentieth century, Campeche's economy relied on natural resource extraction, including logwood (1517–1910), chewing gum (1907–1947), pink shrimp fishing (1947–1994), and oil and natural gas (from 1977) (Vadillo-López, 2008). These industries, driven by global demand, eventually gave way to synthetic substitutes or resource depletion. While they brought jobs, population growth, and wealth to some, they also provoked environmental harm, social inequality, and health risks. The region experienced substantial changes, notably the decline of logwood due to the appearance of chemical dyes and the rise and fall of the chewing gum industry, which was influenced by American investment and synthetic alternatives after World War II (WWII) (Alcalá-Ferráez, 2010; Chablé-Vega et al., 2024; Vadillo-López, 2008).

A. C. Peña-Puch (✉) · E. Rivera-Arriaga
Instituto de Ecología, Pesquerías y Oceanografía del Golfo de México, EPOMEX, Universidad Autónoma de Campeche, Campeche, México
e-mail: angcpena@uacam.mx; evrivera@uacam.mx

J. W. Day
Department of Oceanography and Coastal Sciences, College of the Coast & Environment, Louisiana State University, Baton Rouge, LA, USA
e-mail: johnday@lsu.edu

R. G. Hunter
Comite Resources, Baton Rouge, LA, USA
e-mail: rhunter@comiteres.com

A. Espinoza-Tenorio et al. (eds.), *The Fishing and Oil Industries in Coexistence*,
https://doi.org/10.1007/978-3-032-02621-7_12

The Campeche population relies heavily on coastal-marine ecosystems and marine species, which have been integral to local culture since pre-colonial times. Campeche City, once the primary port of the Yucatan Peninsula, supported a vital fishing industry, which was crucial for both the economy and sustenance (Salazar-Anaya, 2011; SEMAR, 2024). Despite the historical abundance of marine life, overfishing has led to a decline in species and a loss of biodiversity. The shrimp industry, booming since 1946, faced a downturn after years of overexploitation (Gracia, 1995; Ramírez-Rodríguez, 2015; Vadillo-López, 2008; Wakida-Kusunoki et al., 2016).

Understanding the nekton populations and the dynamics of coastal ecosystems in the *Laguna de Términos* is crucial, especially since the impact of oil production on fisheries is rarely quantified (Campos-Flores & Crespo-Guerrero, 2018; Soto et al., 2004). In this region, a diverse array of primary producers exists, with production and migration patterns influenced by tropical climate factors (Salazar-Anaya, 2011; SEMAR, 2024; Vadillo-López, 2008). Primary production in coastal ecosystems exhibits seasonal peaks, with different producers thriving at various times, which ensures year-round high productivity (Rojas-Galaviz et al., 1992; Day et al., 1982). Mangrove litterfall increases during the rainy season, while submerged aquatic vegetation (SAV) is more productive in the dry season due to increased water clarity. Phytoplankton production is higher in the rainy season, fueled by nutrient influx and shifts in river discharge patterns. This productivity is increased by the input of organic matter from the mangroves (Rivera-Monroy et al., 1998). Nekton, including shrimp juveniles and adults, migrate through tropical coastal ecosystems, such as *Laguna de Términos*, following the productivity cycles of different habitats (Yáñez-Arancibia et al., 1988, 1993; Sánchez-Gil et al., 2008). These migrations enable the optimal use of food resources, connecting distant habitats to the Gulf of Mexico's fisheries resources. Consequently, damage to inshore habitats by oil activities can impact offshore fisheries.

The Cantarell oil field, named after the fishers who discovered it near Carmen Island, marked the start of the *Sonda de Campeche* oil era in 1977. This industry shift had significant socio-ecological effects locally, altering Campeche's traditional economic reliance on forestry, agriculture, and fishing (Vadillo-López, 2008). Therefore, the objective of this chapter is to describe the conflicts that have occurred from the second half of the twentieth century to the present between fishing, oil production, and conservation of the coastal area of Campeche, through a review of the literature.

12.2 Fishery Sector

In 1956, the state government encouraged the population of the henequen area to take up fishing through the modification of Municipal limits (from Calkiní, Hecelchakán, and Tenabo) so that they had access to the coast (Lazos-Ruíz & Pinkus-Rendón, 2022). However, the history of fishing and hunting has led to a decline in the populations of particular species, such as manatees (*Trichechus manatus*). Substantial changes in the fishing sector, including boats, fishing equipment, and technologies, occurred in the mid-twentieth century. This was related to changes in public policies, pressures from the shrimp industry, and technological alterations in fishing practices (Fig. 12.1).

12.2.1 Boom and Decline of the Shrimp Fishery (1980–1998)

During WWII, the Japanese discovery of shrimp grounds in the Gulf of Mexico led to a fishery initially dominated by American and Cuban ships, but eventually, the Mexican fleet took precedence (Sierra, 1998; Vadillo-López, 2008). The pink shrimp (*Farfantepenaeus duorarum*) emerged as the primary catch, alongside white shrimp (*Litopenaeus setiferus*) and brown shrimp (*Farfantepenaeus aztecus*), with Campeche's coastal waters and ports becoming central to the industry (Sierra, 1998).

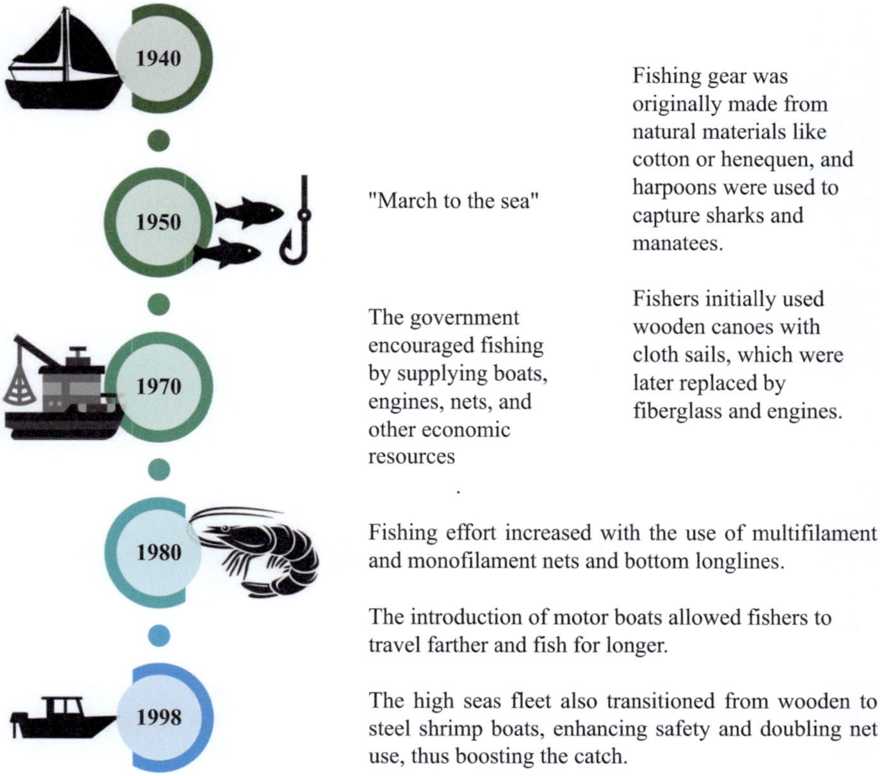

Fig. 12.1 Transition of fishing practices, gear, and equipment and the growth of fishing effort in the 1980s. (Sources: Cahuich-Campos, 2021; Ehuan-Noh et al., 2020; Espinoza-Tenorio et al., 2011; Martínez-Candelas et al., 2020; Vadillo-López, 2008; Villalobos-Zapata & Mendoza, 2010)

Government subsidies for port infrastructure (March to the sea program), technological advancements, and increased fishing efforts culminated in significant industry growth by 1970, marked by an expansive fleet and supporting shipyard and processing jobs, which, despite lower wages, provided vital family income (Ramírez-Rodríguez, 2015). The industry's prosperity led to diverse investments by businesspeople. However, the shift to a cooperative model in 1982 had a significant impact on investment and profitability, affecting both owners and workers (Ramírez-Rodríguez, 2015; Vadillo-López, 2008).

From 1960 to 1980, the shrimp industry saw a significant increase in catch rates, exceeding 50% (Vadillo-López, 2008). However, this rapid growth led to overfishing, posing a threat to species conservation and causing ecological harm due to the methods employed in fishing. In response, shrimp fleets from the USÀ and Cuba were barred from participating in the fishery starting in 1980 to mitigate these effects (Klima et al., 1987). Despite these measures, the industry continued to decline, influenced by a combination of social and environmental factors, including the use of efficient but destructive fishing gear and a lack of localized fishery management plans.

The crisis in the shrimp sector was further exacerbated by decades of resource overexploitation, neglecting the need for species preservation and sustainable harvesting practices (Vadillo-López, 2008). Population growth also played a role, as the demand for coastal land for settlement led to increased deforestation and pollution of environmental conditions in mangrove areas. Illegal fishing, particularly of juvenile shrimp in critical breeding grounds like *Laguna de Términos*, severely impacts shrimp populations by reducing the number of young entering the mature stock. Additionally, the expansion of the oil industry in the area has been associated with further declines in the shrimp fishery (Arreguín-Sánchez et al., 1997). However, the extent to which oil spills have affected the fishery, whether because of human activity or natural seepage from the ocean floor's asphalt deposits, remains unclear (Wakida-Kusunoki & Caballero-Chávez, 2009).

12.2.2 Conflicts Between the Fishery and Oil Sectors

The *Ixtoc*-I oil spill in 1979 represented a significant conflict between sectors, leading to severe socio-ecological damage. Efforts to mitigate the political and social fallout included financial and social aid to the fishing industry. However, this response has been linked to environmental degradation over the past 50 years, including ecological degradation and marine resource depletion due to overfishing, overinvestment, and increased fishing effort (Flores-Hernández & Ramos-Miranda, 2004). Fishing organizations, driven by industry leaders, sought financial aid from PEMEX to offset the impacts of oil exploration and exploitation (Botello-Ruvalcaba et al., 2010).

Investments in the fishing sector over the last half-century have led to instability, with zones established in 2003 for oil extraction impacting the fishing industry's navigation and harvesting (Navarrete del Próo, 2007; Zalik, 2009). The situation worsened following the 2013 energy reform, which significantly altered fishing operations and reduced available fishing areas (Campos-Flores & Crespo-Guerrero, 2018; Espinoza-Tenorio, 2019; Ramos-Muñoz & Espinoza-Tenorio, 2020).

The fishing community has observed a correlation between increased oil activities and hydrocarbon spills in coastal areas, with significant repercussions (Campos-Flores & Crespo-Guerrero, 2018). These spills are believed to deplete fish populations, adversely affecting the catch of fishers (García-Cuéllar et al., 2004) and the health of local communities. In *Laguna de Términos*, fishers have engaged in prolonged discussions with PEMEX regarding compensation for damaged nets, acid rain, and reduced catches (Palacios-Solano & Frutos-Cortés, 2014). They acknowledge the need to monitor oil spills and leaks as a means of managing their disputes with PEMEX (Peña-Puch et al., 2021).

Maritime oil and gas operations pose risks to the coastal ecosystem, navigation, and fishing practices (Rivera-Arriaga et al., 2019). PEMEX's political influence over fishing leaders hinders community protests against environmental damage (Uribe-Iniesta, 2004). The ineffectiveness of strategies to alleviate tension between fishing and oil activities is partly due to insufficient consideration of the region's unique socioeconomic and environmental aspects (Espinoza-Tenorio, 2019; Pérez-Sánchez et al., 2000). In Campeche, the oil sector's social responsibility is limited to benefit distribution, which should be fair across all municipalities, not just coastal ones (Rivera-Arriaga et al., 2019).

12.3 Oil Sector on the Coast of Campeche

Exploration for oil in the SWGoM began in the early 1950s, with exploitation commencing in the 1970s. The SWGoM currently produces 75% of Mexico's oil, which is concentrated in this region (Schifter et al., 2015) (Fig. 12.2). *Cayo Arcas* functions as a key hydrocarbon export port in Campeche, while the Port of Lerma receives fuel to supply the area. The *Laguna Azul* port in *Ciudad del Carmen* supports PEMEX's offshore operations in the *Sonda de Campeche* (Zenteno-Escutia, 2013). The expansion of oil platforms has resulted in the establishment of restricted zones to regulate navigation and fishing activities (Bustamante-Ahumada, 2004). The nautical chart S.M. 840 "Bahía de Campeche," created in May 1981 by PEMEX (the state-owned company Petróleos Mexicanos, by its acronym in Spanish) and INEGI (National Institute of Statistics and Geography), delineates secure maritime routes and zones around the oil fields, aiming to enhance maritime safety, protect marine life, and minimize oil activity risks to fisheries (DOF, 1998).

Updated in March 2007, this chart introduced additional safety measures and restricted airspace to further safeguard human life at sea (Zenteno-Escutia, 2013). A collaborative agreement by the Secretariat of the Navy (SEMAR), Ministry of Communications and Transportation (SCT), PEMEX, and SAGARPA (Ministry of Agriculture and Rural Development, now SADER), documented on September 11,

2003, in the Federal Official Diary (DOF), established stringent security protocols in the *Sonda de Campeche* to protect the marine oil infrastructure.

In 2011, the International Maritime Organization (IMO) formed a subcommittee to review maritime activities near Campeche and *Cayo Arcas*, Mexico (OMI, 2011). Romo (2015) notes that the Cantarell oil field spans over 21,000 km^2, with initial production in 1979 averaging 88 million barrels per day (mbd), increasing to 611 mbd in 1980, and peaking at 1156 mbd in 1981 from just 40 wells. Following 2002, the Cantarell complex, including the Ek-Balam field, underwent expanded oil and gas exploration and exploitation. However, from 2004 to 2009, Cantarell's oil output declined as natural gas production rose (Fig. 12.2). By mid-2014, the complex included fields such as *Nohoch, Chac, Akal, Kutz, Ixtoc, Takín, Ek, Balam, Sihil, Kambesah*, and *Után*, with the latter being non-productive. Cantarell primarily yields heavy crude oil, except for the *Kambesah* and *Ixtoc* fields (Romo, 2015).

The constitutional reforms and amendments concerning energy, published in the Federal Official Diary on December 20, 2013, marked a significant shift in Mexico's energy policy. Following this, the "Ronda Cero" process results were released on August 13, 2014, allocating exploration and extraction fields to PEMEX. The Hydrocarbons Law, specifically Article 29, fraction I, empowers the Secretary of Energy to select contractual areas with the National Hydrocarbon Commission's technical assistance. This guidance was crucial for identifying potential shallow water and heavy oil fields for the "*Ronda Uno*" tender (Comisión Nacional de Hidrocarburos, 2014). The SWGoM saw an increase in offshore drilling, with more than 2000 wells and extensive pipeline networks stretching to the *Sonda de Campeche* (Pulster et al., 2020).

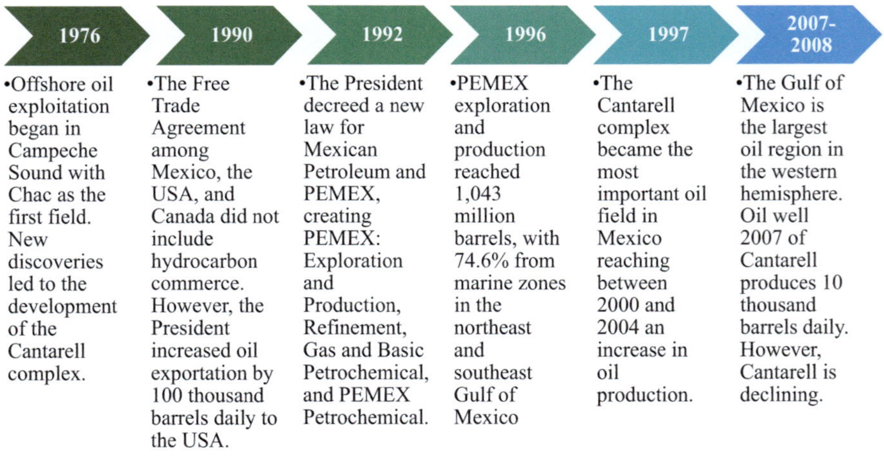

Fig. 12.2 A summary of the history of oil exploitation of the Cantarell complex. Source: (COLMEX, 2008; Palacios-Solano, 1991; Zenteno-Escutia, 2013)

12.3.1 Impacts of Petroleum on Marine Resources and Ecosystems

García-Cuéllar et al. (2004) have highlighted the significant hydrocarbon pollution in the Gulf of Mexico due to PEMEX's extensive activities, including exploration, production, and transportation. This pollution originates from various sources, including oil spills, industrial activities, and natural seepage. The *Sonda de Campeche* produces two types of oil, Maya (heavy) and Itsmo (light), each with distinct properties that affect their environmental impact, including toxicity and the rate of damage.

Elevated temperatures can lead to the evaporation and photooxidation of oil spills, with weather patterns affecting their spread and degradation (Gold-Bouchot, 2004). The impact of oil on marine life is significant, as seen in the macrobenthic community near oil rigs, where higher contaminant levels and reduced macroinfauna abundance were observed (Hernández-Arana et al., 2005). Further research by Pulster et al. (2020) indicates that polycyclic aromatic hydrocarbons (PAH), a toxic component of oil, present long-term ecological risks, with widespread exposure by Gulf of Mexico fish. Additionally, the disposal of produced water from hydrocarbon extraction introduces pollutants such as heavy metals and hypersaline brines into the environment. Improper disposal of this water can lead to soil and groundwater pollution, air pollution from evaporation tanks, and long-term issues such as acid rain and health problems (Bravo, 2007).

Research by Soto et al. (2014) on the *Ixtoc*-I spill indicates that natural attenuation processes such as evaporation and biodegradation were significant in mitigating environmental impacts, a sentiment echoed in a PEMEX CEO's statement from the International Symposium on the *Ixtoc*-I oil spill. However, Guzmán del Próo et al. (1986) found that such spills have a profound impact on marine life, with zooplankton biomass in the SWGM plummeting post-spill although seasonal patterns of biomass remained consistent.

Since 1975, the zooplankton community in *Sonda de Campeche* has experienced a notable decline in stability, with hydrocarbons detected in most samples (Guzmán del Próo et al., 1986). The origin of these hydrocarbons remains uncertain, with possibilities including the *Ixtoc*-1 spill among others (Fig. 12.3). Soto et al. (2014) found that oil-degrading bacteria are widespread in *Sonda de Campeche* water and sediments, indicating a strong potential for natural oil biodegradation. Despite this, no enduring impact from the 1979 *Ixtoc*-I blowout has been observed. Licea reported a significant decrease in phytoplankton biomass, productivity, and diversity post-spill, especially in small phytoflagellates sensitive to oil and dispersants (Licea et al., 1982).

Industrial activities in Campeche's coastal regions, notably oil and gas operations, have raised substantial environmental concerns. The presence of PAHs and heavy metals has been confirmed, resulting in ecological impacts, including

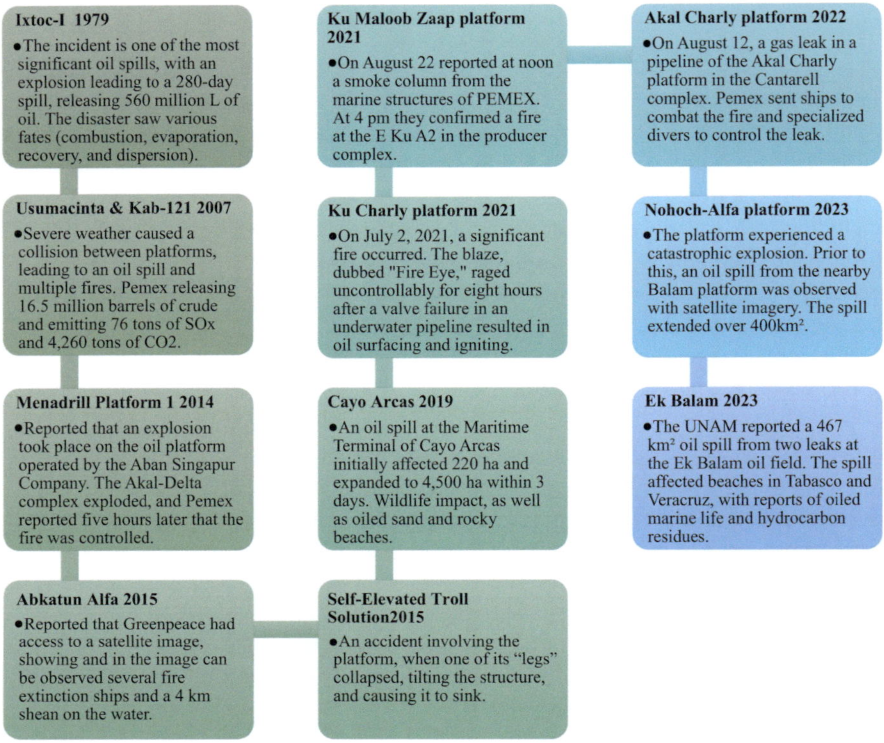

Fig. 12.3 Oil Spills in PEMEX installations in *Sonda de Campeche* 1979–2023. Sources: (CENAPRED, 2019; Comisión de investigación de la Facultad de Ciencias Naturales Universidad Autónoma Del Carmen, 2019; Corona, 2015; Herzka et al., 2021; La Jornada maya, 2015; Latinus, 2023; Miranda, 2014; Myer, 1984; Tapia, 2023)

alterations to aquatic community structures and food chains (Botello et al., 2004; Gold-Bouchot, 2004). Moreover, the bioaccumulation of these pollutants in marine organisms, such as shellfish and shrimp, has been linked to adverse effects, affecting individual species and entire marine ecosystems (Botello et al., 1983; Soto & González, 2009).

12.3.2 Subsidies and Social Aid to Fisher Communities

The aftermath of the *Ixtoc*-I oil spill was marked by significant environmental damage, which sparked social movements, such as the *Pacto Ribereño*, to demand action (Velázquez-Guzmán, 1982). In response, the 1980s saw initiatives such as the National Development Plan and the "Plan Campeche" to support the fishing industry, aiming to enhance employment and improve quality of life in Campeche. The federal government's inclusion of Campeche in the Commission for the Development of Oil Zones in 1986 aimed to diversify the economy with benefits for urban development, fishing, agriculture, and ecological protection (Solano-Palacios & Frutos-Cortés, 2016).

The 1990s marked a legal battle initiated by the National Chamber of the Fishing and Aquaculture Industry (CANAINPESCA) of Campeche against PEMEX for compensation due to reduced shrimp catches, attributed to restricted fishing zones and pollution (Ramos-Muñoz & Espinoza-Tenorio, 2020; Solano-Palacios & Frutos-Cortés, 2016). Persistent conflicts between the fishing and oil industries led to dissatisfaction and political pressure, with PEMEX's compensation measures considered inadequate by the fishing community, resulting in intensified protests and road blockades (Hanson, 2002; Rabelo-Avalos et al., 2021).

The 2000s marked a period of heightened dissatisfaction within the fishing sector, affecting both large-scale shipowners and small-scale cooperatives. This led to negotiations with PEMEX, resulting in proposed subsidies and support initiatives, such as the Trust Fund for the Fishing Reconversion of the Gulf of Mexico (FIFOPESCA) in 2004, aimed at fostering productive projects beneficial to the fishing community (Arias-Rodríguez & Ireta-Guzmán, 2009; Solano-Palacios & Frutos-Cortés, 2016). PEMEX also engaged in coordination agreements with state governments to provide resources and support to the productive sectors, including fishing cooperatives and permit holders (Arias-Rodríguez & Ireta-Guzmán, 2007).

Federal efforts to reduce fishing effort by incentivizing shipowners to retire high-seas vessels were met with limited success, as predominantly obsolete vessels were surrendered, failing to effectively address overfishing (Arias-Rodríguez & Ireta-Guzmán, 2009). The allocation of fishing subsidies led to social tensions, as they were distributed unevenly across regions and social groups, sometimes resulting in conflicts such as platform closures and vessel hijackings (Quist & Nygren, 2015).

Following energy reforms in 2013 that affected fishing zones, PEMEX initiated community development programs in 2014, including the Community and Environmental Support Program (PACMA, by its Spanish acronym), which promotes social development through cooperation between the public and private sectors in oil zone communities (Campos-Flores & Crespo-Guerrero, 2018; Solano-Palacios & Frutos-Cortés, 2016).

The fishing sector has been a significant beneficiary of programs like the instrument that drives Programs, Works, and Actions (PROA, by its Spanish acronym), receiving substantial support in supplies and equipment from PEMEX, particularly in Carmen Municipality. Initiatives have included community development plans, capacity building, and infrastructure feasibility studies, all aimed at enhancing the productivity of fishing communities (Campos-Flores & Crespo-Guerrero, 2018; PEMEX, 2022).

However, the decline in PEMEX's revenues led to reduced subsidies, impacting the economic growth and communities' wealth of communities. The dependency on PEMEX's support has created a perceived social debt among fishers, which remains unresolved despite various social support initiatives, and has cast doubt on the potential for successful collaboration between the oil and fishing sectors (Ramos-Muñoz et al., 2020; Ramos-Muñoz & Espinoza-Tenorio, 2020; Solano-Palacios & Frutos-Cortés, 2016). Additionally, the lack of diversification, support, and administrative and technical training for productive projects is evident, as well as a lack of a long-term vision to materialize tangible benefits for the fishing sector.

12.4 Conservation of the Coast of Campeche

Coastal habitats, including mangroves, seagrass beds, and saline lagoons, are highly susceptible to environmental threats within NPAs (Natural Protected Area). Aguirre-Macedo et al. (2020) highlight their vulnerability, while the International Coral Reef Initiative (ICRI) notes the severe consequences of oil spills on coral reefs, which affect their survival and disrupt vital processes (ICRI, 2021). The condition of the Convention on Wetlands of International Importance (Ramsar) sites is concerning, with mangrove and seabird species requiring increased protection against various threats, including habitat destruction and loss of mangrove cover (Vidal et al., 2015). Furthermore, human activities such as urbanization and oil exploitation are intensifying these impacts (CONANP, 2024), which are compounded by insufficient local cooperation and resources for effective conservation management (Neger & Crespo-Guerrero, 2021).

The Ramsar Convention of 1971 emphasizes the crucial role of water in maintaining biodiversity and regulating environmental conditions within wetlands (Villalobos-Zapata & Mendoza, 2010). Initiated in Mexico in 1986, the Convention recognizes the Yucatan Peninsula's coastal wetlands for their extensive biodiversity and the ecosystem services they provide, which are vital for waterfowl habitats and other species (Correa-Sandoval & Contreras-Balderas, 2008). These areas also contribute to local economies through ecotourism (Vidal et al., 2015). On the Campeche coast, the RAMSAR sites were established until 2004 (Table 12.1).

Table 12.1 Sites designated for conservation in the coastal area of the State of Campeche. Source (INE & SEMARNAP, 1997; SEMARNAT, 2002, 2024; SEMARNAT & CONANP, 2006; Vidal et al., 2015)

Coastal area	NPA	Ramsar	Attributes
Flora and Fauna Protection Area *Laguna de Términos*. This region encompasses 705,016 hectares in the municipalities of El Carmen, Palizada, Escárcega, and Champotón	1994	2004	Includes three river-lagoon systems with a mangrove cover of 127,000 hectares. Moreover, this region is notable for colonies of large aquatic birds such as the Jabiru stork (*Jabiru mycteria*)
Los Petenes Biosphere Reserve. This area covers 282,857 hectares and is distributed across the municipalities of Calkiní, Hecelchakán, Tenabo, and Campeche	1999	2004	It is characterized by "petenes," which are islets of vegetation temporarily flooded, associated with freshwater springs (underground or cenotes), buttonwood mangrove (*Conocarpus erectus*), and seagrasses
The Celestún Biosphere Reserve. This region encompasses 81,482 hectares in the municipality of Calkiní (Campeche) and the municipalities of Celestún and Maxcanú (Yucatán)	2004	2004	Includes significant communities of mangroves, seagrasses, coastal dunes, estuaries, and hypersaline coastal lagoons. Area for migratory birds such as the American Flamingo (*Phoenicopterus ruber*)
Playa Chenkan sanctuary. This area covers 39.55 hectares and is in the municipality of Champotón	2024	2004	It is characterized by sandy beaches, which favor the outstanding nesting of the hawksbill turtle (*Eretmochelys imbricata*) and the green sea turtle (*Chelonia mydas*)

NPA Natural Protected Area

The SWGoM, particularly *Laguna de Términos*, is identified as vulnerable to oil spills, which can affect marine turtles. The State of Campeche's coastal wetlands, including four Ramsar sites, are essential for economic and cultural resources, relying on the conservation of diverse ecosystems, and are protected under the General Law of Ecological Equilibrium and Environmental Protection (LGEEPA) (Vidal et al., 2015).

12.5 Conclusions and Final Thoughts

Competition between hydrocarbon activities and fisheries has impacted the latter beyond environmental and economic effects. Oil spills since the beginning of Cantarell have been increasing, with consequent negative environmental impacts ranging from heavy metals to toxic pollutants and chemicals used to control those spills. In addition, restrictions for accessing large areas around hydrocarbons' infrastructure resulted in higher costs for Fishers and fewer captures due to overfishing.

The environmental and social impacts of the *Ixtoc*-I oil spill led to numerous studies in Mexico and the USA, which tracked the effects on marine and coastal organisms and ecosystems, including both the punctual and cumulative impacts of hydrocarbons. However, in Mexico, the government provided financial support to the fishing industry, without perceiving that this aid inadvertently contributed to social inequality and promoted social dependency on Pemex. Moreover, the *Ixtoc*-I incident resulted in intricate and enduring challenges that arise from the complex interplay between industrial activity and environmental stewardship.

Through time, hydrocarbon activities and oil spill events have significantly impacted coastal ecosystems, particularly causing wetland loss through three primary mechanisms: modification of surface hydrology, subsidence from fluid extraction, and pollution from oil spills and produced water (Day et al., 2022, 2020). These activities alter the natural water flow, leading to issues such as stream capture and saltwater intrusion. Additionally, spoil banks from canal dredging disrupt the exchange of water and other materials in wetlands, which can result in longer flooding durations and lower productivity. Over time, these spoil banks may subside, but they can also provide some protection to the remaining marshlands. All of these highlighted the immediate repercussions on coastal and marine ecosystems, resource depletion, and socioeconomic disparities, exacerbated by policy changes and the expansion of industrial activities, which also set in motion a series of socio-political responses that continue to shape the region's migration and, therefore, the landscape.

The Deepwater Horizon oil spill triggered monitoring programs and a great deal of research on US coastal and marine resources and ecosystems. However, in Mexico, after the Ixtoc-I oil spill, only a few researchers assessed the ecosystem and the impacts. Therefore, assessing damage and planning for ecological restoration after an oil spill is tough due to the limited availability of pre-existing data. The scarcity of ecological data in the Gulf of Mexico hinders the ability to conduct thorough ecotoxicology and vulnerability assessments, making it challenging to ascertain the scope of damage. Recent efforts have been made to enhance knowledge of the Gulf of Mexico through the Gulf of Mexico Research Consortium (CIGOM, by its Spanish acronym), which has implemented oceanographic observation networks to generate scenarios of possible hydrocarbon contingencies. However, it is imperative to prioritize gathering ecological data on resources and ecosystems, including socioeconomic and human health aspects, within a monitoring program, to assess the current health of the Gulf of Mexico (https://cigom.org).

It is evident that while efforts have been made to address the aftermath, the path towards a harmonious coexistence between the oil industry and the coastal communities demands a transdisciplinary approach. This approach must prioritize ecological sustainability, equitable and alternative economic practices, and proactive community engagement to foster resilience and ensure that both natural and human resources are sustainably managed.

In both countries sharing the Gulf of Mexico with Cuba, it is essential to establish an international agenda to effectively address the Cartagena Convention and its Protocol on Oil Spills. At national level, Mexico needs a new marine policy that integrates environmental ethics into hydrocarbons' operations, and a local government that enhances its capacities for monitoring, response, and innovation, and builds collaboration with federal, state and municipal agencies, universities, private sector, NGOs and society, designing a protocol for early warning, mitigation, cleaning and restoring coastal and marine ecosystems.

Acknowledgments This study is part of the project "Advances of vertical and horizontal governance in coastal Social-Ecological Systems with presence of coastal fisheries and community tourism in the NPA of Campeche," of Peña-Puch AC, funded by CONACYT with the postdoctoral research appointments in Mexico 2021–2022 (929608).

During the preparation of this chapter, the authors used COPILOT AI to enhance the English writing process, improving readability and language proficiency. After using this tool/service, authors reviewed and edited the content as needed and take full responsibility for the content of the publication.

References

Aguirre-Macedo, L., Pérez-Brunius, P., & Saldaña-Ruiz, L. (2020). Tomo II Vulnerabilidad ecológica del Golfo de México ante derrames de gran escala, En: P. Pérez-Brunius y M. Aguirre-Macedo (Coord. Gral.) *Regiones, Especies y ecosistemas vulnerables ante derrames de petróleo de gran escala en el Golfo de México*, CIGOM.

Alcalá-Ferráez, C. (2010). La ciudad de Campeche a través de viajeros extranjeros 1834–1849. *Relaciones, Estudios de Historia y Sociedad, XXXI* (122), 201–244.

Arias Rodríguez, J. M., & Ireta Guzmán, H. (2007). El acceso a la información para la contraloría social. *Los donativos y donaciones de PEMEX a Tabasco*. Fundar, Centro de análisis e investigación AC. 1–132.

Arias Rodríguez, J. M., & Ireta Guzmán, H. (2009). *Pesca y Petróleo en el Golfo de México*. Asociación Ecológica Santo Tomás AC. 1–32.

Arreguín-Sánchez, F., Schultz-Ruiz, L., Sánchez, J., & Alarcón, T. (1997). Estado actual y perspectivas de la pesquería de camarón. In D. Flores-Hernández, P. Sánchez-Gil, J. Seijo, &

F. Arreguín-Sánchez (Eds.), *Análisis y diagnóstico de los recursos pesqueros críticos del Golfo de México* (Vol. 7, pp. 185–203). EPOPMEX.
Botello, A., Gofii, J., & Castro, S. (1983). Levels of organic pollution in coastal lagoons of Tabasco state, Mexico; I: Petroleum hydrocarbons. In *Bull. Environ. Contam. Toxicol (31)*. Springer-Verlag New York.
Botello, A., Villanueva, S., & Rosales-Hoz, L. (2004). Distribución y contaminación de metales en el Golfo de México. In M. Caso, I. Pisanty, & E. Ezcurra (Eds.), *Diagnóstico ambiental en el Golfo de México (Vol. 2)*. INE, INECOL, Texas A&M University-Corpus Christi, SEMARNAT.
Botello-Ruvalcaba, M., Villaseñor Talavera, R., & Mezo Villalobos, S. (2010). *Ordenamiento pesquero ribereño marino 2010*. CONAPESCA.
Bravo, E. (2007). Los impactos de la explotación petrolera en ecosistemas tropicales y la biodiversidad. Acción Ecológica. 1–61.
Bustamante-Ahumada, R. (2004). *El desarrollo costero de México*. Sociedad Mexicana de Geografía y Estadística, Ed.
Cahuich-Campos, M. (2021). Memorias del mar: Historia de una cooperativa pesquera campechana. *Glifos, 28*, 28–35.
Campos-Flores, G., & Crespo-Guerrero, J. (2018). Organización espacial de la pesca comercial ribereña en el Área de Protección de Flora y Fauna Laguna de Términos, México. *Investigaciones Geográficas, 96*, 1–21.
CENAPRED. (2019). *Pozo Ixtoc-1, el mayor derrame de petróleo en el mar ocurrido en México*. CENAPRED.
Chablé-Vega, M., García-Hernández, E., Martínez-Heredia, J., Villalpando-Aguilar, J., Arreola-Enríquez, J., López-Rosas, I., & Alatorre-Cobos, F. (2024). The return of natural dyes: The case of logwood tree (Haematoxylum campechianum L.). *Biotechnic & Histochemistry, 99*(5), 269–277. https://doi.org/10.1080/10520295.2024.2367535
COLMEX. (2008). In Centro de Estudios Históricos (Ed.), *Fuentes para la Historia del petróleo en México, 1900–2008*. El Colegio de México.
Comisión de investigación de la Facultad de Ciencias Naturales Universidad Autónoma Del Carmen. (2019). *Comunicado desde la Facultad de Ciencias Naturales de la Universidad Autónoma del Carmen*. UNACAR. 1–3.
Comisión Nacional de Hidrocarburos. (2014). *Documento soporte de decisión para la selección de las áreas de exploración que puedan ser incluidos en la Ronda 1 de licitación*. Aguas someras.
CONANP. (2024). ¿Qué hacemos? CONANP. www.gob.mx/conanp/que-hacemos
Corona, S. (2015). *Cuatro muertos al incendiarse una plataforma de Pemex en México*. El País. www.elpais.com
Correa-Sandoval, J., & Contreras-Balderas, A. (2008). Distribution and abundance of shorebirds in the coastal wetlands of the Yucatan Peninsula, Mexico. *Wader Study Bulletin, 115*(3), 148–156.
Day, J., Day, R., Barreiro, M., Ley-Lou, F., & Madden, C. (1982). Primary production in the Laguna de Terminos, a tropical estuary in the southern Gulf of Mexico. *Oceanologia Acta*. No. SP, 269–276.
Day, J., Clark, H., Chang, C., Hunter, R., & Norman, C. (2020). Life cycle of oil and gas fields in the Mississippi River Delta: A review. *Water, 12*, 1492.
Day, J., Hunter, R., & Clark, H. C. (2022). Energy production in the Mississippi River Delta: Impacts on coastal ecosystems and pathways to restoration. *Springer Lecture Notes in Energy, 43*, 242.
DOF. (1998). *Reglas por las que se establece el sistema de organización del tráfico marítimo en el Golfo de México y en la Terminal de Cayo Arcas, Campeche*.
Ehuan-Noh, R. G., Mariaca, R., Sáenz-Arroyo, A., & Espinoza, A. (2020). Tácticas y saberes: los capitanes de la pesca ante la variabilidad ambiental del mar. *Sociedad y Ambiente, 23*, 1–29. https://doi.org/10.31840/sya.vi23.2199
Espinoza-Tenorio, A. (2019). ¿Mancharse las manos de negro? El dilema ético de la investigación en territorios petrolizados. *Revista Mexicana de Ciencias Políticas y Sociales, 64*(237), 183–210.

Espinoza-Tenorio, A., Espejel, I., & Wolff, M. (2011). Capacity building to achieve sustainable fisheries management in Mexico. *Ocean and Coastal Management, 54*, 731–741. https://doi.org/10.1016/j.ocecoaman.2011.07.001

Flores-Hernández, D., & Ramos-Miranda, J. (2004). In UAC, EPOMEX, & SEMARNAT (Eds.), *Las pesquerías artesanales en el Golfo de México*. El manejo costero en México.

García-Cuéllar, J., Arreguín-Sánchez, F., Hernández-Vázquez, S., & Lluch-Cota, D. (2004). Impacto ecológico de la industria petrolera en la sonda de Campeche, México, tras tres décadas de actividad: una revisión. *Interciencia, 29*(6), 311–319.

Gold-Bouchot, G. (2004). Hidrocarburos en el sur del Golfo de México. In M. Caso, I. Pisanti, & E. Ezcurra (Eds.), *Diagnostico Ambiental del Golfo de México (2,:657–682)*. INE, INECOL, Texas A&M University Corpus Christi, SEMARNAT.

Gracia, A. (1995). Impact of artisanal fishery on production of the pink shrimp Penaeus Farfantepenaeus duorarum Burkenroad, 1939. *Ciencias Marinas, 21*(3), 343–359. https://doi.org/10.7773/cm.v21i3.993

Guzmán del Próo, S., Chávez, E., Alatriste, F., de la Campa, S., De la Cruz, G., Gómez, L., Guadarrama, R., Guerra, A., Mille, S., & Torruco, D. (1986). The impact of the Ixtoc-I oil spill on zooplankton. *Journal of Plankton Research, 8*(3), 557–581.

Hanson, H. D. (2002). *Oiling the system: How activists and the state shaped the politics of petroleum development in Tabasco*. University of California, Davis.

Hernández-Arana, H., Warwick, R., Attrill, M., Rowden, A., & Gold-Bouchot, G. (2005). Assessing the impact of oil-related activities on benthic macroinfauna assemblages of the Campeche shelf, southern Gulf of Mexico. *Marine Ecology Progress Series, 289*, 89–107.

Herzka, S. Z., Zaragoza Álvarez, A., Peters, E. M., & Hernández Cárdenas, G. (2021). *Atlas de línea base ambiental del Golfo de México*. CIGOM-CICESE. 1–122.

ICRI. (2021). *Oil spills and coral reefs*. https://icriforum.org/oil-spills-coral-reefs/

INE, & SEMARNAP. (1997). *Programa de Manejo del Área de Protección de Flora y Fauna Laguna de Términos* (INE, Ed.). http://www.inecc.gob.mx/descargas/publicaciones/171.pdf

Jornada maya, L. (2015). *Colapsa una plataforma de PEMEX*. La Jornada Maya. www.lajornadamaya.mx

Klima, E., Castro-Meléndez, R., Baxter, N., Patella, F., Cody, T., & Sullivan, L. (1987). Mexus-gulf shrimp research 1978–1984. *Marine Fisheries Review, 49*(1), 21–30.

Latinus. (2023). El derrame de petróleo en la Sonda de Campeche alcanzó 467 kilómetros cuadrados, estiman especialistas de la UNAM. Latinus. www.latinus.us

Lazos-Ruíz, A., & Pinkus-Rendón, M. (2022). Historia ambiental de la región de la Reserva de la Biosfera los Petenes en Campeche, México. *Revista Etnobiología, 20*(2), 236–251.

Licea, S., Luna, R., Torres, P., & Trejo, C. (1982). Evaluación de los posibles efectos del derrame del Pozo Ixtoc-I sobre las comunidades. *Informe Final del fitoplancton y la producción primaria en la Sonda de Campeche*. UNAM/Progr. Coord. Est. Ecol. Sonda Campeche, 1–44.

Martínez-Candelas, I. A., Pérez-Jiménez, J. C., Espinoza-Tenorio, A., McClenachan, L., & Méndez-Loeza, I. (2020). Use of historical data to assess changes in the vulnerability of sharks. *Fisheries Research, 226*. https://doi.org/10.1016/j.fishres.2020.105526

Miranda, A. (2014). *Reportan explosión en plataforma de Pemex en Campeche*. Excelsior. www.excelsior.com.mx

Myer, P. G. (1984). *IXTOC I: case study of a major oil spill*. https://digitalcommons.uri.edu/ma_etds/133

Navarrete del Próo, A. (2007). *Análisis de la biología pesquera del camarón rosado (Farfantepenaeus duorarum Burkenroad 1939) en la Sonda de Campeche, Campeche, México [Tesis de Doctorado]*. UNAM.

Neger, C., & Crespo-Guerrero, J. (2021). Problemática de la gestión de las Áreas Naturales Protegidas de México: un análisis en las Reservas de la Biósfera de Los Tuxtlas y Los Petenes. *Anales de Geografía de La Universidad Complutense, 41*(2), 463–481.

OMI. (2011). *Organización del tráfico marítimo, notificaciones para buques y cuestiones conexas incluidas las disposiciones generales sobre organización del tráfico marítimo y el examen de las interpretaciones unificadas de la IACS.*

Palacios-Solano, I. (1991). El petróleo mexicano en la perspectiva del Tratado del libre comercio. *Revista Problemas Del Desarrollo, XXII*(86), 45–64.

Palacios-Solano, E., & Frutos-Cortés, M. (2014). Petróleo y sustentabilidad en la Laguna de Términos: Entre las políticas de ordenamiento territorial y los nuevos esquemas de la industria petrolera regional en la península de Atasta, Carmen, Campeche. In *Problemas del desarrollo regional en la región sur-sureste de México: política, economía y sociedad* (pp. 71–81). UQROO.

PEMEX. (2022). *Informe de sostenibilidad 2022.*

Peña-Puch, A., Pérez-Jiménez, J. C., Espinoza-Tenorio, A., Méndez-Loeza, I., & Williams-Jara, G. (2021). The sustainability of small-scale fisheries in oil-producing sections of the Gulf of Mexico. *The Extractive Industries and Society, 100975.*

Pérez-Sánchez, E., Muir, J., & Lindsay, R. (2000). Social and economic issues in aquaculture development for coastal communities of Tabasco, México. MicrobehaVior and Macroresults: *Proceedings of the Tenth Biennial Conference of the International Institute of Fisheries Economics and Trade*, January 6.

Pulster, E. L., Gracia, A., Armenteros, M., Toro-Farmer, G., Snyder, S., Carr, B., Schwaab, M., Nicholson, T., Mrowicki, J., & Murawski, S. (2020). A first comprehensive baseline of hydrocarbon pollution in Gulf of Mexico fishes. *Scientific Reports, 10*, 1.

Quist, L. M., & Nygren, A. (2015). Contested claims over space and identity between fishers and the oil industry in Mexico. *Geoforum, 63*, 44–54.

Rabelo-Avalos, A., Ramos-Muñoz, D., Díaz-Perera, M., & Mesa-Jurado, M. (2021). El petróleo en Tabasco: propuesta para una periodización. *Revista Mexicana de Sociología, 83*(1), 129–155.

Ramírez-Rodríguez, M. (2015). La Pesquería de camarón en Campeche: Desarrollo histórico y perspectiva. *Ciencia Pesquera, 23*(1), 73–87.

Ramos-Muñoz, D., & Espinoza-Tenorio, A. (2020). ¿Cooperar O No Cooperar? Crónica Del Dilema Por El Espacio Compartido Del Golfo De México. *Cadernos de Estudos Sociais, 35*(2), 1–25.

Ramos-Muñoz, D., Zepeda-Domínguez, J., & Espinoza-Tenorio, A. (2020). La reforma energética mexicana en el marco de la pesca y el petróleo en el Golfo de México. *Revista Costas, 2*(1), 67–80.

Rivera-Arriaga, E., López-Chan, Ó., Williams-Beck, L., León-Olea, R., & Gutiérrez-Rivera, A. (2019). Hydrocarbon governance in the Gulf of Mexico: Building a benefit-sharing system. *Ocean and Coastal Management, 177*, 156–165. https://doi.org/10.1016/j.ocecoaman.2019.04.015

Rivera-Monroy, V., Madden, C., Day, J., Twilley, R., Vera-Herrera, F., & Alvarez-Guillen, H. (1998). Seasonal coupling of a tropical mangrove forest and an estuarine water column: Enhancement of aquatic primary productivity. *Hydrobiologia, 379*, 41–53.

Rojas-Galaviz, J., Yáñez-Arancibia, A., Day, J., & Vera-Herrera, F. (1992). Estuarine primary producers: Laguna de Terminos case study. In U. Seeliger (Ed.), *Coastal plant communities of Latin America* (pp. 141–154). Academic Press.

Romo, D. (2015). The Cantarell oil field and the Mexican economy. *Revista Problemas Del Desarrollo, 46*, 183. http://www.scielo.org.mx/

Salazar-Anaya, D. (2011). Los puertos del inmigrante en México, 1884–1910. *Antropología Boletín Oficial Del INAH, 92*, 21–44.

Sánchez-Gil, P., Yáñez-Arancibia, A., Tapia, M., Day, J., Wilson, C., & Cowan, J. (2008). Ecological and biological strategies of *Etropus crossotus* and *Citharichthys spilopterus* (Pleuronectiformes: Paralichthydae) related to the estuarine plume, Southern Gulf of Mexico. *Journal of Sea Research, 59*, 173–185.

Schifter, I., González-Macías, C., Salazar-Coria, L., Sánchez-Reyna, G., & González-Lozano, C. (2015). Long-term effects of discharges of produced water the marine environment from petroleum-related activities at Sonda de Campeche, gulf of México. *Environmental Monitoring and Assessment, 187*, 11.

SEMAR. (2024). *Catálogo de Cartas y Publicaciones Náuticas.*

SEMARNAT. (2002). *Programa de Manejo de la Reserva de la Biosfera Ría Celestún (RBRC).* In Semarnat. Comisión Nacional de Áreas Naturales Protegidas Dirección Jurídica.
SEMARNAT. (2024). *Declaración del Santuario playa Chenkan como ANP.* In Declaración del Santuario playa Chenkan como ANP.
SEMARNAT, & CONANP. (2006). *Programa de Conservación y Manejo de la Reserva de la Biosfera Los Petenes* (C. Dirección General de Manejo para la Conservación, Ed.).
Sierra, C. (1998). *Breve historia de Campeche México* (FCE, COLMEX, & FHA, Eds.).
Solano-Palacios, E., & Frutos-Cortés, M. (2016). *El impacto de la reforma energética en México: una mirada nacional y regional* (UNACAR & AMECIDER; 1st ed.).
Soto, L., & González, C. (2009). *PEMEX y la salud ambiental de la Sonda de Campeche.* IMP, UNAM, BATTELLE, UAM.
Soto, L. A., Sánchez-García, S., & López-Veneroni, D. (2004). *Ambientes influidos por emanaciones naturales de hidrocarburos y gas en el suroeste del Golfo de México* (pp. 51–58). Universidad y Ciencia. www.ujat.mx/publicaciones/uciencia
Soto, L., Botello, A., Licea-Durán, S., Lizárraga-Partida, M., & Yáñez-Arancibia, A. (2014). The environmental legacy of the Ixtoc-I oil spill in Campeche sound, southwestern Gulf of Mexico. In *Frontiers in marine science (Vol. 1, issue NOV).* Frontiers Media S.A.
Tapia, P. (2023). *Estos son todos los datos en la explosión de la plataforma Nohoch -Alfa de PEMEX.* Forbes. www.forbes.com.mx
Uribe-Iniesta, R. (2004). In UNAM (Ed.), *Petróleo, mayas y trópico: la convivencia de la industria petrolera y los yokt'anob en los pantanos del Golfo de México* (pp. 1–11). Segundo congreso nacional de Historia Económica.
Vadillo-López, C. (2008). *Una historia regional en tres tiempos: Campeche siglos XVIII-XX.* Península, III(2).
Velázquez-Guzmán, M. G. (1982). Afectaciones petroleras en Tabasco: El movimiento del Pacto Ribereño. *Revista Mexicana de Sociología, 44*(1), 167. https://doi.org/10.2307/3540198
Vidal, L., Vallarino, A., Benítez, I., & Correa, J. (2015). Implementación del plan estratégico Ramsar en humedales costeros de la Península de Yucatán: Normativas y regulación. *Latin American Journal of Aquatic Research, 43*(5), 873–887. https://doi.org/10.3856/vol43-issue5-fulltext-7
Villalobos-Zapata, G., & Mendoza, J. (2010). *La Biodiversidad en Campeche: Estudio de Estado.* Comisión Nacional para el Conocimiento y Uso de la Biodiversidad, Gobierno del Estado de Campeche, Universidad Autónoma de Campeche, El Colegio de la Frontera Sur.
Wakida-Kusunoki, A., & Caballero-Chávez, V. (2009). Efectos del derrame de hidrocarburos del pozo Kab sobre la pesca ribereña en el litoral de Campeche y Tabasco, México. *Ciencia Pesquera, 17*(1995), 65–73.
Wakida-Kusunoki, A., Rojas-González, R., Toro-Ramírez, A., Medina-Quijano, H., Cruz-Sánchez, J., Santana-Moreno, L., & Carrillo-Nolasco, I. (2016). Caracterización de la pesca en camarón en la zona costera de Campeche y Yucatán. *Ciencia Pesquera, 1*(24), 3–13.
Yáñez-Arancibia, A., Lara-Domínguez, A., Rojas-Galaviz, J., Sánchez-Gil, P., Day, J., & Madden, C. (1988). Seasonal biomass and diversity of estuarine fishes coupled with tropical habitat heterogeneity (southern Gulf of Mexico). *Journal of Fish Biology, 33*(Suppl. A), 191–200.
Yáñez-Arancibia, A., Lara-Domínguez, A., & Day, J. (1993). Interactions between mangrove and seagrass habitats mediated by estuarine nekton assemblages: Coupling of primary and secondary production. *Hydrobiologia, 264*, 1–12.
Zalik, A. (2009). Zones of exclusion: Offshore extraction, the contestation of space and physical displacement in the Nigerian delta and the Mexican Gulf. *Antipode, 41*, 557–582.
Zenteno-Escutia, P. (2013). Transformación político-territorial, infalibles en el aprovechamiento del petróleo en la Bahía de Campeche. *Revista Geografica, 153*, 153–166.

Chapter 13
The Dispute Between Fishing and Oil Over the Maritime Territorial Space of the Sonda de Campeche

E. Solano-Palacios and M. Frutos-Cortés

13.1 The Disputed Maritime Territorial Space of the Sonda de Campeche and the Geopolitical Context

When we talk about geopolitics, we are not talking about political geography, since they are different concepts. In the case of political geography, according to Sánchez (2001), it is understood that it "is interested in the relationship between the political organization of society and geographic space at all levels and scales" (*op cit.*, p. 18). That is, it deals with the power relations that may exist between individuals, between individuals and institutions, or between institutions or social groups at different territorial scales (global, international, or local), thus constituting a subdiscipline of geography that refers to a field of scientific knowledge whose object of study is linked to the power relations of social groups in a given space.

However, regarding geopolitics, in general terms—although a stricter definition could be adopted—it is defined as the science of war or that of states. However, there are at least three dominant conceptions in geopolitical thought: the first conceives it as a philosophy used to justify territorial expansion, originating from the followers of the German school of geopolitics; the second considers it synonymous with political geography; and the third refers to guiding a country's security policy in terms of its geographic factors (Muñoz, 2001). For the specific case of this paper, the latter conception will be adopted.

From this perspective, it is considered that despite the transformations in geopolitics, what has not changed is the role played by geographic space (García & Ronquillo, 2005), because international actors operate within it, factors that influence the determination of this strategic space: the characteristics of states, the unity

E. Solano-Palacios (✉) · M. Frutos-Cortés
Universidad Autónoma del Carmen (UNACAR), Ciudad del Carmen, Campeche, México
e-mail: esolano@pampano.unacar.mx; mfrutos@pampano.unacar.mx

A. Espinoza-Tenorio et al. (eds.), *The Fishing and Oil Industries in Coexistence*,
https://doi.org/10.1007/978-3-032-02621-7_13

of peoples, the cultural and economic elements, and the actors involved, their hierarchy based on the role they play in geographic space. Continuing with García and Ronquillo, this category has currently been used to explain the behavior and future actions of various international actors, including not only the State but also transnational corporations, global organizations such as financial, political, military, and environmental organizations, as well as other actors (*op cit.*).

For Lacoste (1977), geopolitics is defined as the discipline that studies rivalries over territories, countries, and continents, with the understanding that from the moment we become aware that the articulation of knowledge regarding space is a strategic form of knowledge, it becomes a form of power.

Regarding maritime territorial space, at the Latin American level, Cunill (1995) argues that, beginning in the 1940s, there began to be a sustained emergence of maritime and Antarctic landscapes in relation to jurisdictional rights over the spaces of the American sea, which in universal terms measured from 3 to 14 nautical miles, making up the territorial sea. The cited author points out that the reasons for the valorization of the sea and its resources were the implications of the first formulas for expanding and delimiting underwater maritime space, resulting from the search for underwater oil. With the signing of the Gulf of Paria Treaty in 1942 between Venezuela and Great Britain, signed by its colony of Trinidad, the continental shelf was incorporated into the domain of coastal states.

In the case of Mexico, its extensive coastline gave rise to international claims with the United States and Cuba, in 1972 and 1975, respectively, particularly in relation to fishing activities (Székely, 1991). However, within the geostrategic interests of the Mexico-United States bloc, they signed the Treaty on the Delimitation of the Continental Shelf in the Western Region of the Gulf, beyond 200 nautical miles, in June 2000 in Washington, D.C. This situation exemplifies how, over the last 40 years, nation states have appropriated large swathes of maritime space. Specifically, the application of the 200-mile exclusive economic zone rule entails the annexation of approximately 35 percent of the ocean floor and the resources of the overlying waters. Overall, the global surface area under state control and jurisdiction has almost doubled (Urteaga, 1988).

The exploitation of ocean resources has a geopolitical dimension for the same reasons as the exploitation of almost any land area. Because they are limited resources, and in some cases very limited, and also because their geographical distribution is uneven. What differentiates them from other natural resources is that maritime space has not traditionally had a defined legal status; this understandably reinforces the conflicts surrounding its use and appropriation. This conflict is logical, considering that technological development has made new marine resources accessible and drastically changed the methods of their exploitation, just as nation states' awareness of the limited nature of many natural resources and their strategic value has grown in recent decades.

Within this framework, the Gulf of Mexico is one of the territorial maritime spaces where social relations of an economic, cultural, and political nature have developed at various historical moments, as Mexico shares borders and boundaries with other countries on the continent, such as the United States and Cuba. This geographical location makes it a strategic area of international interest due to its significant fishing and petrochemical potential. Historically, there has been ongoing political tension between the United States and both Latin American nations over the occupation of their territories and natural resources. Within its Exclusive Economic Zone (EEZ), Mexico possesses rich oil deposits that are key maritime territorial spaces for the industrial development of energy in the entire northern region of the American continent.

The Sonda de Campeche is located in the Gulf of Mexico basin. This marine zone became part of the dynamics of oil extraction production in 1975, with the announcement by the state-owned oil company Petróleos Mexicanos (PEMEX) of the presence of hydrocarbon fields 75 km from *Ciudad del Carmen* (Tribuna, August 13, 1975). This initiative was conceived in response to an oil-related economic policy facing a global crisis, and at the national level, it was considered strategic for maintaining the country's position in the oil market.

Under these conditions, at the regional level, the Sonda de Campeche[1] was integrated not only into what some authors called the oil boom in southeastern Mexico (Tudela, 1984), but also into an international and national context of geopolitical significance, thanks to the accelerated exploitation of the main oil regions of Mexico and the world, starting in the 1970s and continuing into the first decades of the twenty-first century.

The Sonda de Campeche is indisputably linked to *Ciudad del Carmen*, considered for many years the oil capital of Mexico due to its location opposite the oil fields of this geological province, and to *Isla del Carmen*, the former shrimp producer on the Gulf of Mexico coast. Both activities formed their own socio-productive systems at different scales and spatiotemporal stages, involving social and economic actors, as well as diverse components and problems. In the case of the shrimp production system, it began in the late 1940s, 30 years before the formation of the oil production system in the same area of Campeche. It subsequently entered into dispute over maritime territorial space due to the spatial displacement of deep-sea fishing, due to the Mexican state's energy policy, which prioritized the exploitation of one of the world's giant oil fields: the Cantarell field. These socio-productive systems have been characterized as part of an extractive model and a regional enclave economy, corresponding to economic development models that prioritize the exploitation of natural resources converted into raw materials and linked to national and international markets tied to geopolitical frameworks.[2]

In the Sonda de Campeche, the territorial configuration under these two socio-productive systems has undergone three stages: one that we call "exploited

---

[1] The Sonda de Campeche is located in the territorial waters of the Gulf of Mexico, in the SWGoM and off the coasts of the states of Tabasco and Campeche (Ángeles, 1996).

[2] It is important to note that shrimp fishing in the Sonda de Campeche began with fishing by American, Cuban, and Japanese vessels. Later, due to Mexican government regulation, it was exploited by Mexican concessionaires. In the case of oil production, the Mexican government has regulated its extraction through PEMEX, and the energy reforms of 2008 and 2013 allowed private capital participation.

Table 13.1 Main components of the dispute process between productive systems (2001–2024)

Fisheries	Oil
– Reduction of fishing zones and delimitation of maritime exclusion zone (Sonda de Campeche in 2003)	– Expansion of maritime exploration and exploitation zone
– Government program for managing fishing effort in the Gulf of Mexico (Gulf of Mexico - 2005)	– International marketing program
– Regulatory framework for temporary extension of closed seasons	– National security policy against terrorism (Intersecretarial Agreement 117)
– Trust Fund Program for productive-social and environmental support (PEMEX-Government)-PACMA	– US energy security policy on the continent
	– Capital investment and privatization of exploration and exploitation areas with the implementation of the 2013–2018 energy reform

Note: Main characteristics of the exclusive territories and disputes between fishing and oil in the Sonda de Campeche (DOF, September 11, 2003)

territories," where shrimp fishing predominated exclusively from the 1940s to approximately the mid-1970s. In the second stage, called "competing territories," not only did fishing activity predominate, but it also marked the beginning of oil exploitation beginning in the 1970s, which has become a predominant activity until the present day, with a lesser presence of deep-sea shrimp fishing; and the third stage, corresponding to the first two decades of the twenty-first century, known as exclusivity and dispute, characterized by the hegemony of oil and the almost total exclusion of fishing within an expanding and deepening oil territory, but with depleted oil fields, in addition to the presence of national and transnational companies with private investment in the sector.

It is in this last stage that the dispute between the two socio-productive systems takes center stage. This is characterized by the fact that oil activity is in a phase of intensive exploitation by PEMEX and the presence of private capital companies, as well as the development of new oil fields that generated tension and conflict with the shrimp fishing sector, not only in *Ciudad del Carmen* but also throughout the Gulf of Mexico fleet due to the advances in the occupation of more spaces in the so-called exclusion zones (Table 13.1).

The reconfiguration of the maritime territory tended to exclude shrimp fishing activity because it required the installation of the physical infrastructure for exploration and intensive exploitation of the Cantarell oil field, which began its decline in 2006 (Fig. 13.1). PEMEX, the former parastatal and now productive company of the Mexican State,[3] predominated in the configuration of the maritime territorial space during the period of exclusivity and dispute.

[3] PEMEX was created by the Lázaro Cárdenas administration in June 1938 (DOF, July 20, 1938), months after the oil expropriation in March. Its transformation into a productive state-owned company, 75 years later, was conceived within the framework of the 2013 energy reform (DOF, August 11, 2014).

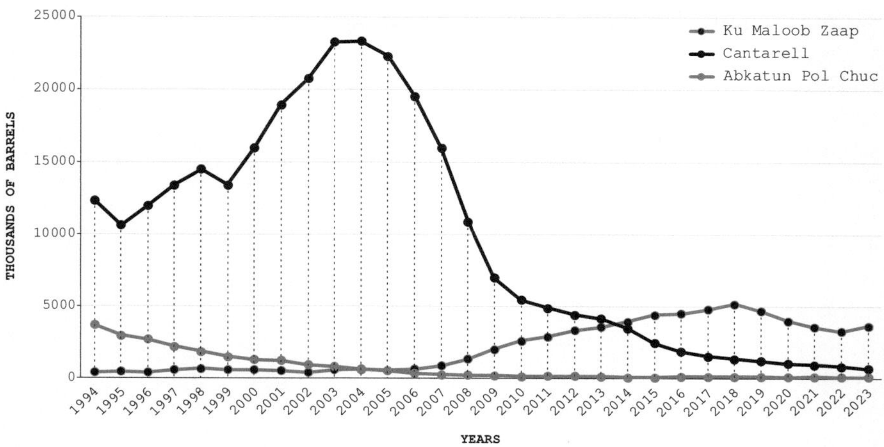

Fig. 13.1 Oil production in the Sonda de Campeche, 1994–2023. Note: The graph shows oil production in the Sonda de Campeche with the exploitation of the main producing fields: Cantarell, Ku Maloob Zaap, and Abkatun Pol Chuc, from the 1990s to the second decade of the twenty-first century. Source: Prepared with data from the National Hydrological Commission (CNH, 2024b)

It is important to note that in 1949, the government of Miguel Alemán decreed that the subsoil of the lands covered by the territorial waters of the Gulf of Mexico, adjacent to the area between the Santecomapan Bar in the state of Veracruz and the Paso Real Bar in the state of Campeche, within a 5-km extension from the low tide line, would be incorporated into PEMEX's assets (DOF, March 11, 1949). In that same presidential decree, PEMEX was granted the concession for inland waters covering the subsoil of the land covered by the waters of Lagunas del Carmen, Machona, and Mecoacán in the state of Tabasco and Laguna de Términos in the state of Campeche, starting from the line marked by the average high water levels in those lagoons.

Subsequently, in the 1970s, the state-owned company expanded its hydrocarbon and gas exploration and production areas, particularly in the Sonda de Campeche with the discovery of the giant Cantarell field. This was due to the geological studies conducted previously, which provided the basis for detecting the presence of hydrocarbons in the sea. Drilling of the Chac 1 well began in June 1974, and its phase as an oil and gas producer ended in 1976 (AMGE, 2020). Regarding the importance of the Campeche coast for the hydrocarbon industry, researcher Javier Meneses stated the following in 1983:

> This Chac 1 well had a depth of approximately 5,000 meters, cutting a sedimentary column from the Recent to the Jurassic. It ultimately produced oil in a Paleocene calcareous breccia at a depth of 3,500 meters, with a daily production of 952 barrels in March 1976. The geological information obtained was very valuable. This well is the one that revealed Mexico's significant marine oil wealth, the most important after the Persian Gulf (Meneses, 1983, p. 64).

Beginning in the 1980s and continuing into the twenty-first century, there was a process of oil expansion by sea and land in the Campeche region, in line with production and the participation of stakeholders involved in this process. However, it wasn't until 1990 that PEMEX initiated a process of economic modernization within the framework of the neoliberal economic policies implemented by the federal government as part of the structural adjustment program. PEMEX's modernization had the following objectives: (1) to guarantee energy sufficiency; (2) to strengthen the energy sector's connection with the economy, society, and environmental protection; and (3) to consolidate a modern and better-integrated energy sector.

The process of exclusion from the deep-sea fishing production system by the oil production system occurred in three phases. The first phase took place on Cayo Arcas Island, a safe harbor for fishers, which, starting in 1978, became the port of arrival for oil tankers exporting oil extracted from the Cantarell field. The second phase of exclusion occurred between 1984 and 1985, primarily due to the expansion of oil activities, including the laying of pipelines from offshore platforms to the coast. This expansion led to the construction of a gas compressor and nitrogen plant located on the Atasta Peninsula, in the municipality of Carmen. A third phase occurred in 2003 due to national security concerns, during which the perimeter of both maritime and air restrictions on all vessels or ships in the Sonda de Campeche was expanded. The first two phases were driven by the need to expand oil activity, and the third was motivated by national security concerns.[4]

The area occupied by oil activities included the Sonda de Campeche, as well as the deep-sea ports of Isla del Carmen, located in *Ciudad del Carmen*, the port of Seybaplaya (located in the current municipality of the same name (formerly belonging to the municipality of Champotón), and the port of Lerma (located in the town of the same name). These ports are currently occupied by oil companies, not only by PEMEX, but also by companies that have contracts derived from the 2013 energy reform.

The Isla del Carmen port is located within the urban area of the Municipality of Carmen. It was established in 1949 under the name *Laguna Azul* to serve as a deep-sea port for shrimp fishing and is part of the Campeche State Port System. It is currently used as a logistics platform for oil activities, with more than 100 companies operating there that provide services to marine facilities in the region. Considered an industrial and fishing port, PEMEX uses it for coastal shipping for its supply, inputs, pipe loading, metal structures, boats, ropes, and personnel movement. With the approval of the energy reform in 2013, this port was considered for expansion of

[4] Interview conducted in Ciudad del Carmen, Campeche, on September 8, 2008, with Engineer Ramón Martínez, aquaculture engineer and environmental consultant in the region. Most informant interviews were conducted during 2023 and 2024, digitally audio-recorded and transcribed using Nvivo software. A total of 20 testimonies from fishers and PEMEX workers were transcribed, with the prior signing of an informed consent form guaranteeing the confidentiality and proper use of personal data.

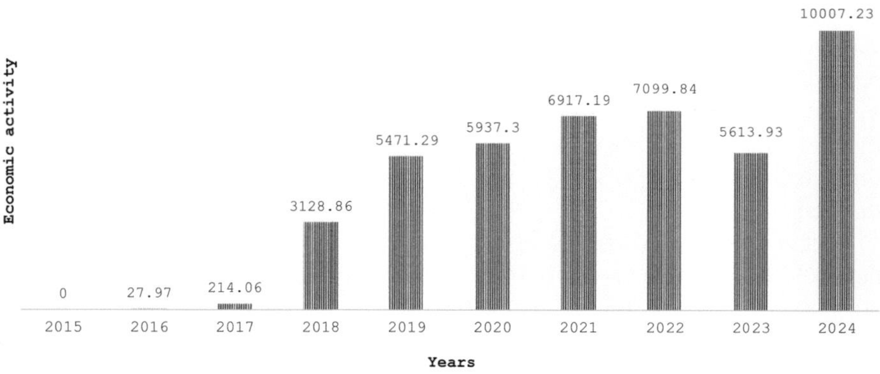

Fig. 13.2 Investments approved by contracts and allocations in Campeche, 2014–2024 in million dollars. Note: The graph shows the significant increase in investment in Mexico's oil industry, despite the drop in production. Prepared with data from the Sistema de Información de Hidrocarburos (2014–2024)

its offshore activities in the Sonda de Campeche and surrounding areas due to the presence of approximately 100 companies linked to the oil sector.[5]

The period of exclusivity and territorial dispute in the Sonda de Campeche has been characterized by the presence of national and transnational companies, as well as PEMEX itself, and a lesser presence of fishing activities and personnel, not only deep-sea shrimp farming but also small-scale fishing, which occurs in the area surrounding Laguna de Términos.

With the energy reform, according to data from the National Hydrocarbons Commission (CNH), the presence of transnational oil companies in Mexico reached 73 companies from 20 countries, directly participating in oil rounds, where 107 exploration and extraction contracts were awarded by the Mexican government. The total investment was $10.445 billion, and in the case of the Sonda de Campeche, companies such as ENI, Lukoil, Repsol (CNH, June 27, 2024a), among others, arrived (Fig. 13.2).

13.2 The Dispute Over Territorial Maritime Space from the Perspective of the Actors and Productive Alternatives

As previously stated, the Sonda de Campeche is the geographic area where natural resources, hydrocarbons, and natural gas are exploited by PEMEX and its service providers. This area also includes marine species such as the brown shrimp

[5]These services include food and hospitality, diving services, fumigation, supply of materials and equipment, and personnel transportation. Supplies, boats, drinking water, fuel and lubricants, maritime support and logistics, among others (API Campeche, 2024).

(*Farfantepenaeus aztecus*), pink shrimp (*Farfantepenaeus duorarum*), and white shrimp (*Litopenaeus setiferus*). These crustaceans are in high demand on the international market and are exploited by Mexican vessels from the states of Tamaulipas, Veracruz, Yucatán, Quintana Roo, and Campeche. Fishers are organized in cooperative fishing production societies, goods, and services, and by private enterprise through business groups. Various actors in both production systems still interact in this area.

Shrimp fishers' maritime territory is known for their sense of direct ownership of nature, which allows it to exist in the local knowledge of some fishers and shipowners. For them, the discovery of oil in the Sonda de Campeche was made by a shrimp fisher from the city, Mr. Rudecindo Cantarell. The tension over maritime territorial space has conflicting manifestations for the different actors involved in this process. On the one hand, there are oil industry workers who view the territory as a temporary stopover, providing them with employment and better living conditions for themselves and their families. They are far from having a closer connection with the local population and the place. The "platformers" who have had no ties to shrimp fishing have found a permanent home in the oil industry within the region; most have decided to settle in Ciudad del Carmen, pursuing their family and work life for short periods or establishing themselves permanently there:

> I am originally from Veracruz, born and raised in the state of Veracruz, in the city of Veracruz and my circumstance of staying in this city is temporary, really only for periods of work specifically to work in the oil industry, I work for a company that is subcontracted by PEMEX for offshore construction... logically, the economic aspect, well yes there are many benefits let's say that practically from the previous company where I was to the company where I am the economic change in income was very important, obviously more, in some way as a family it also allows circumstances of a better quality of life, a better social status or perhaps we at some point acquire a greater number of commitments, maybe if I were not working in this area maybe I would not have the family that I have, my family would be much smaller and I would not have the economic benefits that at some point we tried to channel for the future of the children.[6]

However, for the deep-sea shrimp fishers of *Ciudad del Carmen*, the territory is considered part of their living space, which means employment and permanence in the place, as well as helping their families with the income they earn, even though they consider that they arrived at a time of scarcity in shrimp production. Here is a testimony:

> I'm from Villahermosa, Tabasco, and I came here to work during the shrimp season. My family is here in Ciudad del Carmen, and they told me, "Well, come here, son, there's work here, there's work for everyone, there's lots of work, lots of shrimp," and that's how I came here. I came here when I was 12 and started working as a courier. I started as a courier, then moved on to cooking, then to a winchman, and finally became a truck driver as I got a little older. However, since the product is no longer available, there's nowhere to work. The shrimp business lasted quite a while and that job helped me a lot, because I defended myself, I kept my house and helped my family and everyone learned, my brothers, everyone

[6] Braulio Chamorro, an oil services company worker in the Sonda de Campeche since 2000. Interview conducted in Ciudad del Carmen, June 2024.

> learned, they made a career and when I'm down, they sometimes give me a hand because they recognize that I helped them a lot, well, although I didn't learn anything, but anyway, yes, I was the oldest.[7]

In each activity, the worker's relationship with the territory generates a relatively similar perception of knowledge of it as a lived and productive space. Thus, for fishers, the marine territory where they fish is recognized due to their continuous voyages through the Sonda de Campeche and the harvesting of fish resources. For workers in the oil industry, the territory where they work is associated with areas of oil infrastructure, such as platforms, vessels, and productive activities of hydrocarbon exploration and exploitation, which are technically specialized.

> Look, in the sea we have Cantarell, then we have the Akatun mantle, the Ku Malob Zab field, there is a new field that is Litoral Tabasco that is close to the coast, consider *Atasta* throughout the basin up to Dos Bocas, that is where they are extracting a lot of good quality oil and gas right now, that strip, the marine strip that is close to the coast there is a lot of gas in the entire Gulf of Mexico, because I have gone to work as far as Tampico, all the areas that have been explored there is a lot of gas, we have gas now that is even enough to give away.[8]

However, in the process of territorial exploitation by both activities at different times, confrontation and tension arise, as the territory has been socially transformed by its appropriation by economic agents, transforming it into a transnational social space resulting from human action on natural resources, such as shrimp fishing and hydrocarbon production. Thus, in this phase of confrontation and tension, visions emerge that assume a territorial dispute not from an individual perspective, but rather from the institutions of the Mexican State, which generate the legal resources to exploit the territory, as well as the development model for the region. In this sense, the perception that individuals present regarding this situation is considered a benefit or detriment to their occupational activity, which translates into the polarization of social relations built through processes of exclusion and socioeconomic precariousness in the region (Fig. 13.3).

In addition, the so-called socio-environmental complaint industry was also supported by both the small-scale and deep-sea fishing sectors, leading PEMEX to create a social compensation policy called Mutual Benefit Works (OBM) and the Community and Environmental Support Program (PACMA). These programs are developed within the Corporate Social Responsibility policy that Mexico signed with the United Nations, since it is a company that since its inception has been polluting in oil areas in its exploration, extraction, and production phases (Tampico, Coatzacoalcos, Poza Rica, La Chontalpa (Tabasco), Chiapas, Campeche (Sonda de Campeche). This situation has led to problems with claims and socio-environmental conflicts among different social sectors, especially rural ones, due to the demand for

[7] Interview with Joaquín García, deep-sea fisher from Ciudad del Carmen. Conducted in Ciudad del Carmen, Campeche.

[8] Interview with Manuel Téllez, an oil worker at Petróleos Mexicanos. Conducted in Ciudad del Carmen, Campeche.

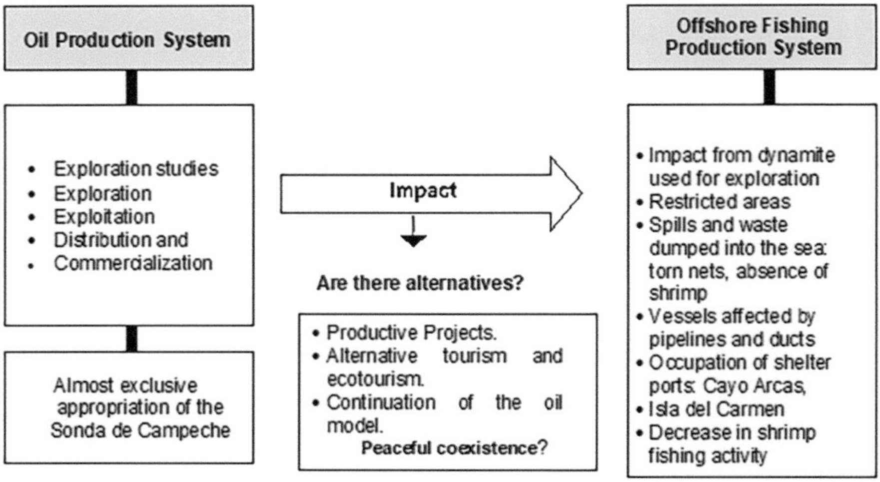

Fig. 13.3 Impact and alternatives of the oil production system on the fishing production system. Note: Own elaboration

environmental damages, such as water pollution, fishing area degradation, pasture conversion to agricultural fields, and acid rain (Toledo et al., 1983; Chenaut, 2017).

However, since 1999 and subsequently with its accession to the UN Global Compact in 2006, PEMEX has promoted and managed the image of a clean, environmentally responsible, and socially responsible industry (La Jornada, January 5, 2006). This type of environmental action, amid a production crisis due to peak oil prices, in addition to its internal crisis (financial, labor, and administrative), has benefited the company in regions such as Tabasco, Veracruz, and Campeche, where socio-environmental demands on rural communities have been addressed by integrating sustainability principles into its production operations, such as: (1) reducing the risk of emissions and discharges into the environment; (2) continuing the development of greenhouse gas emission reduction projects to contribute to combating global climate change; (3) strengthen support for natural areas for the protection of biodiversity in regions where PEMEX operated; and 4) integrate sustainable development concepts across the organization as part of the company's conduct (PEMEX, 2006).

In line with the principle of sustainability, PEMEX's socio-environmental policy is offset by the PACMA, which it began implementing in 2013 in states with oil activities, including Tabasco, Campeche, Veracruz, Chiapas, Tamaulipas, Nuevo León, Oaxaca, Puebla, Guanajuato, Coahuila, San Luis Potosí, and Hidalgo. This type of program is based on Article 4 of the Petroleum Law and requires PEMEX, as well as the national and transnational oil industry bidding for a contract area, to allocate 2 percent of their investment to the population through actions in the areas of education, sports, health, environmental protection, public safety, civil protection, equity, productive projects, and infrastructure. In the case of the state of Campeche, PACMA began to allocate greater resources to the fishing sector and

Table 13.2 PEMEX support to the state of Campeche through PACMA, 2019–2025

Year	Type of support	Beneficiary State	Support (MXN)	Source
2025	In-kind support (fishing gear, engine parts)	Ciudad del Carmen	Not specified (in 2024 it was around 23 million)	PACMA
2024	In-kind support (nets, repairs, other supplies)	Ciudad del Carmen	20,000,000	PACMA
2023	In-kind support (nets, engine parts)	Ciudad del Carmen	34,000,000 (delayed)	PACMA
2021	Social actions (includes support for fishers)	Campeche	1,496,000,000 (total for various actions)	PACMA
2019	Support for the fishing sector	Campeche	16,000,000 (planned)	PACMA

PEMEX (2019–2025). Gerencia de Responsabilidad y Desarrollo Social. Tabular Data. 2019–2025

social actions, which were primarily granted to the municipality of Carmen, mainly in *Ciudad del Carmen* (Table 13.2).

The energy reform promoted by the Peña Nieto administration in 2013 raised expectations of expanding the oil production system in the Sonda de Campeche. This did not materialize, but it did have a negative impact by increasing the state of Campeche's economic dependence on resources sent by the Federation, particularly in Ciudad del Carmen. This impact was not initially accepted by the state government, which stated that the state of Campeche did not depend on the arrival of PEMEX to reactivate the local economy, even though this is an industrial energy sector to which other service companies are linked to the oil company. The government proposed that one of the areas to reactivate the local economy was the deep-sea and coastal port of Seybaplaya, suggesting that when companies arrived to carry out port operations, other services, such as hotels and food services, would be required (Tribuna, March 14, 2020).

More than 40 years after the arrival of oil exploitation in the Sonda de Campeche, local business and government sectors have begun to consider not only the continuity of the oil extraction model in the region, but also new economic alternatives that allow combining PEMEX's presence in the region with a new socio-productive model, such as tourism, which facilitates links with the megaprojects in southwestern Gulf of Mexico (SWGoM) that are priorities for the so-called the Fourth Transformation (4 T) government. It is in this vein that the First Port Industrial Forum was prepared at the beginning of 2024. This initiative was organized by the Business Coordinating Council (CCE, by its Spanish acronym) of Ciudad del Carmen in coordination with the government of the state of Campeche (Secretariat of Economic Development, Secretariat of Tourism, State Energy Agency of the State of Campeche) and the municipality of Carmen (Tourism Coordination) (CCE, January 24, 2024).

The objective of this business and government event was to present the future of the oil and port industry in SWGoM through lectures by energy and oil specialists, as well as the showcase of services from companies in the region's oil and gas sector (Port Industrial Forum, March 13, 2024). Inexplicably, the small-scale and deep-sea fishing sector in Campeche and the Gulf of Mexico were absent, as they were not included in the First Port Industrial Forum to discuss the future of the regional economic model.[9]

Finally, in the case of the business actors organized in the Business Coordinating Council of Ciudad del Carmen, they have proposed not being left out of the oil development of the Sonda de Campeche or SWGoM in the next 20 years. To this end, they have presented proposals for the state or federal government to contribute investment, thereby triggering job creation, services, and integration into the regional oil and rail circuit that was developed during the 2018–2024 6-year term. Some of these proposals are: 1) the Atasta Pueblo-Ciudad del Carmen highway bypass (exceeding MXN 6 billion); 2) the dredging of the navigation channel of the Isla del Carmen industrial port; and 4) the expansion of the Sabancuy-Villahermosa-Escárcega highway. Once again, in this series of proposals, the traditional and deep-sea fishing sector has been relegated; that is, there is no room to include an activity that, due to its sociocultural value and geographic nature, should be a priority for the state of Campeche.

13.3 Conclusions

The Sonda de Campeche is a maritime territorial space of interest to the regional economy of the SWGoM, due to the oil, fishing, and port activities carried out by Mexico and its neighboring countries, such as Cuba and the United States. This makes it a present and future geopolitical frontier at the local, regional, national, and international levels.

The oil production system continues to operate and expand its activities in the Sonda de Campeche. Oil and gas exploration and extraction to supply the domestic and international markets continues with Mexico's trading partners, but with a significant variation added by the Fourth Transformation government, which can be defined as a "sovereignty policy." This initially consisted of suspending auctions for contracts derived from the energy reform approved in 2013. This so-called neoliberal energy reform was considered a viable and necessary public policy by business

[9]Among the service companies that registered for the First Port Industrial Forum are the following: ENGIE Mexico, TotalEnergies, Applus+ Mexico, Lukoil Lubricants Mexico, VIKING Life-Saving Equipment, Arrow Marine AMLS, Naviera Portus Nostrum, Base Maritima Coral, Servicios de Ingenieria Naval Panamá México, Autotab Commercial Vehicles, DNH, Sponyat Jet, Procarga, Cargoweld CCUs & Power Solution, BS GRÚAS, Argo Communications, Corporación Itaipu SA de CV, Mota-Engil Mexico, Cargoweld CCUs & Power Solution, Tormag Mexico, ATE GROUP, Safety Offshore of Mexico, Ocean Divers Mexico, Cornelius Offshore Solutions, Andamios Ring Lock de Mexico, Cargo Lift, Sistemas Visuales 360, Studio 7, Servicios Maritimos y Terrestres CDA, NGlobal Logística, Remagsa SA de CV., Chevrolet de la Era, Casa Briceño, Servicio Cooperativa de Pemex, PACADA, AMPI Ciudad del Carmen, Gestión 360, Ecosolutionsmx, La Jornada Maya, GP Oil Magazine, GE Business Management Magazine, Phoenix Publishing (Port Industrial Forum, March 13, 2024 https://www.facebook.com/FIP.2024

sectors in the SWGoM. However, given the negative results and unmet expectations it left during its implementation, it ultimately "disillusioned" the private sector itself (Barbosa, 2016, p. 16) and was rejected by the government of Andrés Manuel López Obrador and his project, known as 4 T.

The negative impact of the energy reform in the area of influence of the Sonda de Campeche can be summarized as follows: 1) the decline in local employment in PEMEX's exploration and production areas, coupled with the fall in global oil prices in 2014; 2) the oil reform also entailed a reform of PEMEX, which was forced to cut jobs in almost the entire country; The State Productive Enterprise was allocated to stimulate private capital investment and business. One hundred and ten contracts with a total investment of MXN $143 billion were authorized in the oil rounds awarded by the National Hydrocarbons Commission (CNH, 2018, p. 3). The awarding of these oil rounds had very little benefit to peasant communities, *ejidos*, fishers, and population centers near oil fields on land and offshore, despite the commitment to comply with social and environmental policies such as the PACMA (Mexico Oil Rounds), which was established in the contractual areas of the oil sector (Rondas México, n.d.).

More than 10 years after the energy reform and with only 110 contracts awarded, oil activity has increased minimally in the extractive areas of SWGoM. The oil industry has invested more than MXN $27 billion from 2015 to 2023. The investment potential is even greater, with over MXN 47 billion in approved investments throughout the entire life of the contracts, which could span several decades (AMEXHI, 2023).

Meanwhile, the deep-sea fishing production system, which had exploited shrimp farming areas in the region for decades, continues to operate without the fishing grounds currently reserved exclusively for a group of national and foreign oil companies, which hold current contracts in offshore and onshore blocks. Despite the high socioeconomic impact that fisheries continue to have in Campeche, as they are resources with commercial value and are a source of employment and income for thousands of families (Caballero & Morales, 2021), the sector suffers from a series of competitive disadvantages such as capture seasons and closed seasons, lack of regulation in the management of information on their status, intense pressures to travel to more distant fishing areas, lax regulations for extraction and marketing, dependence on intermediaries, frequent thefts of engines and damage to fishing equipment, and finally, the precariousness and aging of fishers who will hardly find generational replacement.

References

Administración Portuaria Integral (API) Campeche. (2024). *Prestadores de servicio puerto Isla del Carmen*. https://apicampeche.com.mx/directorio/prestadores-de-servicio/isla-del-carmen/

Ángeles, F. J. (1996). Estratigrafía del jurásico superior del subsuelo en la Sonda de Campeche (Golfo de México). In *Tesis de maestría en ciencias con especialidad en Geología*. Instituto Politécnico Nacional. https://revistamaya.com/wp-content/uploads/2022/04/Angeles-Aquino-1996.pdf

Asociación Mexicana de Empresas de Hidrocarburos (AMEXHI). (2023). *Convención nacional petrolera 2023: claves para el crecimiento energético de México*. https://www.amexhi.org/comunicados/amexhi-lleva-a-cabo-la-septima-edicion-de-la-convencion-nacional-petrolera-2023-claves-para-el-crecimiento-energetico-de-mexico/

Asociación Mexicana de Geofísicos de Exploración (AMGE). (julio-septiembre, 2020). Complejo Cantarell. Aportaciones referentes de los ingenieros geofísicos de Pemex en el 50 aniversario de la instauración de la carrera de ingeniería de geofísica en México. *Boletín técnico de AMGE, A.C. Boletín técnico* No. 3. Vol. 3: 36–38. https://issuu.com/solucioncuatro.4/docs/boletin_3__vol.3/58

Barbosa, F. (2016). Paradojas de la Reforma Energética. En Solano, E. y Frutos, M. (coeditores), *El impacto de la reforma energética en México: una mirada nacional y regional* (15–40). AMECIDER/UNACAR.

Caballero, V., & Morales, R. G. (2021). *Pesquería de escama marina en el estado de Campeche*. Instituto Nacional de Pesca y Acuacultura.

Chenaut, V. (2017). Impactos sociales y ambientales de la explotación de hidrocarburos en el municipio de Papantla, Veracruz (México) Chenaut, V. y Valladares, L. *Etnicidad, territorio y pueblos indígenas: paradojas y desafíos de las políticas neoliberales*. No. 28. https://doi.org/10.4000/eces.2261. https://journals.openedition.org/eces/2433

Comisión Nacional de Hidrocarburos (CNH). (13 de junio de 2018). CNH ha aprobado más de 12 mil millones de dólares de inversión en exploración y desarrollo de contratos. *Boletín de la CNH*. https://hidrocarburos.gob.mx/media/1146/20180613-cnh-ha-aprobado-m%C3%A1s-de-12-mil-millones-de-d%C3%B3lares-de-inversi%C3%B3n-en-exploraci%C3%B3n-y-desarrollo-de-contratos.pdf

Comisión Nacional de Hidrocarburos (CNH). (2024a). *Rondas petroleras*. Comisión Nacional de Hidrocarburos. https://rondasmexico.gob.mx/media/5447/20240627_cifrasrelevantes.pdf

Comisión Nacional de Hidrocarburos (CNH). (2024b). Mapa de hidrocarburos. https://mapa.hidrocarburos.gob.mx/

Consejo Coordinador Empresarial de Ciudad del Carmen. (24 de enero de 2024). https://www.facebook.com/CCE.Carmen.AC/?locale=es_LA

Cunill, P. (1995). Eclosión de espacios marítimos y antárticos. Transformaciones y movilizaciones de recursos marítimos. Revalorización de espacios insulares. En *Las transformaciones del espacio geo histórico latinoamericano, 1930,-1990* (94–110). Fondo de Cultura Económica/Colegio de México.

DOF. (11 de julio de 1938). Decreto que crea la institución Petróleos Mexicanos. Secretaría de Hacienda y Crédito Público. *Diario Oficial de la Federación*, 20 de julio de 1938.

DOF. (11 de marzo de 1949). Decreto que incorpora al patrimonio de Petróleos Mexicanos, el subsuelo de los terrenos cubiertos por las aguas territoriales del Golfo de México y otros que el mismo específica. Secretaría de Economía. *Diario Oficial de la Federación*. https://www.dof.gob.mx/nota_to_imagen_fs.php?cod_diario=193134&pagina=4&seccion=1

DOF (11 de septiembre 2003). Acuerdo intersecretarial número 117 en materia de pesca en la sonda de Campeche, *Diario Oficial de la Federación*, https://www.dof.gob.mx/nota_detalle.php?codigo=690229&fecha=11/09/2003#gsc.tab=0 (2003)

García, M., & Ronquillo, G. (2005). *Estados Unidos, petróleo y geopolítica las estrategias petroleras como un instrumento de reconfiguración geopolítica*. Plaza y Valdez editores.

Lacoste, Y. (1977). *La geografía un arma para la guerra*. Elementos críticos.

Meneses, J. (1983). *El nuevo petróleo de México. Crónica de un descubrimiento*. Porrúa.

Muñoz, L. (2001). La geopolítica en la relación entre Estados y la proyección de su interés nacional. En *Geopolítica, seguridad nacional y política exterior México y el Caribe en el siglo XIX* (21–41). Instituto de Investigaciones "Dr. José María Luis Mora"/Universidad Michoacana de San Nicolás de Hidalgo.

PEMEX. (5 d enero de 2006). En el pacto mundial de la ONU sobre empresas y globalización. *La Jornada*.

Petróleos Mexicanos (PEMEX). (2006). *Informe de Desarrollo Sustentable*. https://www.pemex.com/etica_y_transparencia/transparencia/informes/Paginas/informes.aspx
Petróleos Mexicanos (PEMEX). (2019–2025). Gerencia de *Responsabilidad y Desarrollo Social*. Datos tabulares. 2019–2025.
Port Industrial Forum. (13 de marzo de 2024). *Presentación*. https://www.facebook.com/FIP.2024
Rondas México. (n.d.). *Administración de contratos*. https://rondasmexico.gob.mx/esp/contratos/
Sánchez, J. E. (2001). Poder, espacio y geografía política. *Geografía política. Síntesis*:17–39.
Sistema de Información de Hidrocarburos. (2014–2024). I*nversiones aprobadas de contratos y asignaciones en Campeche, 2014–2024*. https://sih.hidrocarburos.gob.mx/
Székely, A. (1991). *El derecho del mar*. Instituto de Investigaciones Jurídicas. Universidad Nacional Autónoma de México. https://archivos.juridicas.unam.mx/www/bjv/libros/1/279/2.pdf
Toledo, A., Farrera, H., & Núñez, A. (1983). *Cómo destruir el paraíso: el desastre ecológico del Sureste*. Centro de Ecodesarrollo.
Tribuna (13 de agosto de 1975). Petróleo en Campeche. *Tribuna. Memorias*, 1975–2005.
Tribuna. (14 de marzo de 2020). https://tribunacampeche.com/carmen/2020/03/14/no-solo-de-pemex-dependemos-para-reactivar-economia/
Tudela, F. (1984). *La modernización forzada del trópico: el caso de Tabasco. Proyecto Integrado del Golfo*. El Colegio de México, CINVESTAV IFIAS, UNRISD.
Urteaga, L. (1988). Los recursos naturales y la nueva geografía política del mar. *Geocrítica*. No. 75. mayo de 1988. Universidad de Barcelona. https://www.ub.edu/geocrit/geo75.pdf

Chapter 14
Environmental Ocean and Coast Policy Instruments and Governance in the Gulf of Mexico

L. M. Gama-Campillo, C. J. Pacheco-Figueroa, J. D. Valdez-Leal, E. J. Moguel-Ordoñez, E. E. Mata-Zayas, C. Villanueva-García, and H. M. Díaz-López

14.1 Introduction

Environmental policy planning in Mexico incorporates strategic measures to protect ecosystems, water quality, terrestrial and marine resources, and biodiversity from natural and human-induced threats, such as resource exploitation or oil spills. This holistic approach is influenced by diverse factors based on environmental and economic perspectives. As one of the most biodiverse countries in the world, Mexico faces the dual challenge of promoting economic growth while managing its natural resources and safeguarding its environmental heritage. The importance of this balance cannot be overstated, as it directly impacts the livelihoods of innumerable communities and the ecosystem's health upon which they depend. In Mexico, environmental planning instruments are crucial tools for managing and protecting natural resources, promoting sustainable development, and mitigating environmental impacts. These instruments are designed to align economic growth with natural environmental preservation and are implemented through various laws, policies, and regulations (Anglés-Hérnandez et al., 2021). The primary environmental planning instruments in ocean regions include:

1. A foundational legal framework for environmental policy:

   - The General Law of Ecological Equilibrium and Environmental Protection (LGEEPA, by its Spanish acronym) is the main environmental law. It establishes the principles and policies for environmental protection, conservation, and sustainable use of natural resources.

L. M. Gama-Campillo (✉) · C. J. Pacheco-Figueroa · J. D. Valdez-Leal ·
E. J. Moguel-Ordoñez · E. E. Mata-Zayas · C. Villanueva-García · H. M. Díaz-López
Universidad Juárez Autónoma de Tabasco, Villahermosa, México
e-mail: lilia.gama@ujat.mx; coral.pacheco@ujat.mx; juan.valdez@ujat.mx; emoguel@ujat.mx; ena.mata@ujat.mx; claudia.villanueva@ujat.mx; hilda.diaz@ujat.mx

A. Espinoza-Tenorio et al. (eds.), *The Fishing and Oil Industries in Coexistence*,
https://doi.org/10.1007/978-3-032-02621-7_14

- The General Law on Climate Change (LGCC, by its Spanish acronym) provides the legal basis of climate policy, outlining strategies for reducing greenhouse gas emissions and adapting to climate change.
- The General Wildlife Law (LGVS, by its Spanish acronym) aims to conserve biological diversity through various mechanisms, including the creation and management of natural protected areas.
- The National Water Law (LAN, by its Spanish acronym) provides the guidelines for governing the use, management, and protection of water resources, ensuring their sustainable use.

- These laws are supplemented by specific norms (NOM, by its Spanish acronym) that provide detailed technical standards for various environmental aspects, including air and water quality, as well as hazardous waste management. They are crucial for implementing and enforcing environmental policies, ensuring compliance with international agreements and treaties.

2. Environmental Impact Assessment (EIA) is a process established in the law that, through an analysis of activities, evaluates the potential environmental impacts of proposed projects before they are carried out, to suggest possible measures to prevent, reduce, or compensate for their effects.
3. Urban Development Plans are required for managing land use in urban areas. They integrate environmental considerations into city and town planning, ensuring that development occurs sustainably through the outline of zoning regulations, infrastructure development, and environmental protection measures, especially in coastal zones.
4. Marine Ecological Planning is a strategic instrument that guides the sustainable use of marine resources. It involves analyzing environmental, social, and economic factors to determine the best use and management of resources in ocean regions. This instrument helps prevent conflicts between development and conservation efforts by designating areas for specific activities based on their environmental suitability.
5. Marine Protected Areas (MPAs) are set aside to preserve biodiversity and ecosystems. These areas are protected under federal regulations and are managed to conserve their natural resources.
6. Climate Change Adaptation and Mitigation Plans, developed to address the challenges caused by climate change, with strategies to mitigate greenhouse gas emissions, enhance energy efficiency, promote the use of renewable energy, and search for adaptation measures to global warming impacts.
7. And pollution control programs, established to reduce pollution from industrial, agricultural, and urban activities, limiting emissions, discharges, and waste, and promoting cleaner production techniques, through regulations and monitoring.

The main objective of Environmental Ocean and Coast Planning Policy Instruments, the subject of this chapter, is to emphasize the importance of sustainable planning. Through this chapter, this planning process will be analyzed, existing instruments will be mentioned, and their formalization and usefulness will be discussed. For Mexico, it is crucial due to the country's rich but vulnerable natural resources. This type of instrument aims to reconcile the diverse and often conflicting interests of various stakeholders, including the government, oil companies, and fishing communities (Ramos-Muñoz et al., 2019), throughout the governance process. Regularly, each group has defined interests in how to use natural resources, but their objectives can often be at odds, and the government prioritizes activities according to the country's economic needs.

14.2 Environmental Planning as a Public Policy in Mexico's Marine Zones

This type of planning represents a critical intersection between policy, environmental stewardship, and economic development. Rather than being a mere administrative exercise, the process is a dynamic and complex system that needs to balance the often-competing demands of natural resource exploitation and environmental conservation (Espinoza-Tenorio et al., 2014). Marine spatial planning (MSP), as described by Díaz-Mondragón et al. (2021), is gaining traction worldwide, being integrated into the political agendas and regulatory frameworks of many countries. Organizations such as the Intergovernmental Oceanographic Commission of UNESCO (IOC-UNESCO) are generating support for spatial planning methodologies and providing technical assistance to implement them.

The Marine Ecological Management Programs in Mexico are environmental policy instruments aimed at establishing guidelines and regulations for the preservation, restoration, protection, and sustainable use of natural resources in specific areas within Mexican marine zones, including adjacent federal areas (coastal areas interface), as outlined by the LGEEPA, and its corresponding regulations. Each marine ecological plan defines zones for different activities, based on suitability, through regulations for productive activities to prevent environmental impacts or conflicts with other sectors. This planning should be dynamic and adjusted according to needs and environmental changes. In that sense, to contribute to ocean health, the employment of marine environmental planning policies must take a holistic approach, considering the interconnections among different spatial zones. Particularly, coastal zones are sensitive areas that serve as buffer and transition zones, supporting diverse ecosystems, including mangroves, coral reefs, and estuaries. These areas are also where human activities, such as agriculture, urban development, and tourism, have the most direct impact on marine environments. In addition, it is critical to consider the potential contributions of river discharges to the marine environment. Rivers often transport pollutants, nutrients, and sediments from inland areas to the sea, significantly affecting coastal and marine ecosystems. For example, excessive nutrient runoff from agriculture can lead to eutrophication, creating zones of anoxia (low oxygen levels) in coastal waters (Signoret et al., 2006). These anoxic zones can be detrimental to marine life, causing loss of biodiversity and the collapse of local fisheries. Therefore, effective marine environmental planning must integrate land-based activities and riverine inputs into its framework to prevent such negative impacts.

In this context, marine environmental planning policies must be designed to comprehensively manage and protect these interconnected spatial zones. This requires coordinated efforts across different sectors and governance levels, including local, regional, and national authorities, to ensure that the health of oceans, coastal, and sea zones is preserved, taking into account territorial planning in these regions. By addressing the unique challenges posed by each of these areas—the open ocean, coastal areas, and riverine contributions—marine environmental planning can promote the sustainable use of marine resources, protect biodiversity, and enhance the resilience of marine ecosystems to human activities and climate change impacts. Currently, there are only three Environmental Marine Use Planning Instruments: The Gulf of California, the North Pacific, the Gulf of Mexico, and the Caribbean Sea. However, some coastal states also have territorial planning instruments, important for the sustainable use of marine resources (Table 14.1). As shown in Table 14.1, the dates of the decrees range from 2002 to 2024, with many of them not having been updated, indicating a lack of follow-up actions in some cases, as stipulated in the regulations.

On the other hand, it is evident that due to the large extensions that marine zones cover (e.g., the Gulf of Mexico and the Caribbean Sea), a massive coordination effort was required to carry out this complex process and to review the needs of the users of the different sectors. Moreover, these efforts are strenuous to generate, which is probably the case in the South Pacific zone, where this planning instrument does not yet exist. On the other hand, it stands out that despite international trends (Díaz-Mondragón et al., 2021), most of these processes were carried out in previous administrations. They have not been completed for all that would correspond to national marine zones. Surely, there has been no follow-up, which results in not knowing if the proposed strategies for the different regions are helpful or were even considered, as indicated by the law.

As stated, currently, the entire marine zone of the Gulf of Mexico and the Caribbean Sea is covered by a single planning instrument. In contrast, in the Pacific zone, which is divided into three areas, only two (28% of this zone) have an instrument (the Gulf of California and the North Pacific). Additionally, 10.78% of the marine region of the Gulf of Mexico and Caribbean Sea has been designated as natural protected areas, while in the Pacific zone, 32% is protected.

In Mexico, the first step toward sustainable planning is to have a planning instrument, and there is a legal framework that provides the methodology for constructing it and the support to implement this type of planning. Laws such as the Fisheries Law and the Environmental Law outline the guidelines and requirements of their implementation and monitoring strategies (Anglés-Hérnandez et al., 2021). These laws also empower government agencies to enforce regulations and to establish Marine Protected Areas (MPAs) where necessary. On the other hand, MPAs are so far the most effective tools in marine planning, as they restrict or regulate human activities in designated zones to protect biodiversity and preserve critical habitats,

Table 14.1 National, marine, and coastal (states or municipalities) Environmental Land or Marine Use Planning, existing official instruments, with their type, the authority responsible for their instrumentation, the decree date, and a link to the official web to access them

Planning instrument	Type	Authority Level	Date	Link
National (Mexico)	Territorial General	Federal	2012	https://gisviewer.semarnat.gob.mx/aplicaciones/uga_oe2/#
California Gulf	Marine	Federal	2006	https://www.semarnat.gob.mx/archivosanteriores/temas/ordenamientoecologico/Documents/documentos_golfo/zip/decreto_presidencial.pdf
Gulf of Mexico and Caribbean Sea	Marine	Federal	2012	https://dof.gob.mx/nota_detalle.php?codigo=5279084&fecha=24/11/2012
North Pacific	Marine	Federal	2023	https://www.dof.gob.mx/nota_detalle.php?codigo=5698272&fecha=11/08/2023
Baja California	Regional	State	2014	https://www.bajacalifornia.gob.mx/Documentos/sest/desarrollo_sustentable/Programas_Proyectos/OrdenamientoEcologico/PROGRAMA_DE_ORDENAMIENTO_ECOLOGICO_DEL_ESTADO_POEBC_2014/DOCUMENTO_COMPLETO_POEBC_2014.pdf
Campeche	Local	Municipality	2014	https://www.semabicce.campeche.gob.mx/wp-content/uploads/2025/03/Campeche.zip
Champoton	Local	Municipality	2018	https://www.semabicce.campeche.gob.mx/wp-content/uploads/2025/03/Champoton.zip
Hecelchakan	Local	Municipality	2014	https://www.semabicce.campeche.gob.mx/wp-content/uploads/2025/03/Hecelchakan.zip
Colima	Regional	State	2025	https://admiweb.col.gob.mx/archivos_prensa/banco_img/file_68151a0e3d95b_POET_2025.pdf
Lagoon of Cuyutlan, Colima	Regional	State	2025	https://admiweb.col.gob.mx/archivos_prensa/banco_img/file_68151a0e3d95b_POET_2025.pdf
Manzanillo	Local	Municipality	2016	https://admiweb.col.gob.mx/archivos_prensa/banco_img/file_68152ab05ea86_POETManzanillo_2016.pdf

continued

Table 14.1 (continued)

Planning instrument	Type	Authority Level	Date	Link
Chiapas	Regional	State	2012	https://semahn.chiapas.gob.mx/portal/descargas/ord_territorial/Estado_Chiapas/decreto.pdf
Sub-basin of the Coapa River, Chiapas	Regional	State	2012	https://semahn.chiapas.gob.mx/portal/descargas/ord_territorial/Coapa/decreto.pdf
Jalisco	Regional	State	2018	http://siga.jalisco.gob.mx/moet/
Cabo Corrientes, Jalisco	Local	Municipality	2011	https://sigat.semadet.jalisco.gob.mx/ordenamiento/bitacora.html
Costa Alegre, Jalisco	Local	Municipality	2021	https://periodicooficial.jalisco.gob.mx/sites/periodicooficial.jalisco.gob.mx/files/09-11-2i-vi.pdf
Michoacan	Regional	State	2011	https://drive.google.com/drive/u/0/folders/1X3qc5wesohirczmn3B9x1UTsNiz2YemV
Coastal mountains of Michoacan	Regional	State	2010	http://dsiappsdev.semarnat.gob.mx/datos/portal/poet/2023/d_sc_191003.pdf
Lazaro Cardenas; Michoacán	Local	Municipality	2019	https://periodicooficial.michoacan.gob.mx/download/2019/agosto/14/6a-1519.pdf
Oaxaca	Regional	State	2016	https://www.oaxaca.gob.mx/medioambiente/programas-de-ordenamiento-ecologicos/
South coast of the territory of Oaxaca	Regional	State	2024	https://www.oaxaca.gob.mx/medioambiente/wp-content/uploads/sites/59/2024/11/Anexo-del-Decreto.-RESUMEN-EJECUTIVO_POERTR-SS-C.pdf
Santa María Tonameca, Oaxaca	Local	Municipality	2021	https://apps1.semarnat.gob.mx:8443/dgiraDocs/documentos/oax/estudios/2021/20OA2021HD023.pdf
Cancun-Tulum, Quintana Roo	Regional	State	2001	http://bitacora-ambiental.semaqroo.gob.mx/images/decretos/decreto_poet_cancun_tulum_-1.pdf
Mayan Coast of Quintana Roo	Regional	State	2006	http://bitacora-ambiental.semaqroo.gob.mx/images/decretos/decreto_costa_maya_31_octubre.pdf
Coastal zone of Sian Ka'an Biosphere Reserve, Quintana Roo	Regional	State	2002	http://bitacora-ambiental.semaqroo.gob.mx/images/decretos/periodico_oficial_oet_sian_kaan%20(1).pdf

Benito Juarez, Quintana Roo	Local	Municipality	2014	http://bitacora-ambiental.semaqroo.gob.mx/images/decretos/Decreto_POEL_MBJ_2014.pdf
Cozumel, Quintana Roo	Local	Municipality	2008	http://sema.qroo.gob.mx/bitacora/index.php/decretos/municipio-de-cozumel?download=10:decreto-cozumel
Isla Mujeres, Quintana Roo	Local	Municipality	2008	http://sema.qroo.gob.mx/bitacora/index.php/decretos/municipio-de-isla-mujeres?download=15:decreto-poel-isla-mujeres
Othon P. Blanco, Quintana Roo	Local	Municipality	2015	http://sema.qroo.gob.mx/bitacora/index.php/decretos/municipio-de-othon-p-blanco?download=19:decreto-poel-othon-p-blanco
Solidaridad, Quintana Roo	Local	Municipality	2009	http://sema.qroo.gob.mx/bitacora/index.php/decretos/municipio-de-solidaridad?download=17:decreto-poel-solidaridad
Coastal area of El Rosario, Sinaloa	Local	Municipality	2024	https://media.transparencia.sinaloa.gob.mx/uploads/files/11633/Tomo%20I%20PMOTDU%202023%202050_25014%20Rosario,%20Sinaloa_compressed.pdf
Coastal region of Sonora	Regional	State	2009	https://gisviewer.semarnat.gob.mx/aplicaciones/uga_oe2/
Tabasco	Regional	State	2024	https://tabasco.gob.mx/bitacora-programa-de-ordenamiento-ecologico-regional-del-estado-de-tabasco
Burgos-Tamaulipas Basin	Regional	State	2011	https://www.dof.gob.mx/nota_detalle_popup.php?codigo=5234595
Bobos and Solteros River basin, Veracruz	Regional	State	2008	https://docs.google.com/viewer?a=v&pid=explorer&chrome=true&srcid=0B87bmmJLaLIgM2EyN2Q5YjAtYjg0ZC00Y2IxLTllOTctZTEyYWJkNGM4MDUy&hl=es
Coatzacoalcos River basin, Veracruz	Regional	State	2008	https://sursureste.org.mx/wp-content/uploads/2022/09/Programa-de-Ordenamiento-Ecologico-de-la-cuenca-Baja-del-Rio-Coatzacoalcos-Veracruz.pdf
Tuxpan River basin, Veracruz	Regional	State	2012	https://repositorio.veracruz.gob.mx/medioambiente/wp-content/uploads/sites/9/2022/11/GOE_242_20_julio_2012_POE-Tuxpan.pdf
Yucatan	Regional	State	2023	https://bitacoraordenamiento.yucatan.gob.mx/
Coastal region of Yucatan	Regional	State	2007	https://bitacoraordenamiento.yucatan.gob.mx/documentos/index.php?IdOrdenamiento=6#

based on planning instruments when available. However, the implementation of these instruments, plans, and policies regularly faces significant challenges. One of the primary challenges is ensuring compliance with regulations. Although the laws are clear, enforcement can be inconsistent, especially in remote areas where monitoring is challenging. Furthermore, economic pressures, such as the demand for oil and the livelihoods of fishing communities, can lead to resistance against restrictions (Solo-Gabriele et al., 2021), even when they are necessary for long-term sustainability.

Mexico has increasingly relied on economic instruments to address these challenges as part of its environmental planning policy. These instruments include financial incentives for sustainable practices, such as subsidies for eco-friendly fishing gear or tax breaks for companies that invest in environmental protection. By aligning economic incentives with environmental goals, these tools help to mitigate the resistance to conservation measures and promote more sustainable behavior among stakeholders. Environmental planning in Mexico's marine zones is a multifaceted process that integrates scientific research, stakeholder engagement, legal frameworks, and economic instruments. It is considered a crucial tool for balancing the needs of economic development with the imperative of conserving marine ecosystems; in this sense, reconciling governance and governability becomes imperative.

14.3 Governance and Governability, as the Support for Planning Instruments

Although governance and governability are related, their intents differ (Brenner & Vargas del Río, 2010). While governance is concerned with the processes, structures, and mechanisms through which decisions are made and policies are implemented in society, governability refers to the government's ability to exercise its authority, to implement policies, and maintain order and stability in society. In that sense. Environmental planning for marine areas in Mexico is not only a policy tool but also a strategic approach that has significant implications for various productive activities. The success of its construction depends on governance and its implementation with governability. In that sense, this overall planning relies on the ability to mediate between the often-conflicting demands of economic growth and environmental conservation, ensuring that Mexico's marine resources continue to provide ecological, economic, and social benefits. Marine areas are home to a wide range of species and ecosystems that are vital not only to the environment but also to the economy, which is undergoing increased exploitation. The Gulf of Mexico is a prime example of the pressure exerted by human activities related to resource extraction. That is why the challenge is to find a balance between the need for economic development and the imperative of conserving these fragile ecosystems through governance. If marine resources are overexploited or degraded, the consequences could be severe, affecting the livelihoods of millions of people who depend on these resources. For instance, overfishing can lead to the collapse of fish populations, threatening food security and the income of fishing communities. Similarly, oil spills and other forms of pollution from extractive industries have had devastating impacts on marine habitats, affecting biodiversity, tourism, and even human health.

For the implementation of this type of instrument, zoning must be carefully delimited to ensure that economic activities are carried out in a manner that minimizes their impacts. The implications of the analysis for zoning and social action are significant, and their success depends on the detailed specification of the sustainability concept (Huppes & Simonies, 2009). To navigate these conflicting interests, these instruments are developed based on governance, serving as a bridge between policy formulation and its practical implementation. Their effectiveness, however, hinges on the specific context in which they are applied, involving a complex web of public and private organizations, as well as numerous individuals, each contributing their own expertise, roles, and perspectives (Jordan & Benson, 2013). Despite their legally binding nature in Mexico, these instruments often struggle to produce the desired behavioral changes, indicating a gap between policy intent and actual outcomes.

Due to the challenges in their implementation, in many cases, the actions require complex and multifaceted management. For example, spatial planning for the establishment of marine protected areas employed different strategies to address pressures on marine ecosystems, such as land-use changes adjacent to watersheds, coastal development, and overfishing. These pressures are often exacerbated by human activities, making the need for robust and adaptive planning for critical situations. The process of developing these instruments is often lengthy and involves multiple stages (Díaz-Mondragón et al., 2021), requiring extensive research and data collection to establish a factual basis for decision-making. This information is then synthesized during a deliberative phase, where it is combined into a comprehensive plan that integrates various types of knowledge and data. Importantly, this process is not linear; instead, it is an interactive one, with new insights being incorporated throughout the planning process. The ultimate goal is to construct a map that would have the competing interests and value preferences of the different stakeholders involved, considering alternative futures.

There is no doubt that the complex interplay between oil extraction and fishing in regions like Tabasco underscores the urgent need for a more integrated, transparent, and inclusive environmental planning. While environmental instruments and zoning initiatives represent important steps toward balancing economic development with ecological preservation, their effectiveness ultimately depends on equitable governance and the genuine inclusion of local communities in the decision-making process.

14.4 The Gulf of Mexico and the Case of Tabasco, Mexico

Two of the most economically significant activities in Mexico, particularly in the Gulf of Mexico and in Tabasco, are oil extraction and fishing. These industries are crucial to the country's economy, providing employment and contributing to the national income. However, their environmental impacts are profound, especially if they are not managed with sustainability in mind. Environmental policy seeks to address these challenges by employing a range of policy instruments and governance frameworks to mitigate environmental impacts. As mentioned above, they depend on clearly delineating areas where these activities can be carried out and areas where they are prohibited. In this sense, protected areas, fisheries refuges, and fisheries management plans are all outlined in legislation, such as fisheries and environmental laws.

In this sense, the hydrocarbon industries, which encompass both oil and natural gas, have taken a leading role in some of the country's environmental reform processes. This leadership is partly driven by a combination of factors, including municipal and federal environmental regulations, increased enforcement of these regulations, and the development of local and cross-border markets for Mexican gas. These elements provide incentives for the hydrocarbon industries to adopt better environmental practices. However, a review of the zoning shows how the spatial configuration of the oil industry is often shielded by political influence and makes it challenging for communities living on the Tabasco coast areas, making it difficult to understand how coastal communities navigate their daily lives in the context of offshore operations (Ramos-Muñoz et al., 2019). Therefore, it is a challenge to integrate these needs into planning instruments, as this involves disputes over sea access between fishing and oil governance.

The Mexican government recognizes the importance of these instruments in influencing decisions of both industries and citizens. By incorporating environmental considerations into economic activities, such as oil extraction and fishing, these tools should help ensure that the environment is not overlooked. The process of environmental planning has to take into account the realities of financial constraints and political power dynamics. These factors, including citizens' demands for a healthy environment and technical viability, must be considered to ensure that these environmental plans are both effective and feasible. It is important to point out that their elaboration requires a specialist group analyzing the information and constructing the target map, and a committee evaluating the entire process. As a personal experience (first author), I was invited to participate in the committee on the construction of the zoning plan for the Gulf of Mexico and the Caribbean Sea, as a researcher representative for Tabasco. The Committee is divided into an executive body, responsible for decision-making and implementing the planning process, and a technical part, responsible for providing technical inputs and considering sector needs. The committee was formed by one researcher and one authority from each of the coastal states in this region (Tamaulipas, Veracruz, Tabasco, Campeche, Yucatán, and Quintana Roo), as well as 11 representatives from the federal government. It took 3–4 years to construct it (2007–2010) and was decreed in 2012, after a public consultation. The entire extension covers an area of 995,486.2 km^2, corresponding to 168,462.4 km^2 of the coastal component and 827,023.8 km^2 of the Marine component, with a length of the coastline comprising approximately 2900 km. During this process, the importance of considering the risk associated with natural events, including climate change, was discussed. This was due to an increase in the magnitude, duration, and extension of some hydrometeorological events, generating significant problems for productive activities (Rosete-Vergés et al., 2013). However, analyzing certain impacts on different sectors, such as fishing, was crucial, as environmental territorial planning plays a key role in ensuring the sustainability of fish stocks. In this context, implementing measures such as catch limits, seasonal closures, and the establishment of fisheries refuges can help prevent the overexploitation of marine resources.

Despite the clear benefits of environmental planning, significant challenges also need to be addressed. One of the primary challenges is the need for enhanced coordination between different levels of government and across various sectors. Environmental issues often transcend jurisdictional boundaries, and effective planning necessitates collaboration among federal, state, and local authorities, as well as among various industries and interest groups. Another challenge is the need for more comprehensive data and better tools for monitoring and enforcement. Without accurate and up-to-date information, it is difficult to make informed decisions about how to manage marine resources. Advances in technology, such as satellite monitoring and geographic information systems (GIS), offer new opportunities for improving the effectiveness of environmental territorial planning. However, these tools require investment and capacity-building to be fully utilized (Díaz-Mondragón et al., 2021).

The oil-producing regions of the Gulf of Mexico, especially in Tabasco and Campeche, are not only of strategic economic importance to Mexico but also highlight significant geopolitical and governance challenges at both local and regional levels. This type of instrument was expected to propose a delimitation for their development. Historically, these areas had been sites of competing interests, where national energy needs, local community rights, environmental concerns, and the influence of multinational corporations intersect. The geopolitics of these regions involves the distribution of power and resources between the federal government, state governments, and local communities, all of which have different stakes in the management of oil resources. One key geopolitical issue in these zones is the tension between economic development and environmental protection. The exploitation of oil resources in the Gulf of Mexico has led to significant economic benefits for the country. However, it has also raised concerns about environmental degradation and the displacement of traditional livelihoods, such as fishing (Ramos-Muñoz et al., 2019). Local communities, particularly those heavily reliant on fishing, often find themselves marginalized in decision-making processes, resulting in conflicts over resource access and utilization. Effective governance in these regions requires balancing these competing interests through inclusive and transparent decision-making that considers both the economic and environmental impacts of oil production. Additionally, the governance of oil production in the Gulf of Mexico is complicated by the involvement of powerful multinational corporations. These entities bring in vital investment and technology, but their interests may not always align with those of local communities or environmental priorities. These dynamic highlights the need for robust regulatory frameworks and effective oversight mechanisms to ensure that oil extraction benefits all stakeholders, not just a select few. The federal government must navigate these geopolitical challenges by enforcing regulations that protect the environment and local communities while also fostering economic growth.

Tabasco has a long history of conflict with the oil industry, particularly in the agricultural, livestock, and fishing sectors, stemming from an erosion of trust in public institutions. However, local authorities have implemented various strategies to address these conflicts, and the effort has been more local than federal, with inconsistent success. A historical review reveals how legislation has been adjusted, and even planning instruments, such as environmental policy, have been used to validate binding processes that privilege certain activities, including generating fishing exclusion areas to favor oil activities (Quist, 2019), leaving other sectors unprotected, notably the fishing sector, which affects hundreds of families. Addressing the asymmetries of power and influence, particularly the dominance of the oil industry, remains a central challenge. For environmental planning to fulfill its role in ensuring sustainability and social justice, it must move beyond technical exercises and become a participatory process that reflects the diverse interests, rights, and knowledge of those who live and depend on these territories.

14.5 Conclusion

We must conclude by mentioning that environmental planning in Mexico's marine zones is a vital component of the country's broader environmental management strategy. By providing a framework for balancing economic growth with environmental conservation, this planning process should ensure that Mexico's rich marine resources are used sustainably and equitably. The impact of this planning on productive activities such as oil extraction, fishing, and tourism should be regulated, and how these industries operate and contribute to the national economy should be considered. While challenges remain, the opportunities for enhancing environmental planning through improved data, strengthened enforcement, and increased stakeholder engagement are substantial. As Mexico continues to develop its marine resources, the importance of environmental planning is expected to grow, playing a crucial role in shaping the future of the country's marine ecosystems and the communities that depend on them. As global environmental changes intensify, the management of marine zones in Mexico faces a critical crossroads. The interplay between local and global factors has profound implications for the future of Mexico's marine ecosystems. Climate change, ocean acidification, sea level rise, and the increasing intensity of human activities, such as overfishing and coastal development, are all exerting unprecedented pressures on these ecosystems in ways that are difficult to predict. As a result, the future of marine zone management in Mexico will require innovative approaches that are adaptable, science-based, and inclusive of all stakeholders.

Given these challenges, the future of marine zone management in Mexico will depend on the development and implementation of resilient management strategies. Traditional approaches to environmental management, which often rely on fixed regulations and static boundaries, are increasingly inadequate in the face of dynamic and interconnected global changes. Instead, there is a need for management strategies that are flexible and can respond to changing environmental conditions, using models and scenarios to anticipate future changes to adjust management plans accordingly. For example, MPAs may need to be re-evaluated and potentially expanded or relocated to ensure they continue to protect critical habitats as conditions change. Similarly, fisheries management plans must be updated to reflect shifting species distributions and the potential for increased vulnerability of certain stocks to overfishing under changing climate conditions.

Building resilience is another crucial aspect because it enhances the ability of marine ecosystems, various sectors that utilize natural resources, and coastal communities to absorb shocks and maintain their essential functions. This can be achieved through various measures, including the restoration of degraded habitats, the protection of biodiversity, and the implementation of sustainable resource management practices.

Advances in science and technology will play an important role in shaping the future of marine zone management in Mexico. Actions enhancing monitoring strategies and data collection are essential for updating these instruments and their management strategies. Moreover, predictive modeling and scenario analysis can help managers anticipate the effects of different management actions under various future conditions. In addition to scientific and technological advancements, there is a growing recognition of the importance of incorporating traditional and local knowledge into marine zone management. Indigenous and local communities often possess a deep understanding of their environments, gained through generations of interaction with the land and sea. Integrating this knowledge with scientific research can lead to more effective and culturally appropriate management strategies that are better suited to the specific contexts of different marine zones in Mexico. The future of marine zone management in Mexico will also depend on the ability to maintain and increase governance by engaging a wider range of stakeholders in the decision-making process. Each group brings valuable perspectives and resources to the table, and their involvement is essential for developing management plans that are both effective and equitable. One of the key challenges in stakeholder engagement is balancing the interests of different groups, particularly when conflicts arise between economic development and environmental conservation. The future of marine zone management in Mexico will be shaped by the authority's ability to implement governance and an interest in responding to the complex and interconnected challenges today and in the future.

Faced with this dilemma, the future of the marine zone in Mexico is not only the responsibility of technical committees or government institutions, but also a matter of democratic governance, social justice, and environmental responsibility. As Mexico debates the balance between development and conservation, decision-making must reflect the interests of all stakeholders, particularly coastal communities, indigenous peoples, and small-scale fishers, who are often sidelined despite being the most affected. The government must implement a long-term vision that prioritizes a collective interest over short-term economic gain, ensuring that marine ecosystems are protected as the foundation of livelihoods, food security, and cultural identity. There is no doubt that strengthening marine governance requires a strong political will, effective intersectoral coordination, and a commitment to equity that prioritizes sustainability over development. In this sense, this will be the only way to ensure an inclusive approach to addressing the environmental challenges of our time and charting a resilient future for the marine environment.

Acknowledgments AI was used to search for suggestions to improve English writing. We would like to extend special thanks to the book editors and chapter reviewers for their insightful suggestions, which significantly improved this chapter.

References

Anglés-Hérnandez, M., Rovalo Otero, M., & Tejado Gallegos, M. (2021). *Manual de Derecho Ambiental Mexicano*. Instituto de Investigaciones Jurídicas. Serie Doctrina Jurídica 915. UNAM. https://biblio.juridicas.unam.mx/bjv/detalle-libro/6429-manual-de-derecho-ambiental-mexicano 915.

Brenner, L., & Vargas del Río, D. (2010). Gobernabilidad y gobernanza ambiental en México. La experiencia de la Reserva de la Biosfera Sian Ka'an. *Polis: Investigación y Análisis Sociopolítico y Psicosocial, 6*(2), 115–154.

Díaz-Mondragón, S., Pedroza-Páez, D., Bojórquez-Tapia, L. A., & Díaz de-León, A. J. (2021). Contribution of marine spatial planning in Mexico to marine and coastal management. *Revista Costas, Esp. 2*, 75–90. https://doi.org/10.26359/costas.e0421

Espinoza-Tenorio, A., Moreno-Báez, M., Pech, D., Villalobos-Zapata, G. J., Vidal-Hernández, L., Ramos-Miranda, J., Mendoza-Carranza, M., Zepeda-Domínguez, J. C., Alcalá-Moya, G., Pérez-Jiménez, J. C., Rosete, F., León, C., & Espejel, I. (2014). El ordenamiento ecológico marino en México: un reto y una invitación al quehacer científico. *Latin American Journal of Aquatic Research, 42*(3), 386–400. https://doi.org/10.3856/vol42-issue3-fulltext-1

Huppes, G., & Simonies, U. E. (2009). Environmental policy instruments. In J. J. Boersema & L. Reijnders (Eds.), *Principles of environmental sciences*. Springer. https://link.springer.com/chapter/10.1007/978-1-4020-9158-2_13

Jordan, S., & Benson, W. (2013). Governance and the Gulf of Mexico coast: How are current policies contributing to sustainability? *Sustainability, 5*, 4688–4705. https://doi.org/10.3390/su5114688

Quist, L. M. (2019). Fishers' knowledge and scientific indeterminacy: Contested oil impacts in Mexico's sacrifice zone. *Maritime Studies, 18*, 65–76. https://doi.org/10.1007/s40152-018-0123-7

Ramos-Muñoz, D., Ramos-Reyes, R., Zamora-Cornelio, L. F., Hernández-De la Cruz, A., & Espinoza-Tenorio. (2019). Exclusión en el Golfo de México: una visión desde los pescadores sobre la industria petrolera en Tabasco. *Cuadernos de Geografía: Revista Colombiana de Geografía, 28*(2), 357–372. https://doi.org/10.15446/rcdg.v28n2.73511

Rosete-Vergés, F. A., Enríquez Hernández, G., & Aguirre von Wobeser, E. (2013). El componente del riesgo en el Ordenamiento Ecológico del Territorio: el caso del Ordenamiento Ecológico Regional y Marino del Golfo de México y Mar Caribe. *Investigaciones Geográficas, 80*, 7–20. https://doi.org/10.14350/rig.36393

Signoret, M., Monreal-Gómez, M. A., Aldeco, J., & Salas-de-León, D. A. (2006). Hydrography, oxygen saturation, suspended particulate matter, and chlorophyll-a fluorescence in an oceanic region under freshwater influence. *Estuarine, Coastal and Shelf Science, 69*(1–2), 153–164. https://doi.org/10.1016/j.ecss.2006.04.011

Solo-Gabriele, H. M., Fiddaman, T., Mauritzen, C., Ainsworth, C., Abramson, D. M., Berenshtein, I., Chassignet, E. P., Chen, S. S., Conmy, R. N., Court, C. D., Dewar, W. K., Farrington, J. W., Feldman, M. G., Ferguson, A. C., Fetherston-Resch, E., French-McCay, D., Hale, C., Ruoying, H., Kourafalou, V. H. … Yoskowitz, D. (2021). Towards integrated modeling of the long-term impacts of oil spills. *Marine Policy, 131*, 1–7. https://doi.org/10.1016/j.marpol.2021.104554

Chapter 15
Structure of the Governance System of the Red Snapper Fishery Improvement Project in Campeche, Mexico

P. Soria-Gómez, J. A. Zepeda-Domínguez, L. Malpica-Cruz, E. Coronado, and C. M. Fumero-Andreu

15.1 Introduction

The sea is a common resource managed as a public good, where every member of society has equal access to and the ability to appropriate its natural resources. Open access often leads to resource overexploitation and eventual collapse (Defeo, 2015). Until the nineteenth century, natural resources, including fisheries, were widely considered to be infinite. However, as some fisheries became depleted, it became clear that these resources were finite (Sims & Southward, 2006). This realization marked a fundamental shift: fisheries transitioned from being considered public resources, where use by one individual does not hinder use by others, to

P. Soria-Gómez · J. A. Zepeda-Domínguez (✉)
Facultad de Ciencias Marinas, Universidad Autónoma de Baja California,
Ensenada, B.C., México
e-mail: pauleth.soria@uabc.edu.mx; zepeda.jose@uabc.edu.mx

L. Malpica-Cruz
Instituto de Investigaciones Oceanológicas, Universidad Autónoma de Baja California,
Ensenada, B.C., México
e-mail: lmalpica@uabc.edu.mx

E. Coronado
Escuela Nacional de Estudios Superiores, Unidad Mérida, Universidad Nacional Autónoma de México (ENES, Mérida, UNAM), Merida, Yucatán, México

Ecomorphoses, Consultoría, Mérida, México
e-mail: eva.coronado@enesmerida.unam.mx

C. M. Fumero-Andreu
Centro Interdiciplinario de Ciencias Marinas, Instituto Politécnico Nacional,
La Paz, B.C.S., México

SATOUMI REEF, Consultoría, La Paz, México
e-mail: cfumeroa1600@alumno.ipn.mx

A. Espinoza-Tenorio et al. (eds.), *The Fishing and Oil Industries in Coexistence*,
https://doi.org/10.1007/978-3-032-02621-7_15

common-pool resources, where their use is rivalrous, and exclusion is challenging due to the high costs associated with regulation. In 1968, Hardin analyzed the inevitable environmental degradation expected when many individuals simultaneously use a scarce resource without regulation. Since then, and recognizing the limitations of his model, the need to explore alternative approaches to natural resource management has become evident.

Given the current degradation of Social-Ecological Systems (SES), exploring models that better manage common-pool resources is crucial. The next major paradigm shift occurred with the development of Collective Action Theory by Ostrom (1990), which outlines general principles for achieving sustainability in the use of common-pool resources. Ostrom (1990) built on the collective action logic proposed by Olson (2012), who argued that individuals face challenges in prioritizing the collective good over individual interests to prevent collective ruin. Ostrom posited that engaging a group of individuals requires shared interests, which in turn implies collective actions to achieve mutually beneficial outcomes (Gordillo, 2014). This underscores the need to balance human progress with environmental preservation and social justice, enabling informed policymaking and improved governance models.

Governance as a system aims to ensure human well-being, social equity, and ecosystem functionality through balanced decision-making (Ostrom, 1990). In practice, governance is expressed as a system of institutions and individuals interacting through their representatives to achieve balanced agreements (Zetina et al., 2020). Since the 1990s, a paradigm shift has occurred, moving the focus of fisheries management from centralized government to governance frameworks that facilitate fisheries co-management. New management schemes have been pursued to ensure the sustainable use of marine resources.

In Mexico, this paradigm has advanced, establishing a foundation in the General Law of Sustainable Fisheries and Aquaculture (LGPAS) (United Mexican States, 2007), which regulates mechanisms for public participation in national fisheries management. The LGPAS identifies four key sectors (Governmental, Productive, Social, and Academic). Furthermore, considering the relevance of these actors, studies on public participation in Mexican fisheries management have shown that environmental NGOs are significant actors in fisheries management, despite not being legally recognized in management legislation (Zepeda-Domínguez et al., 2017).

Fishing in Mexico operates under two main modalities: small-scale, also known as artisanal, and industrial. The small-scale fisheries employ 95% of fishers nationwide. It is considered small-scale due to its limited infrastructure, inputs, safety measures, and technology regulation (Martínez & González, 2016). This type of fishing is characterized by "...small vessels, low tonnage, traditional fishing methods, and operations near the coast…" (LGPAS, 2023). Among the primary fishery resources captured by small-scale fisheries in Mexico are groupers, snappers, sharks, croakers, shrimp, and others (Arreguín-Sánchez & Arcos-Huitrón, 2011).

Campeche is one of the leading fishing states in Mexico's marine zone. It ranks among the top ten states in fishery production (Peña et al., 2021) and has a population of 10,479 fishers. The most valuable species in terms of production include octopus, jack mackerel, conch, and shrimp (CONAPESCA, 2021). Fishing activity takes place along 523 km of the state's coastline and within 7905 km^2 of the adjacent exclusive economic zone (Ruvalcaba et al., 2010). Campeche has 3269 active small-scale fishing vessels and 26 fish processing plants (CONAPESCA, 2021). In Campeche, the governance system of many fisheries, such as snappers, has been poorly assessed, and information gaps are evident.

15.2 The Red Snapper Fishery in Campeche

The red snapper (*Lutjanus* spp.) fishery is one of the most important for the Yucatán Peninsula, where it is known as "*pargo rojo*" or "*pargo mexicano de Campeche*." This resource significantly contributes to the economy of Campeche. This species, exploited in the Gulf of Mexico, is distributed in marine waters off the coasts of Tamaulipas, Veracruz, Tabasco, Campeche, and Yucatán (DOF, 2023). In 1984, the initial biomass was estimated at 32,957 tons, which decreased to 16,877 tons by 1999. However, over the past 16 years, the decline has further accelerated by 51.2% (Monroy et al., 2002). Mexican states along the Gulf of Mexico and Caribbean Sea reported 36.2% of the national red snapper fishery production in 2013, with Campeche accounting for 21.5% of the total catch (Caballero-Chavez, 2018). Due to its high-quality meat, red snapper is exported to the United States, where it is marketed in various forms (filleted, whole fresh, or frozen).

In the *Sonda de Campeche*, there are no specific regulatory measures targeting the red snapper fishery (such as catch quotas, minimum sizes, or seasonal closures). Fishers are only required to hold a finfish permit issued by the National Commission of Aquaculture and Fisheries (CONAPESCA, by its Spanish acronym), which covers approximately 95 species, with management strategies focused on red grouper and based on technical guidelines from the Mexican Institute of Fisheries and Sustainable Aquaculture (IMIPAS; Monroy et al., 2002). Studies indicate that the red snapper fishery of the Gulf of Mexico is currently exploited at its maximum sustainable yield, with signs of deterioration (Pérez, 2022; DOF, 2023). As part of efforts to recover the fishery and leverage the environmental pressure exerted by export markets, a Fishery Improvement Project (FIP) for this species was initiated in Campeche (Fishery Progress, 2024b). FIPs are market-driven processes, rather than government-led initiatives, involving producers through economic incentives rather than traditional state-regulated, coercive approaches. This approach aims to address the challenge of non-excludability, a characteristic of common-pool resources, while fostering improvements in the social fabric (Alliance for Conservation of Marine Solutions, 2022).

The social fishing network of the Campeche Red Snapper Fishery participating in the FIP (hereafter, red snapper FIP) includes government institutions, academia, cooperatives, and permit holders. However, each of these actors has different perspectives on how to manage the resource. Therefore, the collaborative participation of the various stakeholders in the fishery socio-ecological system in this region is essential for implementing management plans that promote the sustainable use of the resource. This study mapped red snapper FIP stakeholders, identifying the main sectors and key actors, and analyzed their interactions to generate information that can improve the management of the red snapper fishery.

15.3 Method

A stakeholder mapping was conducted to identify the leading actors in the red snapper FIP and their roles. The analysis was based on a normative approach, considering those stakeholders recognized under the General Law on Sustainable Fisheries and Aquaculture (LGPAS, 2023). Stakeholders were categorized according to sectors identified by the LGPAS as follows: federal government, state government, and social and productive organizations. Additionally, research institutions (academia) and civil society organizations participating in the red snapper FIP were included. The following definitions were used: Federal Government: All federal government agencies. State Government: All state government agencies. Social and Productive Organizations: Productive Sector: Actors engaged in extraction or capture, i.e., those who fish and sell to intermediaries but not directly to final consumers or exporters. Commercialization Sector: Actors involved in at least two of the following actions: fishing, purchasing from small producers for intermediary purposes, national or export trade, or retailing. Academia: Institutions that produce information to inform decision-makers. Non-Governmental Organizations (NGOs): Their social objectives are not focused on supporting the productive sector, but rather on participating in fisheries resource management (e.g., environmental NGOs).

Through a consensus among co-authors, fisheries academics, and review of the red-snapper list of members (Fishery Progress, 2024a), 15 stakeholders representing six sectors involved in the red snapper FIP were identified. The identification of actors was focused exclusively on groups within the red snapper FIP context.

Structured interviews were conducted to determine the roles and interactions of these actors within the red snapper FIP. The interviews focused on identifying the existence and intensity of factors recognized as influencing the development and consolidation of cooperation between actors and sectors (e.g., communication frequency and type).

The data collected in the interviews were used to construct relational matrices, representing the 21 interviewees in the columns and the 15 relevant actors mentioned by them in the rows. A value of 1 was assigned when an actor was mentioned and 0 when not mentioned. This matrix included two elements: interviewees and relevant governance system actors (known as an incidence matrix).

To focus on relevant actors, the incidence matrix was converted into a square adjacency matrix reflecting only the relationships between these actors. This was done by multiplying the original matrix (incidence matrix) by its transpose (a process in matrix algebra where rows and columns are exchanged). The resulting square matrix had both rows and columns representing the relevant actors. The Igraph package (Gábor et al., 2023) in the R environment (version 4.3.1, R Core Team, 2023) was used to model the network of relevant actors. Only undirected links were considered in this study, meaning that while communication between actors was identified, the direction of information flow was not analyzed, limiting the inferences regarding communication effectiveness.

To determine each actor's role within the system, centrality indicators were calculated using the methodology proposed by Velazquez and Aguilar (2005). Degree Centrality: Represents the number of direct connections a node (actor) has with other nodes, measuring each actor's level of connectivity. Closeness Centrality: Measures the distance between a specific node and all others, reflecting each actor's level of collaboration. Betweenness Centrality: Measures how often an actor appears on the shortest paths between two other actors. Prestige Centrality: Determined by the sum of connections from nodes linked to a specific node, reflecting not only the number of connections an actor has but also their quality. A node is more central if it is connected to other influential or well-connected nodes. These centrality indices were estimated using the Igraph package (Gábor et al., 2023) in R (version 4.3.1, R Core Team, 2023).

An adjacent matrix was formed to represent the 15 actors, and actors and institutions in the system were categorized in accordance with national fisheries regulations (e.g., LGPAS). The network indicators were estimated using Igraph. The presence of a link between actors was represented by a 1, and its absence by a 0. The 15 actors were transposed with the sectors, and both matrices were multiplied to group the data into a quadratic matrix derived from the adjacency matrix. Only undirected links were used, and the directionality of relationships was not analyzed, limiting inferences about communication effectiveness.

Based on the results of numerical analyses, strategic recommendations were developed to enhance the participation and collaboration of key actors, interpreted within the framework of applicable legislation. This was done to strengthen the FIP and promote sustainable resource management.

15.4 Results and Discussion

15.4.1 Main Actors in the Socio-Ecological System of the Red Snapper Fishery in the Sonda de Campeche

This study found that the key actors in the red snapper fishery (red snapper FIP) in the *Sonda de Campeche* are within the governmental and commercial sectors. Two hierarchical levels were identified among government actors: federal and state. At the federal level, the technical body, the National Institute of Fisheries and Aquaculture (INAPESCA, now IMIPAS), and the executive bodies, the CONAPESCA and the Secretariat of the Navy (SEMAR, by its Spanish acronym), were present. According to interviewees, SEMAR is considered part of this system due to its responsibility for maritime surveillance and its relationship with the fishery sector, including the red snapper fishery. However, SEMAR is not officially part of the red snapper FIP.

The Fisheries and Aquaculture Institute of Campeche (INPESCA, by its Spanish acronym), a state-level fisheries government agency, is one of the actors most frequently mentioned, which may be attributed to the role of its executive body. According to interviewees, this institution provides support to fishers in the form of financial assistance and work equipment.

Another of the most relevant sectors with the highest number of links between nodes in this fishery is the commercial sector, which includes producers and resource exporters. In the specific case of the red snapper fishery in Campeche, exporters are in Puerto Progreso, Yucatán. During interviews, export companies were categorized as a single entity ("exporters"), without being considered as separate actors. Export plants play a crucial role in the fishery's value chain. Their presence facilitates the distribution and export of red snapper to major markets, such as the United States.

As part of the productive sector, in Campeche, six actors were identified from the localities of *Nuevo Progreso* (*Nuevo Campechito*), *Isla Aguada*, and *Sabancuy*. These actors are partners representing cooperatives or permit holders, taking on leadership roles in the capture segment. However, some of them are not officially part of the red snapper FIP. This segment focuses on selling at the beach, regional, or national levels but does not engage in exports, supermarket chains, or final retail and consumer markets. In this system, two productive actors play central roles in decision-making and coordination with the government and NGOs on socio-environmental issues (Productive_4 and Productive_9; Table 15.1). Other productive systems have observed similar trends (Wade et al., 2023). Specifically, one productive actor is the most recognized, who, in addition to being a producer, also acts as an intermediary, purchasing fishery products from producers and exporting them (Productive_4; Table 15.1). Intermediaries, such as permit holders and cooperative members, also belong to the productive sector but represent a distinct segment. Unlike producers, intermediaries focus on export activities and maintain relationships with both producers and international trading companies, a sector typically inaccessible to producers. Consequently, they play a more significant role in commercial systems (Bodin et al., 2006). The implications of power imbalances within the supply network have been addressed in various studies (Bodin et al., 2006; Rodríguez-Fuentes et al., 2025). However, these were beyond the scope of this study.

In the academic sector, identified institutions include *El Colegio de la Frontera Sur*, Campeche Unit (ECOSUR), *Universidad Autónoma del Carmen* (UNACAR), and the Gulf of Mexico Institute of Ecology, Fisheries, and Oceanography-*Universidad Autónoma de Campeche* (EPOMEX). Although ECOSUR was among the most mentioned actors during interviews, its significance may have been overestimated due to the interviews being facilitated by members of this institution. Nonetheless, the general validity of the results is trusted, as interviewees recognized the importance and connection of some projects that ECOSUR has conducted in the

Table 15.1 Structural properties of the governance network in the red snapper fishery system

Actor code	Actor	Degree	Betweenness	Closeness	Closeness Prestige
GF_2	CONAPESCA	162	0,2033	0,0714	10,000
GF_1	IMIPAS	136	0,0762	0,0714	0,8942
Academia_1	ECOSUR	136	0,1016	0,0714	0,8929
Commercialization_1	Exporters	126	0,0847	0,0714	0,8214
GE_1	INPESCA	125	0,1016	0,0714	0,8318
GF_3	SEMAR	103	0,0508	0,0714	0,6995
Productive_4	Fisher	100	0,1355	0,0714	0,6609
Productive_9	Fisher	98	0,1016	0,0714	0,6507
Productive_1	Fisher	79	0,0169	0,0714	0,5311
Productive_6	Fisher	79	0,0508	0,0714	0,5378
Academia_3	EPOMEX	79	0,0169	0,0714	0,5393
NGO_1	COBI	77	0,0254	0,0714	0,5222
Academia_2	UNACAR	76	0,0338	0,0714	0,5277
Productive_7	Fisher	56	0,0000	0,0666	0,3888
Productive_8	Fisher	42	0,0000	0,0666	0,2930

Note: Centrality measures of a network include degree, betweenness, closeness, and closeness prestige. The identified sectors are as follows: Federal Government (GF), State Government (GE), Resource Exporters (Commercialization), License Holders and Cooperative Representatives (Productive), Research Institutions (Academia), and Non-Governmental Organizations (NGOs)

region, even though these are not directly associated with the red snapper fishery but involve shared participating actors.

In the NGOs sector, *Comunidad y Biodiversidad* A.C. (COBI) was identified as the only representative actor. Although COBI is an environmental NGO, it plays the role of implementer in the red snapper FIP. Its activities include training producers involved in the FIP on fishery monitoring data collection. The low representation of environmental NGOs in this fishery contrasts with other fisheries, especially in the Pacific, where such organizations are more actively involved. However, as in this case, most FIPs in Latin America are led by NGOs (Gómez-Gómez et al., 2024).

The types of actors analyzed in this network are listed in the General Law on Sustainable Fisheries and Aquaculture (2023) and belong to the government sector, commercial sector, social organizations, and academia. Together, they form part of public participation. In the case of the red snapper fishery system, all four sectors were identified, but only two stand out in the social network analysis: the governmental and commercial sectors. The actors with the most social links, such as IMIPAS, CONAPESCA, exporters, and ECOSUR, hold central positions that enable them to exercise leadership roles. These positions enhance their capacity to facilitate collective action within the network. This centrality not only enhances activity coordination but also fosters shared knowledge and perceptions, facilitating joint decision-making (Bodin, 2017). In fact, similar patterns of actor integration and convergence of fishery management goals have been observed in other FIPs in Latin America and Mexico (Gómez-Gómez et al., 2024).

The structure of the red snapper FIP network exhibits strong connectivity, making it a resilient network less vulnerable to potential impacts, such as the loss of a highly central node or reduced communication channels. The management of any natural resource is sensitive to the norms and practices agreed upon by the agents involved. These agreements aim to enhance coordination in resource use, conflict resolution, compensation, and the sharing of knowledge management (Folke et al., 2005). Bodin (2009) emphasizes that developing internal knowledge within a complex system, such as socio-ecological systems, benefits from a high degree of information exchange among specialized agents. Studies have shown how well-connected groups of fishers, characterized using specialized networks, leverage their ties to develop a management system based on shared knowledge (Bodin, 2009).

15.4.2 Role of Actors in the Red Snapper Fishery

Centrality measures revealed a high level of agreement, meaning that actors with the highest degree centrality also exhibit high betweenness and closeness centrality. Regarding closeness centrality, most actors in this system exhibit similar values, indicating that they all possess a comparable capacity to connect with others in the network. The actors with the highest number of connections (degree centrality) are the two federal-level institutions: CONAPESCA and IMIPAS, with CONAPESCA having the most links. However, both CONAPESCA and IMIPAS serve as central nodes in the network, surrounded by the same number of nearby nodes (see Fig. 15.1). An actor situated between many others in the network has high betweenness centrality, meaning they can act as a bridge between those actors (with connections functioning as bridge ties) (Bodin, 2009). Both CONAPESCA and IMIPAS

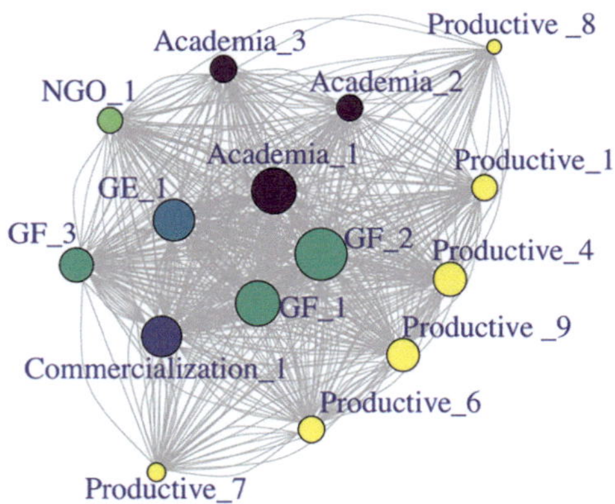

Fig. 15.1 Governance network of the red snapper fishery (*Lutjanus campechanus*) in the Sonda de Campeche, Mexico. In the network, the nodes represent the system's stakeholders. The links indicate the number of connections a node has with other nodes. The color of each node corresponds to the identified sector to which it belongs within the system

also have the highest closeness prestige scores (see Table 15.1), which indicates that their value increases with the frequency and intensity of their interactions with other actors. This is relevant given that communication is considered a proxy for coordination and influence capacity among actors (Vivas, 2001).

On the other hand, IMIPAS and ECOSUR have slightly lower betweenness centrality than CONAPESCA, suggesting they may be less efficient as information disseminators within the system or less recognized by the interviewed actors. While IMIPAS and CONAPESCA are key decision-makers in the fishery system, ECOSUR serves as an academic actor, according to the perceptions of the stakeholders interviewed. However, it is essential to note that their decisions may not necessarily be linked to the red snapper FIP. The influence of the commercial sector, represented by exporters, is noteworthy as they act as intermediaries between producers and consumers. Regarding closeness prestige, CONAPESCA scored the highest, followed by INAPESCA and ECOSUR as the top three actors in the network. The actors with the lowest prestige scores were Productive_7 and Productive_8.

According to the interviewees, INPESCA, as a fishery manager at the state level, is recognized for the support it provides to the fisheries sector. However, as previously mentioned, its influence does not stem from direct contributions to actions within the red snapper FIP, but rather from providing incentives that directly favor fishers and their fishing activities.

It is worth noting that the interviewees mentioned the functions of SEMAR, including activities such as marine surveillance, marine environmental protection, marine rescue and salvage, as well as collaboration with local authorities. While the actors recognize the importance of SEMAR, it is not a part of the red snapper FIP. In the productive sector, the actors who occupy leadership roles within the network, such as cooperative presidents (e.g., Productive_4 and Productive_9; Table 15.1), are highlighted and recognized for their participation in the red snapper FIP. On the other hand, those actors whose participation is limited to operational activities but not related to decision-making within fishing groups are considered actors with limited influence (e.g., Productive_7 and Productive_8; Table 15.1).

Similarly, the low influence of the Non-Governmental Organizations (NGOs) sector stands out, which was represented only by COBI, recognized as the implementers of the FIP but not as representatives of the environmental sector. The low values in intermediation and degree indicate that participation in the red snapper FIP is more influenced by other actors (probably from the productive sector—producers and marketers) than by the influence of NGOs in the system. Although the leadership of NGOs in FIPs is expected (see Gómez-Gómez et al., 2024), they are also expected to serve as bridges that facilitate coordination between the different sectors. Without their participation, it is expected that some red snapper FIP outcomes may be hindered.

The actors mentioned by some interviewees but less connected to others were the Gulf of Mexico Institute of Ecology, Fisheries, and Oceanography (EPOMEX) and the Autonomous University of Carmen (UNACAR). According to the website where FIP progress is reported (fisheryprogress.org, accessed October 2023), these actors are not involved in the project, which explains their low connectivity observed in the analyses.

The results of this study align with those reported by Wade et al. (2023), which indicate that a centralized network consists of a few actors with high centrality scores, while most other actors have low scores (Fig. 15.1). In this way, within the network structure, it is possible to identify that the central actors are those linked to academia, commercialization, and production (Table 15.1). In contrast, the remaining actors are distant and have low centrality values. This limits the diffusion of information within the network, as interactions are concentrated in a few individuals (Wade et al., 2023). This type of highly centralized structure may not be suitable for governing social-ecological systems over time, as it is considered less capable of solving complex tasks (Frank et al., 2007).

Based on the results obtained from the network structure, it is possible to infer that the network is centralized. However, it is recognized that not all actors directly involved in the FIP were included in the interviews, so expanding the sample size in future analysis is recommended.

Defeo (2015) notes that a sectorized management scheme, i.e., one whose decisions are based on a single sector—contrasting with a holistic or systemic view—minimizes ecosystemic, socio-economic, and governance aspects, resulting in losses or declines in natural resources. In Mexico, decentralized management with a holistic approach is sought, based on an ecosystemic approach that considers social, technological, productive, biological, and environmental aspects (e.g., LGPAS) of fishing communities or coastal regions. This approach has been addressed in recent studies for the Campeche region (Coronado et al., 2020; Peña et al., 2021).

It is worth noting that network structures can evolve, and the content transferred through the links can also change over time (Bodin, 2009). Thus, as more NGOs or institutions linked to the environment are involved in the fishery decision-making, the network structure will have the opportunity to change. It is also important to emphasize that the presence of more actors within a network does not necessarily indicate better efficiency or a more balanced power structure.

Promoting and including sectors with low representation (i.e., NGOs and academia), especially through information generation, could provide more support and balance to the decisions made, thus improving governance and, consequently, the sustainability of the red snapper fishery system. In FIPs in Latin America, academia mainly focuses on generating and transferring information, which is one of the objectives of these management efforts and helps establish communication and collaboration networks (Gómez-Gómez et al., 2024; Crona et al., 2019).

The collaborative participation of the different actors that make up the fishing social-ecological system in this region is essential for implementing management plans that benefit the sustainable use of the resource. While the red snapper fishery in the *Sonda de Campeche* is involved in a FIP, there is limited representation and apparent participation from actors (e.g., permit holders, cooperative representatives, NGOs, academia, among others), possibly due to a lack of inclusion mechanisms or information outreach. This would indicate certain critical aspects that need to be addressed to improve the impacts on the management of the red snapper resource, as other FIPs in Latin America have been successful in influencing fisheries management policies and practices beyond the FIP implemented and only the sector involved (Gómez-Gómez et al., 2024).

15.4.3 Improving the Participation and Collaboration

Based on the comparison of topological indicators with mid-range theories recommending public participation, intersectoral dialogues, community agreements, and other measures, the following guidelines are proposed to strengthen the governance of the SES:

- Improve governance: Decentralize decision-making in this fishery by increasing the participation of state governments (Campeche and Yucatán) and producers.
- Productive Sector Involvement: Encourage the participation of the productive sector in the FIP.
- Involvement of Academia and NGOs: Promote the inclusion of NGOs and academia to support government agencies and accomplish specific aims, such as surveillance, biological monitoring, and socioeconomic monitoring of the fishery.
- Research continuity: Replicate the study's approach by including a greater number of key actors.
- Alternative Methodologies: Conduct further studies using other methodologies, such as the interactive governance analytical approach, governance analysis, and actor influence within value chains, among others.

15.5 Conclusions

The actor network of the red snapper fishery participating in the FIP in the Sonda de Campeche identified in this study consists of 21 prominent participants representing six sectors, of which 15 can be considered key actors. The network has a high density, with most actors interconnected. Its structure is centralized, with a few leaders from the federal government, academia, and commercial sectors. The concentration of power in a few actors suggests low governance, underscoring the need to improve the participation of actors in the system.

Acknowledgments NPSG and JAZD acknowledge the CONACYT Project 1564454 "The Challenging Coexistence of Coupled Socio-Ecological Systems: The Fishing and Oil Industries in the Sonda de (PePe 2.0)." The authors thank Pedro Luis Córdova Argüello for editorial support.

References

Alliance for Conservation of Marine Solutions. (2022). *Lineamientos de Apoyo para Proyectos de Mejora Pesquera.* https://solutionsforseafood.org/wp-content/uploads/2023/04/FIP-Guidelines-Spanish-Apr2023.pdf.

Arreguín-Sánchez, F., & Arcos-Huitrón, E. (2011). Fisheries in Mexico: Status of exploitation and ecosystem use. *HIDROBIOLÓGICA, 21*, 3. https://hidrobiologica.izt.uam.mx/index.php/revHidro/article/view/773

Bodin, Ö. (2009). Ecological topology and networks. In R. Meyers (Ed.), *Encyclopedia of complexity and systems science*. Springer. https://doi.org/10.1007/978-0-387-30440-3_164

Bodin, Ö. (2017). Collaborative environmental governance: Achieving collective action in social-ecological systems. *Science, 357*, 6352.

Bodin, Ö., Crona, B., & Ernstson, H. (2006). Social networks in natural resource management: What is there to learn from a structural perspective? *Ecology and Society, 11*, 2.

Caballero-Chavez, V. (2018). Population parameters and fisheries indices of the red snapper *Lutjanus campechanus* in southern Campeche. *Ciencia Pesquera, 26*(2), 29–36.

CONAPESCA (National Commission of Aquaculture and Fisheries). (2021). *Statistical yearbook of aquaculture and fisheries.*

Coronado, E., Salas, S., Torres-Irineo, E., & Chuenpagdee, R. (2020). Disentangling the complexity of small-scale fisheries in coastal communities through a typology approach: The case study of the Yucatan Peninsula, Mexico. *Regional Studies in Marine Science, 36*, 101312. https://doi.org/10.1016/j.rsma.2020.101312

Crona, B., Käll, S., & van Holt, T. (2019). Fishery improvement projects as a governance tool for fisheries sustainability: A global comparative analysis. *PLoS One, 14*(10), e0223054. https://doi.org/10.1371/JOURNAL.PONE.0223054

Defeo, O. (2015). *Ecosystem approach to fisheries: Fundamental concepts and its application in small-scale fisheries in Latin America.* Food and Agriculture Organization (FAO). FAO Fisheries and Aquaculture Technical Paper No. 592. Rome, Italy.

DOF (Official Gazette of the Federation). (2023). *National fisheries charter.* https://www.gob.mx/cms/uploads/attachment/file/892410/CNP_2023.pdf

Fishery Progress. (2024a). *Base de datos y herramientas de seguimiento del progreso del proyecto de mejora pesquera.* https://fisheryprogress.org/

Fishery Progress. (2024b). *Mexico Campeche red snapper—vertical and bottom longline.* https://fisheryprogress.org/fip-profile/mexico-campeche-red-snapper-vertical-and-bottom-longline

Folke, C., Hahn, T., Olsson, P., & Norberg, J. (2005). Gobernanza adaptativa de sistemas socio-ecológicos. *Annual Review of Environment and Resources, 30*(1), 441–473.

Frank, K., Mueller, K., Krause, A., Taylor, W., & Leonard, N. (2007). In W. W. Taylor, M. G. Schechter, & L. G. Wolfson (Eds.), *Globalization; effects on fisheries resources* (pp. 385–423). Cambridge University Press.

Gábor C., Tamás N., Vincent T., Szabolcs H., Fabio Z., Daniel N., & Kirill M. (2023). *igraph (interfaz R).* https://r.igraph.org/articles/igraph.html

Gómez-Gómez, A., Malpica-Cruz, L., Montaño-Moctezuma, C. G., Cisneros-Montemayor, A. M., Salomon, A. K., & Seingier, G. (2024). Diversity, challenges, and opportunities of fishery improvement projects in the Latin America and the Caribbean region. *Marine Policy, 163*, 106116. https://doi.org/10.1016/J.MARPOL.2024.106116

Gordillo, G. (2014). Estado, mercado, comunidad: Ostrom, la controversial. *Revista Mexicana de Sociología, 76*(SPE), 227–258.

Hardin, G. (1968). The tragedy of the commons. *Science*, 1243–1248.

LGPAS. (2023). *Acuerdo por el que se establece la Zona de Refugio Pesquero en aguas marinas adyacentes a San Quintín, B.C. Diario Oficial de la Federación.* https://dof.gob.mx/nota_detalle.php?codigo=5696337&fecha=21/07/2023

Martínez, S. González, F. (2016). La construcción de la política pesquera en México: Una mirada desde el campo geográfico. *Atlantic Review of Economics*, 2.

Monroy-García, C., Garduño-Andrade, M., Espinosa, J.C. (2002). Análisis de la pesquería de huachinango (Lutjanus campechanus) en el Banco de Campeche. *Proc. Gulf Caribb. Fish. Inst. 53*, 507–515.

Olson, M. (2012). La lógica de la acción colectiva [1965]. *Teoría sociológica contemporánea, 124*, 62–63.

Ostrom, E. (1990). *Governing the commons: The evolution of institutions for collective action*. Cambridge University Press.

Peña, A., Pérez, J., Munguía, A., & Espinoza, A. (2021). Sistemas socio-ecológicos como unidad de manejo: el caso de las pesquerías de Campeche, México. *Economía, Sociedad y Territorio, 21*(65), 113–145.

Pérez, B. E. E. (2022). Importancia cultural de los arrecifes del Corredor Arrecifal del Suroeste del Golfo de México en comunidades pesqueras. Tesis para obtener el grado de Doctora en Ecología y Pesquerías Universidad Veracruzana. Veracruz, Mexico. 104 pp.

R Core Team. (2023). *R: A language and environment for statistical computing*. R Foundation for Statistical Computing. https://www.R-project.org/

Rodríguez-Fuentes, M., Zepeda-Domínguez, J. A., Kluger, L. C., Fumero-Andreu, C. M., Ponce-Díaz, G., & Zetina-Rejón, M. J. (2025). Supply networks of fisheries social-ecological systems: A systematic review of the network approach. *Ecological Economics, 227*, 108394.

Ruvalcaba, M., Villaseñor Talavera, R., & Mezo Villalobos, S. (eds.) (2010). *Ordenamiento pesquero ribereño marino: informe de ejecución Botello 2010, Ciudad*. https://www.academia.edu/16610330/Ordenamiento_Pesquero_Ribere%C3%B1o_Marino_Informe_de_Ejecuci%C3%B3n_2010

Sims, D. W., & Southward, A. J. (2006). Dwindling fish numbers already of concern in 1883. *Nature, 439*(7077), 660–660. https://www.nature.com/articles/439660c

United Mexican States. (2007). General law of sustainable fisheries and aquaculture. *Official Journal of the Federation*. Retrieved from https://www.diputados.gob.mx/LeyesBiblio/pdf/LGPAS.pdf

Velázquez-Álvarez y Aguilar-Gallegos, 2005. O.A. Velázquez-Álvarez, N. Aguilar-Gallegos. Manual introductorio al análisis de redes sociales. Ejemplos prácticos con UCINET 6.85 y NETDRAW 1.48. Medidas de centralidad. Universidad Autónoma del Estado de México, Toluca.

Vivas, J. (2001). Análisis de redes sociales y procesos de influencia en la toma de decisión grupal. *Interdisciplinaria, 18*(1), 87–113.

Wade, E., Alexander, S., Gerkey, D., & Biedenweg, K. (2023). Exploring the relationship between fishing actors and network prominence in information-sharing networks in Jamaican small-scale fisheries. *Human Ecology, 51*(5), 877–889.

Zepeda-Domínguez, J. A., Ponce-Díaz, G., & Vergara-Solana, F. J. (2017). El mapeo de actores y la conformación de cuerpos de participación pesquera. *Región y Sociedad, 29*(68), 259–277.

Zetina, M., Zepeda, J., Rodríguez, M., & Fumero, C. (2020). Stakeholder diversity correlates with governance network performance in two artisanal fisheries in Northwest Mexico. *Ocean & Coastal Management, 196*, 105313.

Part IV
Pathways to Sustainability

Chapter 16
Food from the Sea: Valuing the Provisioning Ecosystem Service of Small-Scale Fisheries in Tabasco, Mexico

M. A. Mesa-Jurado, J. M. Palacios-Hernández, A. Espinoza-Tenorio, and S. M. Suárez-García

16.1 Introduction

Marine-coastal ecosystems provide essential benefits for human well-being, notably the provision of food through fishing. However, their value extends far beyond their nutritional benefits. These ecosystems also offer regulating services, such as climate regulation through carbon sequestration in mangroves and seagrasses (Alongi, 2012; Fourqurean et al., 2012), coastal protection against storms and erosion (Barbier et al., 2011), and waste purification through estuarine filtering systems (Costanza et al., 1997). Additionally, they supply cultural services, including recreation, tourism, spiritual significance, and the maintenance of traditional knowledge and practices within coastal communities (Pascual et al., 2017). Supporting services, such as nutrient cycling and habitat provision for juvenile fish and other marine organisms, underpin the productivity of the entire marine food web (Beaumont et al., 2008; Barbier et al., 2011).

Small-scale fisheries (SSF) play a dual role: they supply local markets and meet subsistence needs, ensuring both food security and livelihoods for millions,

M. A. Mesa-Jurado (✉)
El Colegio de la Frontera Sur, Unidad Villahermosa, Villahermosa, Tabasco, México
e-mail: mmesa@ecosur.mx

J. M. Palacios-Hernández · S. M. Suárez-García
Instituto Tecnológico Superior de Centla, Frontera, Tabasco, México
e-mail: sandra.suarez@centla.tecnm.mx

A. Espinoza-Tenorio
El Colegio de la Frontera Sur, Unidad Campeche, Lerma, Campeche, México
e-mail: aespinoza@ecosur.mx

A. Espinoza-Tenorio et al. (eds.), *The Fishing and Oil Industries in Coexistence*,
https://doi.org/10.1007/978-3-032-02621-7_16

especially in socially and economically marginalized coastal areas. Globally, SSF contribute at least 40% (37.3 million tonnes) of total fisheries catches and provide 2.3 billion people, primarily in coastal regions, with key dietary micronutrients (Basurto et al., 2025). Specifically, marine SSF supply 32% of all seafood-derived nutrients worldwide, including zinc, iron, calcium, DHA, EPA, and vitamins A and B12 (Viana et al., 2023). In the Americas, this contribution often exceeds 50%, highlighting their vital role in regional nutrition (Basurto et al., 2025). Despite this importance, fish and fisheries remain largely invisible in strategies to achieve Sustainable Development Goal 2 (Zero Hunger), reflecting a critical gap in both global and regional food security agendas.

At the same time, the ecosystems that support SSF face mounting threats. Coastal and marine environments are increasingly degraded by overfishing, habitat fragmentation, wastewater discharge, and intensive oil and gas extraction. These pressures disproportionately impact fishing communities, whose livelihoods are intricately tied to the ecological health of these environments (Espinoza-Tenorio & Mendoza-Carranza, 2015). The expansion of the oil industry, in particular, has led to both direct impacts, such as pollution and gear loss, and indirect effects, including ecosystem disruption and declining fish productivity (Soto et al., 2014; Andrews et al., 2021). Additional pressures include the displacement of fishers due to increasing coastal traffic and infrastructure, as well as the catastrophic consequences of oil spills on fisheries and coastal livelihoods (Andrews et al., 2021).

In the Southwestern Gulf of Mexico (SWGoM), and especially in the state of Tabasco, these dynamics are particularly pronounced. Offshore oil extraction has turned vast marine spaces into what Quist (2019) describes as *sacrifice zones*, where environmental and social damage is justified in the name of national development. Through processes of enclosure and securitization, access to traditional fishing grounds is restricted, oversight is limited, and harm is externalized to local communities, particularly small-scale fishers. This not only undermines food security but also erodes livelihoods and deepens economic vulnerability. Fishers in Tabasco often travel up to 200 km offshore, navigating waters now dominated by extractive operations, further intensifying the conflict between industrial development and sustainable fishing economies (Quist, 2019). The consequences are evident. In 2023, Tabasco reported a total seafood catch of 28,601 tonnes, supporting approximately 14,000 fishers and over 4000 small-scale vessels (CONAPESCA, 2023). However, this marks a 38% decline from the 46,169 tonnes recorded in 2014, a loss of over 17,500 tonnes in less than a decade.

In this context, recognizing the value of food from the sea as a critical provisioning ecosystem service is essential for building sustainable coastal futures. Beyond simply acknowledging its contribution to food security and livelihoods, sustainability requires the integration of fishers' knowledge, the protection of traditional fishing territories, and robust environmental regulations to mitigate the damage caused by extraction. Routes toward sustainability must also include inclusive governance mechanisms, fair compensation for environmental harms, and investment in community-led adaptation (Hernández García, 2024). Moreover, it supports environmental justice by recognizing the rights of coastal communities that depend directly on the sea for sustenance.

This study aimed to estimate the economic value of the food provisioning ecosystem service derived from small-scale fishing in *Frontera* and *Barra de San Pedro*, two communities in the municipality of Centla, Tabasco. Both face severe environmental pressures from hydrocarbon activities, marine pollution, and climate change, as well as entrenched social challenges, including poverty, limited educational access, and scarce employment opportunities. Despite these constraints, small-scale fishing remains central to survival, resilience, and identity (Hernández García, 2024). By focusing on fishers operating small and medium vessels within cooperatives and under official permits, this work seeks to highlight the multiple benefits of small-scale fisheries and inform more just, effective, and sustainable resource governance.

16.2 Case Study Description

Coastal Tabasco features a diverse landscape comprising mangrove forests, coconut plantations, fishing communities, and oil infrastructure that stretches along a 200-km coastline, where fishing remains a central livelihood, despite the growing presence of the oil industry (Quist & Nygren, 2015). The municipality of Centla is the largest in Tabasco by land area and holds strategic ecological and economic importance. Bordered to the north by the Gulf of Mexico, Centla features vast wetlands, river systems, and coastal plains, and encompasses part of the *Pantanos de Centla* Biosphere Reserve, one of the most significant wetland ecosystems in Mesoamerica (One Earth, 2020). The region's hydrology is defined by the confluence of the *Grijalva* and *Usumacinta* rivers at *Tres Brazos*, where they form a complex deltaic system that includes distributary channels such as *San Pedrito*. This network ultimately drains into the Gulf of Mexico near the town of *Frontera*, at a site known as *La Barra*. According to Mexico's 2020 Population and Housing Census (INEGI, 2020), Centla has a population of 107,731, with women accounting for 50.7% and men 49.3%. The local economy relies primarily on fishing and agriculture although oil and gas activities have become increasingly prominent in recent years (Secretaría de Economía, 2024).

Frontera, the municipal seat of Centla, was founded in 1816 approximately five kilometers inland from the mouth of the *Grijalva* River, and currently has a population of 23,024 (INEGI, 2020). Its economy is primarily based on small-scale and commercial fishing, local trade and services, and activities linked to the oil sector, given its proximity to both offshore and onshore PEMEX facilities. The port of *Frontera*, although limited in size and draft, serves as a strategic hub for regional fishing, oil and gas transport, and coastal commerce. To the east lies *Barra de San Pedro*, a small rural fishing community situated along the *San Pedro* and *San Pablo* rivers, which serve as the natural boundary with the state of Campeche (Fig. 16.1). The community has a population of 571 and is characterized by high levels of

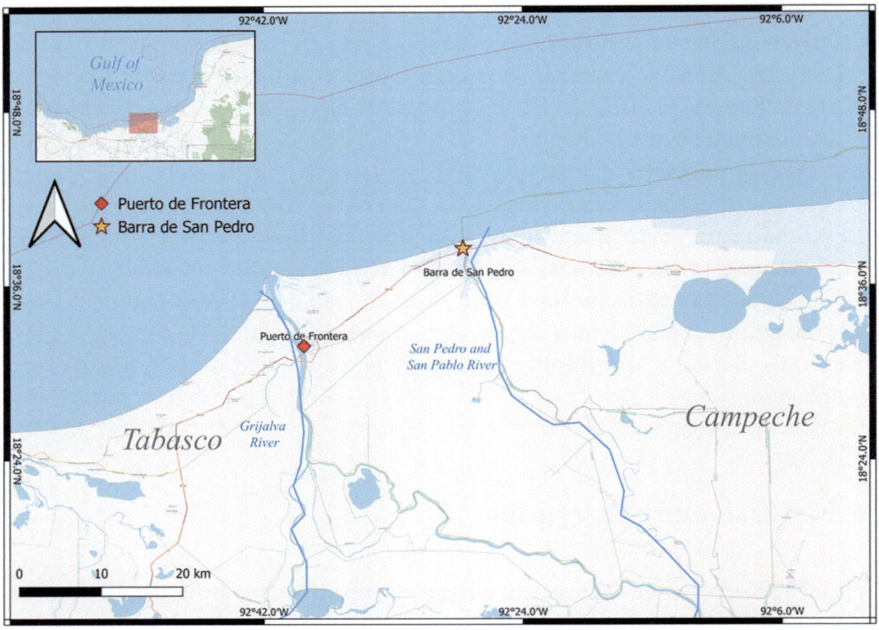

Fig. 16.1 Location of the two communities: *Frontera* and *Barra de San Pedro* in Centla, Tabasco, México

marginalization. Its primary economic activity is small-scale fishing, and many residents are migrants from other states, particularly Veracruz, with significant migration flows recorded since the 1980s (Muñoz-Sánchez & Cruz-Burguete, 2013). While both locations offer basic services such as potable water, electricity, and drainage, the broader municipality faces persistent social vulnerability, with 41.6% of the population living in moderate poverty and 35.2% in extreme poverty (INEGI, 2020). These challenges, compounded by territorial pressures from the oil sector, limit sustainable economic development and equitable access to fishing resources.

The most important species caught by the small-scale fleet in the area include gafftopsail catfish (*Bagre marinus*), southern stingray (*Dasyatis americana*), and red snapper (*Lutjanus campechanus*), which together account for over 68% of total catch weight. Traditional fishing gear is used aboard small and medium-sized vessels. The choice of fishing methods, vessel type, and trip duration depends on the target species and its location (Mendoza-Carranza et al., 2018). Other species exploited are marine fish such as white snook (*Centropomus undecimalis*), black snook (*Centropomus poeyi*), fat snook (*Centropomus parallelus*), liseta and mullet (*Mugil* spp.), lane snapper (*Lutjanus synagris*), tarpon (*Megalops atlanticus*), school shark (*Galeorhinus galeus*), greater amberjack (*Seriola dumerili*), cod (*Gadus morhua*; *Rachycentron canadum*), Spanish mackerel (*Scomberomorus maculatus*), and barracuda (*Sphyraena barracuda*).

Fishing and oil extraction have overlapped in Tabasco since 1949, when onshore oil operations began following the nationalization of the oil industry in 1938. Offshore expansion accelerated between 1977 and 1980, prompting the creation of the *Pacto Ribereño* (1976–1983), a temporary agreement between PEMEX (the Spanish Acronym for Petróleos Mexicanos) and fishing cooperatives. This agreement aimed to compensate for environmental damage and allow both sectors to operate in parallel (Velázquez Guzmán, 1982). However, the coexistence was never free of conflict. By the 1980s, significant environmental impacts, particularly the *Ixtoc-I* blowout in 1979, underscored the tensions between the two activities. These tensions intensified in the 1990s as the number of oil platforms increased, thereby reducing fishing areas and raising concerns over pollution and the degradation of marine habitats. In 2003, the federal government established a 15,907-km^2 offshore exclusion zone (DOF, 2003), prohibiting all non-oil activities, including fishing, around hydrocarbon infrastructure. While justified under national security concerns in the wake of 9/11, the measure also served to safeguard uninterrupted oil operations. The exclusion zones, initially 5 km in radius, combined with recurring oil spills, significantly limited artisanal fishers' access to traditional grounds, damaging fishing gear, and affecting their economic viability (Espinoza-Tenorio & Mendoza-Carranza, 2015; Toledo-Ocampo, 2005).

Between 2014 and 2017, several legal reforms modified these restrictions: the exclusion radius was reduced to 2.5 km but extended to all marine oil infrastructure, while roughly 10,000 km^2 of previously restricted fishing areas were reopened as part of economic reactivation efforts for Tabasco and Campeche (Salazar-De la Cruz et al., 2020). Simultaneously, the 2013 energy reform introduced public bidding rounds that, by 2018, had granted exploration rights over 70,433 km^2 of marine space in the Gulf of Mexico (Ramos-Muñoz et al., 2019), intensifying spatial and environmental conflicts between oil operations and coastal communities.

16.3 Methods

16.3.1 Market Price Method for Environmental Valuation

To estimate the economic value of the food provisioning ecosystem service, the market price approach was used (Azqueta & Pérez, 1997). The market price method is a widely used economic valuation technique that estimates the value of ecosystem services based on observed market prices of goods such as fish, timber, or other natural resources. It calculates the economic value by multiplying the quantity of goods extracted or consumed by their prevailing market prices. Its key strength lies in the use of readily available and verifiable market data, which makes it a transparent and straightforward tool for valuation. In the case of subsistence fishing, the market price method provides a conservative yet robust estimate of the economic contribution of self-consumed catch, shedding light on its significance for household food security and local livelihoods.

Where VFPS represents the *economic Value of the Food Provisioning Service*, $MCpc_i$ is the *monthly* per capita *consumption* of species *i*, and AMP_i is the *average monthly price* of species *i*. The value of the self-consumed fishery for species *i* is calculated by multiplying the per capita consumption of that species by its corresponding average monthly price.

$$\mathrm{VFPS} = \Sigma_{i=0}^{n}\left(\mathrm{MCpc}_i \times \mathrm{AMP}_i\right)$$

To incorporate the self-consumption volumes of fish based on species preferences and accessibility, a *Weighting Index* (WI) was developed for each selected self-consumption species. To calculate the *WI* for each species *i* in each community, the number of self-consumption observations for species *i* (O_i) was quantified relative to the total number of observations recorded in that community's logbook (TCO, for *Total Community Observations*). The Weighting Index (WI_i) was then obtained by dividing the number of observations for species *i* by the total number of observations. This index reflects the relative importance or frequency of each species in self-consumption practices:

$$\mathrm{WI}_i = \mathrm{O}_i \,/\, \mathrm{TCO}$$

To estimate the *Weighted Average of Monthly Trips* (WAMT) per community, interview data were used, in which fishers reported both the duration and number of fishing trips conducted each month. Trip frequency varied significantly among small-scale vessels (which made multiple trips per month), in contrast to medium-sized vessels, which generally carried out only one trip per month.

The WAMT was calculated by multiplying the number of trips of each type by their corresponding weighting index, as follows:

$$\mathrm{WAMT} = \Sigma_{i=0}^{n}\left(\mathrm{MT}_i \times \mathrm{WIT}_i\right)$$

Where MT_i is the number of trips of type *i* per month, and WIT_i is the weighting index for that trip type.

16.3.2 Data Collection Process

Data were collected through semi-structured interviews, self-consumption logbooks, and price monitoring targeting free fishers, permit holders, cooperative managers, and local beachside vendors and fish stalls in local markets. This information enabled the calculation of the quantity of fish allocated for self-consumption, the average frequency of fishing trips, and the most accessible species for self-consumption throughout the year, based on catch volumes. The study employed a mixed-methods approach, combining qualitative and quantitative tools to assess the economic value of food provision comprehensively (Fig. 16.2). Triangulating surveys, direct records by logbooks, and market monitoring data enabled a robust estimate of the economic contribution of fish directly consumed by fishing households, highlighting its role in family well-being and food security.

The study focused on marine finfish fishers from the communities of *Frontera* and *Barra de San Pedro*, who operate small- and medium-sized vessels. In the initial stage, estimating the total fisher population presented challenges due to discrepancies between official CONAPESCA (Spanish acronym for National Commission of Aquaculture and Fisheries) records, mainly in *Frontera*, and the actual number of active, unregistered fishers. To overcome this, in-person visits were conducted at

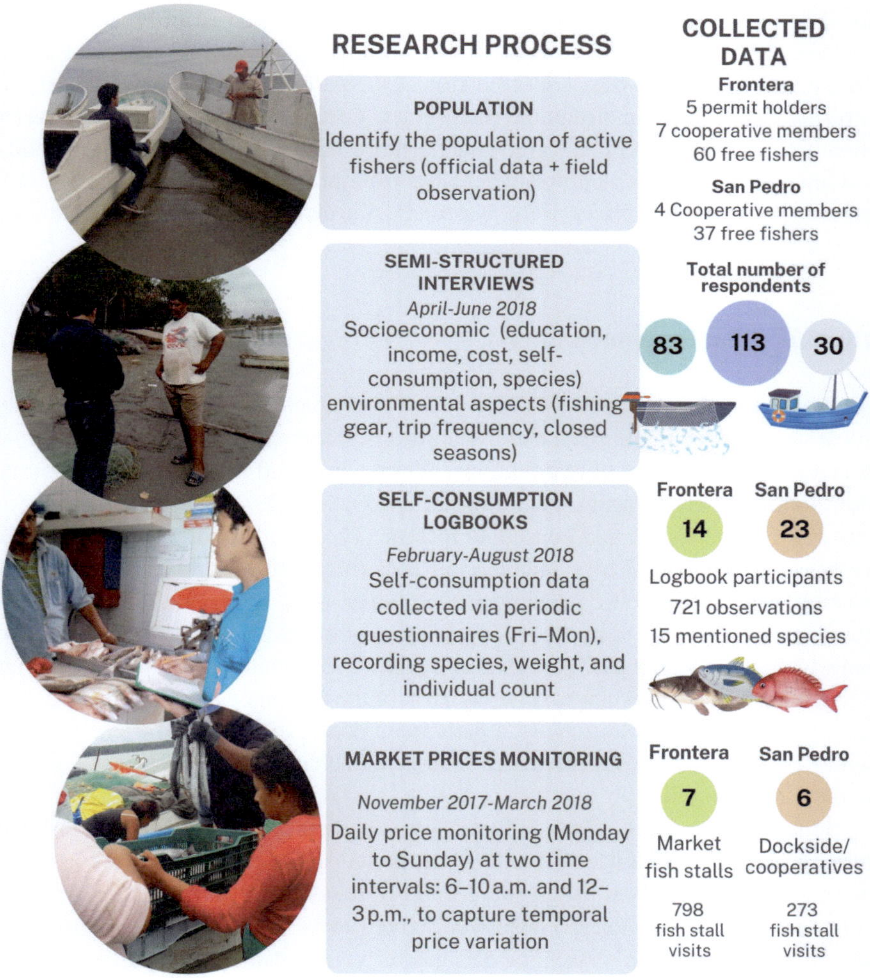

Fig. 16.2 Synthesis of the research process and data collection

cooperatives and landing sites to obtain direct estimates of the number of active fishers.

An initial approach was made to cooperative leaders and permit holders in *Barra de San Pedro* and *Frontera*. The objectives of the study, the estimated duration of visits, and the frequency of interviews with fishers at their establishments were explained in advance to request authorization for conducting field activities. A snowball sampling strategy was employed, beginning with key informants and expanding through referrals from initial participants. Willingness to participate and active involvement in fishing activities were the main criteria for selection. This approach facilitated trust-building and helped identify experienced fishers with relevant knowledge and expertise.

In *Frontera*, three cooperatives and four permit holders agreed to participate in the study, accounting for 192 active fishers out of the 320 officially registered in the locality. In *Barra de San Pedro*, five cooperatives and one permit holder collaborated, representing 142 active fishers from a total of 180 officially registered in the community (based on data collected from CONAPESCA offices). Based on this information, 113 semi-structured interviews were conducted with cooperative members, permit holders, and free fishers (those without a formal fishing permit). Additionally, 37 fishers recorded their self-consumption periodically over a 6-month period. Local market prices were systematically monitored in both communities over a 5-month period.

16.3.2.1 Semi-Structured Interview

A semi-structured interview was developed to characterize the food provisioning ecosystem service in the coastal communities of *Barra de San Pedro* and *Frontera*. The interview consisted of two sections: the first collected socioeconomic information about the respondents, including their education level, economic activities, number of dependents, and household income. The second addressed fishing gear, vessel size, number of trips, income per trip, frequently caught species, and the amount retained for self-consumption. Moreover, other questions related to environmental topics, including knowledge about fishing bans, were asked. The interview was administered between April and June 2018. A total of 113 interviews were conducted, representing approximately 10% of the estimated active fishing population in each locality. Additionally, a participant observation method was employed, with field notes taken. This approach helped to characterize the small-scale fishing in these communities.

Before participation, respondents were given a brief explanation of the research purpose and the scheduled data collection period. Each interview lasted approximately 10–20 min. Most fishers were initially reluctant to participate, even when confidentiality was assured. Those who agreed to be interviewed tended to hold leadership positions on vessels or had greater confidence in answering, typically due to having more years of experience in the fishing activity.

16.3.2.2 Self-Consumption Logbook

The logbook recorded the volume of catch intended for self-consumption after each fishing trip. Key data included the species profile (weight in kilograms and number of individuals), vessel size, type of fishing gear used, and basic information about the informant (name and the cooperative or permit holder for whom they worked at that time). The self-consumption logbook was applied to fishers operating both small vessels (outboard motorboats) and medium-sized boats in the communities of *Frontera* and *Barra de San Pedro*. Data collection occurred periodically (Friday to Monday) between February and August 2018.

16.3.2.3 Market Price Monitoring

A standardized guide was developed for collecting and systematizing market prices in *Frontera* and beachside markets in *Barra de San Pedro* over a 5-month period (November 2017 to March 2018). Data were collected from fish stalls 7 days a week during two time intervals, morning (6:00 a.m. to 10:00 a.m.) and afternoon (12:00 p.m. to 3:00 p.m.) to capture potential price fluctuations by time of day and day of the week.

For systematic monitoring, each fish stall in *Frontera* was labeled with letters A through G, and those in *Barra de San Pedro* with letters H through M, resulting in a total of seven monitored stalls in *Frontera.* Six stalls were selected in *Barra de San Pedro.*

In *Frontera*, market stands were visited on a rotating weekly schedule. During the first week, stands A, C, E, and G were visited on Mondays, Wednesdays, Fridays, and Sundays, while stands B, D, and F were visited on Tuesdays, Thursdays, and Saturdays. This schedule was reversed in the following week and repeated throughout the study, with each visit occurring twice a day. In *Barra de San Pedro*, data were collected exclusively on Saturdays and Sundays, as these were the usual landing days for small-scale fishing vessels, weather permitting. Price observations at cooperatives and beachside vendors were conducted 2 to 3 times per week.

16.4 Results

The following findings are based on data collected through 113 semi-structured interviews with fishers from the communities of *Frontera* and *Barra de San Pedro.* These interviews provide insight into the demographic, social, and economic characteristics of the fishing population in the region, as well as key aspects of their fishing livelihoods. The data cover topics such as fishing effort, organizational structures, vessel use, fishing gear, trip duration, income patterns, fishing grounds, and practices of self-consumption.

16.4.1 Sociodemographic Characteristics

Fishers ranged in age from 15 to 68 years, with the majority (75%) falling between 15 and 41 years of age, followed by 56% between 42 and 63 years, and only 3% being over 63 years old. Fishing is a deeply rooted tradition, with most respondents starting before the age of 15, reflecting early entry into the activity as a cultural norm. In terms of education, over half of the participants had not completed secondary school. A significant proportion only reached primary formal education, and a few reported being functionally illiterate. This educational profile reflects limited access to formal schooling and possibly an early entry into the fishing activity. Regarding family structure, only 3% had no children, while 41% had one to two children, 45% had three to five children, and 12% had six or more children. Furthermore, 44% of fishers reported having at least one child under the age of 18. Most fishers originated from Tabasco although approximately 30% had migrated from other states, mainly Veracruz, Campeche, Guerrero, and Yucatán. This reflects a trend of interstate mobility among fishers in the region, contributing to a culturally diverse fishing workforce in both *Frontera* and *Barra de San Pedro*.

16.4.2 Fishing Organization and Gear Use

The fishing establishments visited revealed two primary organizational structures. In one of them, the owner or permit holder of multiple vessels leases equipment to one or more fishers and later divides the profits evenly, with half allocated to the fishers and the other half to the owner. The second form is the cooperative model, in which affiliated fishers deliver their catch to the cooperative, and revenues are periodically distributed after a financial tally. Most respondents (94.7%) identified themselves as full-time fishers, while 5.3% engaged in other income-generating activities, such as construction, agriculture, aquaculture, or security work, primarily during periods of low fishing productivity.

Fishers used traditional gear types depending on the target species, including bottom longlines (*palangre*), vertical lines (*ristra*), and gillnets (*red de enmalle*) (Fig. 16.3). The most common species included gafftopsail catfish, small sharks, jack, grouper, and *bala*. In *Barra de San Pedro*, all respondents operated small-scale vessels with outboard motors. In contrast, fishers in *Frontera* used both small (58%) and medium-sized vessels (42%). Labor needs to be varied accordingly: small vessels typically required 2–3 fishers, whereas medium-sized boats needed 5–6.

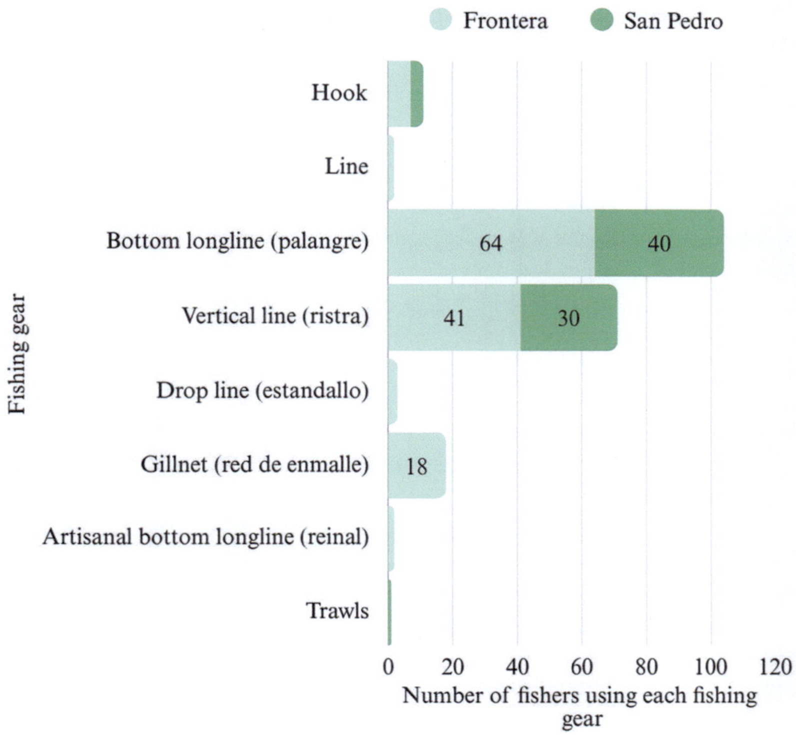

Fig. 16.3 Most common fishing gears in small-scale and medium-sized fleets in the communities of *Frontera* and *Barra de San Pedro*, Tabasco, Mexico

16.4.3 Fishing Effort and Self-Consumption Patterns

Fishing activity in the interviewed communities reveals three distinct operational patterns shaped by vessel type, trip duration, distance to fishing grounds, and income dynamics: one-day trips, medium-duration trips (3–4 days), and extended offshore trips (14–20 days).

One-day trips, primarily undertaken with small-scale vessels equipped with outboard motors, targeted nearshore areas located between 15 and 150 nautical miles. These were the most frequent type of trip, occurring on average 14 times per month, and generated modest but steady income, averaging USD 29.16[1] per outing, with a

[1] The average USD to Mexican Peso exchange rate for the available 2018 data was approximately 18.95 MXN/USD (Gobierno de México, 2018).

wide range between USD 7.91 and 68.60, depending on the catch. Fishers retained an average of 6.95 kg of fish per trip for household consumption, reflecting both food security needs and the flexibility offered by short, low-cost operations (under USD 10.5 per trip). The most frequently retained species included: *curvina* (weakfish), *jurel* (crevalle jack), *cintilla* (Atlantic cutlassfish), *sábalo* (tarpon), *huachinango* (red snapper), *peto* (wahoo), *coruco* (gray snapper), *bacalao* (cod), *besugo* (vermilion snapper), *cojinuda* (pompano), and *bonito* (little tunny). These species, commonly found in coastal waters, served both dietary and economic functions.

Medium-duration trips (3–4 days) were also common among small-scale fishers, especially those venturing slightly farther offshore (around 90–100 nautical miles). These trips occurred with a frequency of about 6–10 times per month, yielding an average income of USD 41.34 per trip and 6.91 kg of fish retained for self-consumption. Despite slightly higher operational demands, they remained accessible and offered a reliable blend of income generation and food provisioning. Commonly reported species for these outings included: *bandera* (Gafftopsail catfish), gray snapper, *curvina* (weakfish), *jurel* (crevalle jack), red snapper, wahoo, *bacalao* (cod), *besugo* (vermilion snapper), *sierra* (Atlantic Spanish mackerel), *villajaiba* (lane snapper), *balá* (American stingray), *cojinuda* (blue runner), *negrillo* (lesser amberjack), and *medregal* (banded rudderfish). The diversity in species reflects opportunistic harvesting across productive mid-shelf grounds.

Extended trips, lasting 14–20 days, were conducted using medium-sized vessels and aimed at distant offshore fishing zones (averaging 221 nautical miles from shore). These trips were infrequent, typically occurring once a month, but represented a high-investment strategy, with income ranging between USD 158.31 and USD 633.25, and an average of USD 363.41 per trip. Operational costs were significantly higher (USD 158.31–237.46), yet fishers retained approximately 8.5 kg of fish per trip for household use, indicating substantial contributions despite lower trip frequency. The most commonly retained species during these expeditions were the *besugo* (vermilion snapper), *medregal* (banded rudderfish), *jurel* (crevalle jack), *villajaiba* (lane snapper), with occasional inclusion of the *balá* (American stingray), *mero* (grouper), parguete, *bacalao* (cod), *cazón* (small shark), and *trucha* (trout). The dominance of large pelagic and demersal species in these catches reflects the productivity and depth of the fishing grounds visited.

In summary, shorter trips allowed fishers to regularly access species of dietary and economic value, contributing to household provisioning with high frequency. In contrast, longer trips involved greater investment and operational complexity, but yielded higher per-trip volumes and access to larger, commercially valuable species. These patterns illustrate how vessel type, trip duration, and fishing range jointly shape both livelihood strategies and food security outcomes, with fishers adapting their practices to optimize the balance between income generation and household needs.

16.4.4 Economic Valuation of the Food Provisioning Ecosystem Service

The economic valuation of food provisioning service was based on a robust triangulation of primary data sources. This included 721 self-consumption records from 37 fishers, systematically collected between February and August 2018 through detailed logbooks maintained after frequent fishing trips. In parallel, an extensive price monitoring exercise was conducted at points of sale, comprising 798 visits to local fish stalls in *Frontera* and 273 visits to beachside landing sites in *Barra de San Pedro*, between November 2017 and March 2018. This triangulated approach, linking firsthand records of fish quantities kept for self-consumption with real-time market and landing prices, ensures a reliable and context-specific estimation of the economic value of fish reserved for household consumption.

According to the self-consumption logbook, in *Barra de San Pedro*, the six species contributing the highest volumes per fisher were *bandera* (Gafftopsail catfish), *besugo* (vermilion snapper), villajaiba (lane snapper), *balá* (American stingray), *coruco* (gray snapper), and *huachinango* (red snapper). In contrast, for small-scale vessels in *Frontera*, the leading species were *bandera, huachinango, villajaiba, coruco, jurel,* and *besugo*. Among medium-sized vessels in *Frontera*, the most retained species were *jurel*, *picuda*, *besugo*, *medregal* (banded rudderfish), *cazón* (small shark), and *bacalao* (cod) (Table 16.1).

Table 16.1 Average per capita weight of fish retained for self-consumption per trip in the communities of *Frontera* and *Barra de San Pedro*, Tabasco, Mexico. The three main species for each fishing fleet are highlighted in bold

Specie	Per capita weight (kg)	
	Frontera	Barra de San Pedro
Cherna	**11.4**	–
Banded rudderfish	**9.4**	–
American stingray	**9**	**7.4**
Small shark	8.9	**8.0**
Crevalle jack	8.5	–
Picuda	7.9	–
Cod	7.1	6.0
Vermillion snapper	6.8	5.6
Gafftopsail catfish	6.4	**6.5**
Red snapper	5.7	5.5
Lane snapper	5.5	5.5
Gray snapper	4.7	4.9
Weakfish	4.1	3.3

Source: Own elaboration from logbooks

Table 16.2 Average market and dockside prices of the main species selected for self-consumption in the communities of *Frontera* and *Barra de San Pedro,* Tabasco, Mexico. The top three species for each fishing fleet are highlighted in bold

Species	Market prices	Max	Min	Dockside prices	Max	Min
	USD					
Red snapper	**5.82**	6.33	4.75	**5.46**	6.33	4.75
Banded rudderfish	**4.07**	4.75	3.69	**2.53**	3.17	1.58
Cod	**4.18**	5.01	3.69	2.13	3.17	0.79
Lane snapper	3.06	3.43	2.37	**2.18**	2.90	1.58
Small shark	3.03	3.43	2.64	1.30	2.37	1.06
Gafftopsail catfish	1.93	2.64	1.58	0.96	1.32	0.53
Gray snapper	0.37			0.37		
Picuda	0.37			0.37		

Source: Own elaboration from market price monitoring

In the case of species selected for self-consumption, both market and dockside prices were collected (Table 16.2). However, some species, such as *coruco* (gray snapper) and *picuda* were not commercially traded, so their prices had to be estimated using the lowest recorded market price as a reference.

Analyzing market versus dockside prices reveals significant disparities that highlight the economic importance of self-consumption practices among fishing households. Across the main species kept for household consumption, the average price difference between market and dockside points was around 62.3%, indicating significant commercial margins within the supply chain. *Huachinango* (red snapper) exhibited a minimal difference of just 6.6%, whereas other species, such as gafftopsail catfish, cod, and small shark, showed the highest margins at 100.5%, 96.6%, and 132.9%, respectively. Moderate gaps were found for species such as *medregal* (banded rudderfish) (60.9%) and *villajaiba* (lane snapper) (40.4%). Even among high-value species, price differences regularly exceeded 30%, illustrating a consistent trend across market categories. These findings suggest that retaining fish directly from the catch could lead to significant economic savings, particularly for species with substantial retail markups. In this context, self-provisioning enhances food security and functions as a key strategy for supplementing income, particularly in low-income coastal communities where purchasing fish at market prices would significantly impact household budgets.

Table 16.3 illustrates the contribution of the food provision service (VFPS) to household income across two coastal communities and fishing fleet types. In *Frontera*, fishers operating small-scale vessels report an average Monthly Economic Value of Food Provision Service (VFPS) of USD 169.59, derived from frequent trips (average of 10.11 per month). This amount represents 42% of the average monthly per capita income (AMCI) in the community (USD 404.57), highlighting the critical economic value of self-consumed fish as a household support mechanism.

In *Barra de San Pedro*, small-scale vessels average 5.17 trips per month, resulting in a VFPS of USD 79.72, which also constitutes a substantial 43% of the local AMCI (USD 180.21). This suggests that, despite lower fishing frequency,

Table 16.3 Comparison of monthly per capita income and economic value of the food provisioning service by vessel type in two coastal communities of Tabasco, Mexico

Community	Small-scale vessel					Medium-scale vessel				
	Average monthly per capita income (AMCI)	Max	Min	Monthly value of the food provision service (VFPS)	% VFPS of AMCI	Average monthly per capita income (AMCI)	Max	Min	Monthly value of the food provision service (VFPS)	% VFPS of AMCI
	USD					USD				
Frontera	404.57	1266.49	63.32	169.59	42%	357.26	791.56	131.93	21.91	6%
Barra de San Pedro	180.21	791.56	47.49	79.72	43%		–	–	–	–

Source: Own elaboration

self-consumption remains a vital contributor to food security, especially for lower income families.

For medium-scale vessels in *Frontera*, the VFPS is notably lower at USD 21.91 per month, representing just 6% of the AMCI for this group (USD 357.26). Although catches per trip are generally higher, these vessels typically make only one trip per month, which limits the total volume of fish available for self-consumption.

16.5 Discussion

This study provides a critical baseline of the food provisioning value of small-scale fisheries before the construction of the *Dos Bocas* refinery, which started in August 2019 (Secretaría de Energía (SENER), 2019). As such, it captures the socio-ecological dynamics and household-level benefits of self-consumption fishing before the most recent industrial transformations in the region. The provisioning service of food from small-scale fisheries emerges as both a nutritional backbone and an economic lifeline for coastal households in Tabasco. Our findings reveal that the value of self-consumed fish can represent up to 44% of monthly per capita income, underscoring its importance in regions marked by environmental vulnerability and limited economic diversification. This aligns with broader global evidence highlighting how self-provisioning contributes not just to food intake but to livelihood resilience, particularly among populations with restricted market access (McClanahan et al., 2015).

In communities like *Frontera* and *Barra de San Pedro*, fishers rely heavily on 1-day or medium-duration trips to secure marine protein. These trips, averaging 6 to 10 times per month, translate into steady access to key nutrients that are often unavailable through other affordable sources. Recent global analyses, including those by Basurto et al. (2025) in *Nature*, estimate that small-scale fisheries contribute at least 44% of the total landed value of global fisheries and provide essential micronutrients to approximately 2.3 billion people. This further validates the relevance of our local observations within a planetary context.

Although not the primary focus of this study, the cultural dimension of self-consumption proved to be a critical component of small-scale fisheries. Fishing knowledge is transmitted intergenerationally, and most interviewees reported the beginning of fishing during childhood, largely due to the absence of viable alternatives linked to limited education and scarce employment opportunities. This finding aligns with observations by Isaac and de Almeida (2011) and Vasco and Sirén (2018), who highlight that fishing in Latin American communities is not only a source of food but also a deeply rooted cultural practice. Coastal populations consistently consume more fish than inland groups and maintain strong emotional and symbolic ties to fishing. However, this cultural reliance is increasingly at risk. Industrial expansion, particularly from oil infrastructure, has resulted in access restrictions, environmental degradation, and gear losses, all of which directly undermine fishers' ability to engage in subsistence practices. These dynamics reinforce patterns of environmental injustice, where the burdens of development fall disproportionately on those most dependent on ecosystem services (Quist, 2019).

Adding to this challenge are the sharp disparities in market structures. Our study found that market prices for key species are, on average, 62.3% higher than dockside values. In some cases, such as small sharks and gafftopsail catfish, this markup exceeds 100%. Such price differentials are indicative of structural inequality along the value chain, where intermediaries capture a disproportionate share of value. For fishing households, retaining part of the catch thus becomes a cost-saving strategy as much as a subsistence practice. The variability in savings across vessel types also matters. In *Frontera*, where small-scale fishers make frequent trips, the economic value of self-consumption reaches USD 169.55 per month, or 42% of the local per capita income. In contrast, medium-scale vessels report lower savings due to infrequent outings. This pattern reveals how trip frequency, not just catch volume, shapes food provisioning outcomes.

In comparison, studies such as Hernández-Flores (2015) estimate wide ranges in the annual value of self-consumed fish in Mexico (from USD 434.78 to 5685.62[2] per cooperative), reflecting how effort, access, and community structure modulate the benefits of this service. Our data contribute to this growing body of work by providing grounded, quantitative estimates of how self-consumption supports household economies in resource-constrained settings.

In summary, self-provisioning in small-scale fisheries, though often undervalued in national policy, emerges as a pillar of food sovereignty, cultural continuity, and economic stability in Tabasco. As Basurto et al. (2025) emphasize, recognizing the role of small-scale fisheries is fundamental to achieving global sustainability goals, particularly SDG 2 (Zero Hunger). In contexts facing industrial encroachment and ecosystem degradation, safeguarding the right to fish and the capacity to self-consume becomes both an equity issue and a strategic imperative for long-term resilience.

16.6 Conclusion

The findings of this study point to the urgent need for sustainability pathways that center small-scale fishers as key actors in coastal food systems. Protecting the provisioning service of self-consumed fish requires not only environmental conservation but also social recognition and institutional support. As oil and infrastructure development continue to expand in Tabasco, alternative approaches must prioritize rights-based access to marine resources, establish effective compensation mechanisms for affected communities, and integrate traditional ecological knowledge into policy frameworks. Strengthening co-management systems, investing in local value chains, and formally recognizing self-provisioning as a pillar of food security can help reduce the vulnerability of fishing households. In doing so, Mexico can chart a

[2] The average USD to Mexican Peso exchange rate for the available 2016 data was approximately 14.97 MXN/USD (Gobierno de México, 2016).

course toward a more just and ecologically resilient coastal development model—one that aligns local needs with global sustainability commitments.

Acknowledgments The original data presented in this study derive from the undergraduate thesis of Juan Manuel Palacios Hernández: *Estimación del servicio ecosistémico de provisión de alimento en la costa de Tabasco. El caso del municipio de Centla, Tabasco*. This research was supported by the project "*La desafiante coexistencia de los sistemas socio-ecológicos acoplados; las industrias de la pesca y petróleo en la Sonda de Campeche*" (Project CF-2019-1564454, CONACyT). Special thanks are extended to J.B. Cruz-Pech for the design of the maps and to Magdalena Hernández for her editorial input and thorough review of the manuscript. During the preparation of this chapter, the authors used two AI-assisted tools solely to improve the quality of English language expression: ChatGPT (OpenAI, GPT-4, May 2024 version) and Grammarly (Grammarly Inc., 2025). ChatGPT was employed to enhance textual clarity, ensure terminological consistency, and suggest stylistic refinements. Grammarly was used to support grammar, spelling, and sentence structure checking. All content generated with these tools was critically assessed, manually revised, and validated by the authors to ensure accuracy, originality, and scholarly integrity.

Competing Interests The authors have no conflicts of interest to declare that are relevant to the content of this chapter.

References

Alongi, D. M. (2012). Carbon sequestration in mangrove forests. *Carbon Management, 3*(3), 313–322. https://doi.org/10.4155/cmt.12.20

Andrews, N., Bennett, N. J., Le Billon, P., Green, S. J., Cisneros-Montemayor, A. M., Amongin, S., Gray, N. J., & Sumaila, U. R. (2021). Oil, fisheries and coastal communities: A review of impacts on the environment, livelihoods, space and governance. *Energy Research & Social Science, 75*, 102009. https://doi.org/10.1016/j.erss.2021.102009

Azqueta, D., & Pérez, L. (1997). *El valor económico de los servicios recreativos de los espacios naturales*. McGraw-Hill.

Barbier, E. B., Hacker, S. D., Kennedy, C., Koch, E. W., Stier, A. C., & Silliman, B. R. (2011). The value of estuarine and coastal ecosystem services. *Ecological Monographs, 81*(2), 169–193. https://doi.org/10.1890/10-1510.1

Basurto, X., Gutierrez, N. L., Franz, N., Mancha-Cisneros, M. M., Gorelli, G., Aguión, A., Funge-Smith, S., Harper, S., Mills, D. J., Nico, G., Tilley, A., Vannuccini, S., Virdin, J., Westlund, L., Allison, E. H., Anderson, C. M., Baio, A., Cinner, J., Fabinyi, M. … Thilsted, S. H. (2025). Illuminating the multidimensional contributions of small-scale fisheries. *Nature*, *637*, 875–884. https://doi.org/10.1038/s41586-024-08448-z.

Beaumont, N. J., Austen, M. C., Mangi, S. C., & Townsend, M. (2008). Economic valuation for the conservation of marine biodiversity. *Marine Pollution Bulletin, 56*(3), 386–396. https://doi.org/10.1016/j.marpolbul.2007.11.013

CONAPESCA. (2023). *Anuario estadístico de acuacultura y pesca 2023*. Comisión Nacional de Acuacultura y Pesca. https://www.gob.mx/conapesca

Costanza, R., d'Arge, R., de Groot, R., Farber, S., Grasso, M., Hannon, B., Limburg, K., Naeem, S., O'Neill, R. V., Paruelo, J., Raskin, R. G., Sutton, P., & van Den Belt, M. (1997). The value of the world's ecosystem services and natural capital. *Nature, 387*, 253–260. https://doi.org/10.1038/387253a0

DOF (Diario Oficial de la Federación) (2003). "ACUERDO mediante el cual las secretarías de Marina, de Comunicaciones y Transportes, y de Agricultura, Ganadería, Desarrollo Rural, Pesca y Alimentación, establecen medidas de seguridad en la Sonda de Campeche." Diario Oficial de la Federación. 11 de septiembre, 2003. http://dof.gob.mx/nota_detalle.php?codigo=690229&fecha=11/09/2003.

Espinoza-Tenorio, A., & Mendoza-Carranza, M. (2015). Pesquerías costeras de Tabasco: problemáticas actuales y escenarios futuros ante el cambio climático. In F. Paz Pellat, J. Wong González, & R. Torres Alamilla (Eds.), *Estado actual del conocimiento del ciclo de carbono y sus interacciones en México: Síntesis a 2015* (pp. 553–559). Programa Mexicano del Carbono, Centro del Cambio Global y la Sustentabilidad en el Sureste A.C., y Universidad Juárez Autónoma de Tabasco, Texcoco, Estado de México.

Fourqurean, J., Duarte, C. M., Kennedy, H., Marbà, N., Holmer, M., Mateo, M. A., Apostolaki, E. T., Kendrick, G. A., Krause-Jensen, D., McGlathery, K. J., & Serrano, O. (2012). Seagrass ecosystems as a globally significant carbon stock. *Nature Geoscience, 5*, 505–509. https://doi.org/10.1038/ngeo1477

Gobierno de México. (2016). *Indicadores económicos—Tipo de cambio peso-dólar estadounidense (22/01/2015–22/01/2016)*. Diario Oficial de la Federación. https://dof.gob.mx/indicadores_detalle.php?cod_tipo_indicador=158&dfecha=22%2F01%2F2015&hfecha=22%2F01%2F2016

Gobierno de México. (2018). *Indicadores económicos—Tipo de cambio peso-dólar estadounidense (01/01/2018–12/06/2018)*. Diario Oficial de la Federación. https://dof.gob.mx/indicadores_detalle.php?cod_tipo_indicador=158&hfecha=12/06/2018&dfecha=01/01/2018&accionI=imprimir

Hernández García, B. (2024). *Cambios en los valores de la naturaleza en la comunidad pesquera de pequeña escala en Paraíso, Tabasco*. El Colegio de la Frontera Sur.

Hernández-Flores, E. (2015). *Valoración económica de la pesca de autoconsumo como servicio ecosistémico de provisión de alimento en la Reserva de la Biosfera La*. Universidad Autónoma de Chiapas, Instituto de Biociencias.

Instituto Nacional de Estadística y Geografía (INEGI). (2020). *Panorama sociodemográfico de Tabasco. Censo de Población y Vivienda 2020*. INEGI. https://www.inegi.org.mx/programas/ccpv/2020

Isaac, V. J., & de Almeida, M. C. (2011). *El consumo de pescado en la Amazonía brasileña* (COPESCAALC Documento Ocasional No. 13). Organización de las Naciones Unidas para la Alimentación y la Agricultura. Rome, Italy, FAO.

McClanahan, T., Allison, E. H., & Cinner, J. E. (2015). Managing fisheries for human and food security. *Fish and Fisheries, 16*, 78–103. https://doi.org/10.1111/faf.12045

Mendoza-Carranza, M., Ejarque, E., & Nagelkerke, L. A. J. (2018). Disentangling the complexity of tropical small-scale fisheries dynamics using supervised self-organizing maps. *PLoS One, 13*(5), e0196991. https://doi.org/10.1371/journal.pone.0196991

Muñoz-Sánchez, P., & Cruz-Burguete, J. L. (2013). Identities and migratory 75 tendencies from fishing in Chiapas and Tabasco [Identidades y tendencias migratorias desde la Pesca en Chiapas y Tabasco]. *Convergencia, 20*(63), 231–257.

One Earth. (2020, September 23). *Pantanos de Centla*. https://www.oneearth.org/ecoregions/pantanos-de-centla/

Pascual, U., Balvanera, P., Díaz, S., Pataki, G., Roth, E., Stenseke, M., Watson, R. T., Dessane, E. B., Islar, M., Kelemen, E., Maris, V., Quaas, M., Subramanian, S. M., Wittmer, H., Adlan, A., Ahn, S., Al-Hafedh, Y. S., Amankwah, E., Asah, S. T. … Yagi, N. (2017). Valuing nature's contributions to people: The IPBES approach. *Current Opinion in Environmental Sustainability, 26–27*, 7–16. https://doi.org/10.1016/j.cosust.2016.12.006

Quist, L. M. (2019). Fishers' knowledge and scientific indeterminacy: Contested oil impacts in Mexico's sacrifice zone. *Maritime Studies, 18*, 65–76. https://doi.org/10.1007/s40152-018-0123-7

Quist, L.-M., & Nygren, A. (2015). Contested claims over space and identity between fishers and the oil industry in Mexico. *Geoforum, 63*, 44–54. https://doi.org/10.1016/j.geoforum.2015.05.015

Ramos-Muñoz, D. E., Ramos-Reyes, R., Zamora-Cornelio, L. F., Hernández-De la Cruz, A., & Espinoza-Tenorio, A. (2019). Exclusión en el Golfo de México: una visión desde los pescadores sobre la industria petrolera en Tabasco. *Cuadernos de Geografía: Revista Colombiana de Geografía, 28*(2), 357–372.

Salazar-De la Cruz, C. C., Zepeda-Domínguez, J. A., Espinoza-Tenorio, A., & Ramos-Muñoz, D. E. (2020). Governance networks in marine spaces where fisheries and oil coexist: Tabasco, México. *The Extractive Industries and Society, 7*(2), 676–685. https://doi.org/10.1016/j.exis.2020.03.012

Secretaría de Economía. (2024). *Centla: Economic profile*. Data México. https://www.economia.gob.mx/datamexico/en/profile/geo/centla

Secretaría de Energía (SENER). (2019, agosto 5). *En marcha la construcción de una nueva refinería con capacidad de procesamiento de 340 mil barriles de petróleo en Dos Bocas, Tabasco*. Gobierno de México. https://www.gob.mx/sener/articulos/en-marcha-la-construccion-de-una-nueva-refineria-con-capacidad-de-procesamiento-de-340-mil-barriles-de-petroleo-en-dos-bocas-tabasco

Soto, L. A., Botello, A. V., Licea-Durán, S., Lizárraga-Partida, M. L., & Yáñez-Arancibia, A. (2014). The environmental legacy of the Ixtoc-I oil spill in Campeche Sound, southwestern Gulf of Mexico. *Frontiers in Marine Science, 1*, Article 57. https://doi.org/10.3389/fmars.2014.00057

Toledo-Ocampo, A. (2005). Marco Conceptual: Caracterización Ambiental del Golfo de México. In A. V. Botello, J. Rendón von Osten, G. Gold-Bouchot, & C. A. Hernández (Eds.), *Golfo de México Contaminacion e Impacto Ambiental: Diagnostico y Tendencias* (2nd ed., pp. 25–52). Universidad Autonoma de Campeche, Universidad Nacional Autonoma de México, Instituto Nacional de Ecología.

Vasco, C., & Sirén, A. (2018). Determinants of wild fish consumption in indigenous communities in the Ecuadorian Amazon. *Society & Natural Resources, 32*(1), 21–33. https://doi.org/10.1080/08941920.2018.1475587

Velázquez Guzmán, M. G. (1982). Afectaciones petroleras en Tabasco: El movimiento del Pacto Ribereño. *Revista Mexicana de Sociología, 44*(1), 167–187.

Viana, D. F., Zamborain-Mason, J., Gaines, S. D., Schmindhuber, J., & Golden, C. D. (2023). Nutrient supply from marine small-scale fisheries. *Scientific Reports, 13*, 11357. https://doi.org/10.1038/s41598-023-37338-z

Chapter 17
Small-Scale Fisheries and Oil Marine Infrastructure Interactions: Challenges for Marine Sustainability and Territorial Planning in the Southern Gulf of Mexico

R. E. Lara-Mendoza, M. Vásquez-Ortíz, S. Caña-Hernández, and M. T. Zarate-Herrera

17.1 Introduction

Oil is a fundamental element in the world's energy generation nowadays. A quarter of it is extracted from offshore fields, and up to 32% of global energy came from this resource before the COVID-19 pandemic. However, a continued increase in overall demand is expected (Andrews et al., 2021; Mackenzie, 2017). Historically, in Mexico, most of the national oil comes from the "*Cuencas del Sureste* Oil Province," where the states of Veracruz, Tabasco, and Campeche contribute 92% of the total national volume, with Tabasco and Campeche being the largest crude oil producers (CNH, 2024). This ranks Mexico as the fourth-largest producer in the Americas and the 13th worldwide, highlighting the country's reliance on oil (Cerón, 2024). On the other hand, the environmental conditions in the southwestern Gulf of Mexico (sGoM) make the region a highly productive area due to the input of nutrients from the Grijalva-Usumacinta system, mangroves, seagrass areas, shallow and deep reefs that propitiate conditions that favor commercial fishing (Uribe-Martínez et al.,

R. E. Lara-Mendoza (✉) · M. Vásquez-Ortíz
Dirección de Investigación Pesquera en el Atlántico, Instituto Mexicano de Investigación en Pesca y Acuacultura Sustentables. Project: Implementation of the Strategic Action Program of the Gulf of Mexico Large Marine Ecosystem, Mexico City, Mexico
e-mail: proyectogom.coordinacion@imipas.gob.mx; mario.vasquez@imipas.gob.mx

S. Caña-Hernández
Centro Regional de Investigación Acuícola y Pesquera en Lerma, Instituto Mexicano de Investigación en Pesca y Acuacultura Sustentables. Campeche, Campeche, México
e-mail: sebastian.cana@imipas.gob.mx

M. T. Zarate-Herrera
Doctorado en Ciencias en Ecología y Desarrollo Sustentable, El Colegio de la Frontera Sur, Campeche, México
e-mail: mitzy.zarate@posgrado.ecosur.mx

A. Espinoza-Tenorio et al. (eds.), *The Fishing and Oil Industries in Coexistence*,
https://doi.org/10.1007/978-3-032-02621-7_17

2019) so that along the coasts of both states a high number of fishing communities have settled that interact with the semi-industrial fleet as well as with the oil industry, in addition to the fact that some settlements are established within protected natural areas such as the *Pantanos de Centla* Biosphere Reserve or the *Laguna de Términos* Flora and Fauna Protection Natural Area, adding complexity to this dynamic.

Due to the high productivity of the area, more than 20 species of elasmobranchs (sharks and rays) and around 100 species of bony fishes are caught artisanally in directed and incidental multi-species fisheries with fishing gear such as nets, longlines, hand lines, trolling lines, among others (Coronado et al., 2022; Lara-Mendoza, 2020), making small-scale fishing an important activity in the country that represents a source of direct and indirect employment, fight poverty and marginalization in coastal communities, and guarantees food security by providing low-cost quality protein to the population (CONAPESCA, 2021). This coexistence between two activities of significant relevance to the country presents a challenge to environmental and fisheries sustainability, as well as to effective territorial planning, since sustainable development depends on the performance and balance of the institutions involved. However, the Mexican governmental system prioritizes economic development and infrastructure, while giving less attention to the fisheries sector (Coronado et al., 2024).

This situation has caused many fishers to move to restricted areas for offshore fishing (exclusion zones), which were decreed as security measures derived from the terrorist events of September 11, 2001, zones that ended up displacing them from their fishing areas and which are attractive to them since the structures of the oil platforms usually works as artificial reefs, aggregating a large amount of fish to their surroundings and consequently to the fishers, who defy the restrictions imposed by the government to obtain profits from the activity they carry out (DOF, 2003; Liina-Maija, 2019). There is also a constant discussion where fishers accuse the oil industry and its extractive activities, spills and seismic studies of being responsible for the decrease in catches because they displace or kill the species on which they depend, while the government and the oil industry blame them for the overexploitation of the species, as well as violating safety regulations that have been imposed, leading to conflicts with fishers, disputes over territory as on most occasions communities and fishers are displaced by the industry and ultimately social consequences for the communities (Liina-Maija, 2019; Arbo & Thanh Thủy, 2016).

Understanding the interactions that occur in the sea allows making decisions based on governance, especially in an area as complex as the sGoM where the traditional approach to fisheries management in which decisions are made in a unidirectional way and where the social context is not often taken into consideration have not managed to stop or reverse the decline of fisheries (Peña-Puch et al., 2021) and where there is also a lack of information about the real interaction between fishers and the oil industry. In this sense, co-management and improved governance, where the participation of all actors involved in the use and management of space and resources is encouraged, emerge as a more effective option in the decision-making process. Therefore, we made an exhaustive effort to identify and determine the potential interactions and the degree of overlap between marine oil infrastructure and small-scale fisheries, as well as the challenges this presents for effective territorial planning and resource sustainability.

17.2 Material and Methods

The spatial information used in this study comes from three sources, mainly obtained through small-scale fisheries research projects during 2016 to 2024, covering the main fishing localities and fisheries in the states of Tabasco and Campeche (Fig. 17.1). The first data source originates from information provided by the fishers, recorded on their navigation equipment (e.g., Furuno GPS 32 and 33) during each fishing trip. They log the geographic position of each trip and fishing set, as well as the depth and date. The second data source comes from fishing trips where

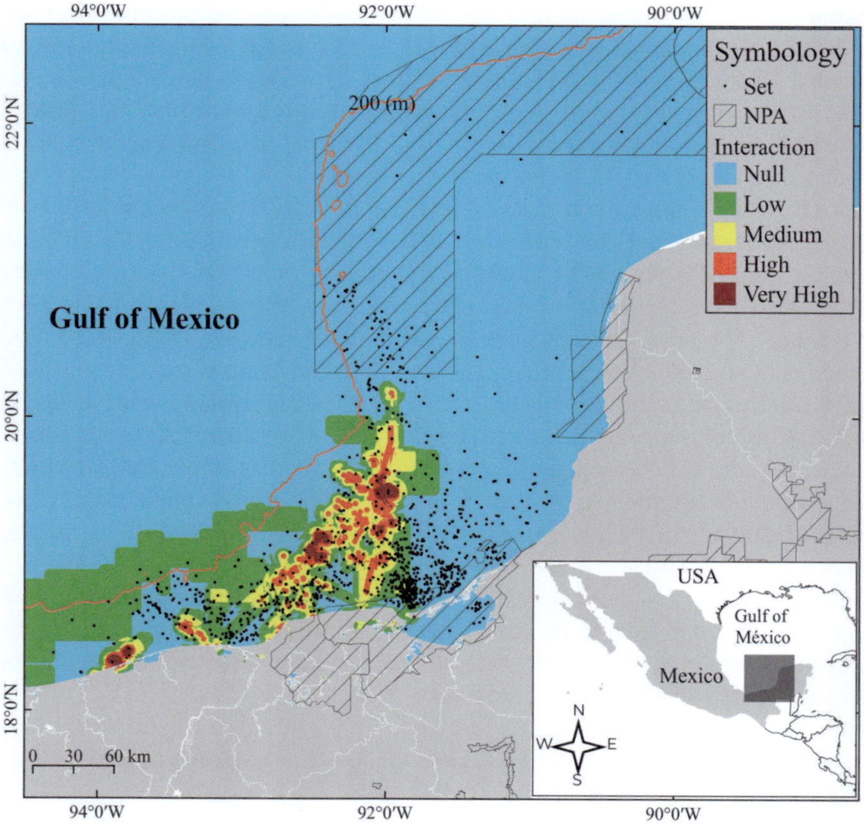

Fig. 17.1 Potential interaction between small-scale fisheries and oil marine infrastructure in the southern Gulf of Mexico

a technician or researcher was on board the vessel to record several biological and fishery information; among other data, they collected the geographic position of each trip and fishing set, as well as the depth and date, using a handheld GPS device (Garmin eTrex 10). The third and most important source of information was obtained through interviews with fishers after each fishing trip, where was recorded information about the fishing area, distance from the coast (in kilometers), geographic heading, depth (in meters), fishing target (species), characteristics of fishing gear, and dynamics of the fishing activity.

For the analysis of the interaction between small-scale fishing and oil infrastructure, it was necessary to identify all small-scale fisheries that could potentially have some type of interaction, which are described in Table 17.1. On the other hand, it is important to mention that there are other small-scale fisheries in the region. However, they do not have any interaction because they are conducted further north in the region and farther from the oil infrastructure, like octopus (*O. maya* and *O. americanus*), crabs (*Callinectes* spp.), and snail fisheries.

Otherwise, the geographical information about the exclusion zones, oil platforms, and other installations (fixed, semi-fixed, and pipelines) installed in the coastal-marine waters from sGoM during the period covered by this study was extracted from official sources such as the Official Journal of the Federation and the National Hydrocarbons Commission in Mexico (CNH, 2024) (https://mapa.hidrocarburos.gob.mx/); and additionally, the update information about the Natural Protected Areas (NPA) was obtained from National Commission of Natural Protected Areas (CONANP, by its Spanish acronym).

A GIS-based multicriteria decision analysis (GIS-MCDA) was carried out to understand the potential interactions between the small-scale fisheries, the oil infrastructure, and NPA the sGoM according to Laaribi et al. (1996) and Malczewski (2006). First, the set contains thematic layers associated with oil infrastructure, such as pipelines, facilities, anchorage areas, and production zones (CNH, 2024). The second data set pertains to fisheries. The data were converted into a vector format, and to conduct the analysis, raster layers with 50 by 50 m pixel size were created. In total, five raster layers were generated showing the Euclidean distance to each infrastructure to zone the interaction with the small-scale fisheries. Thereafter, each layer was reclassified according to the Euclidean distance: (1) Null (>10,000 m), (2) Low (6000–10,000 m), (3) Medium (4000–6000 m), (4) High (2000–4000 m), and (5) Very high (0–2000 m).

Each layer was assigned a weight (Table 17.2) according to its susceptibility to interaction with the fishery (I). Each reclassified layer (L) was multiplied by the value of its respective weight (W), then summed and divided by the number (n) of layers according to the following equation:

$$I = \frac{1}{n} * \sum_{k=1}^{k=n} \left(L * W \right)$$

Table 17.1 Fisheries with potential interaction with oil and marine infrastructure in the southern Gulf of Mexico

Fishery	Fishing target species	Fishing gear	Fishing operation zone
Snook	*Centropomus* spp.	Gillnet	Fishing on the bottom and mid-water at distances from the coast of 13–30 km and depths between 4 and 27 m
Gafftopsail sea catfish	*Bagre marinus*	Bottom longline	Fishing on the bottom at distances from the coast of 5–155 km and depths between 4 and 86 m
Hardhead sea catfish	*Ariopsis felis*	Bottom longline	Fishing on the bottom at distances from the coast of 2–14 km and depths between 3 and 18 m
Spanish mackerel	*Scomberomorus maculatus*	Gillnet	Fishing on the bottom and mid-water at distances from the coast of 3–81 km and depths between 3 and 36 m
King mackerel	*Scomberomorus cavalla*	Trolling line	Operates towed near the surface, or a certain distance below at distances from 28 to 57 km from the coast
Red snapper	*Lutjanus campechanus*	Bottom longline	Fishing on the bottom at distances from the coast of 27–240 km and depths between 9 and 135 m
		Vertical longline	Fishing on the bottom at distances from the coast of 28–185 km and depths between 51 and 114 m
Vermilion snapper	*Rhomboplites aurorubens*	Vertical longline	Fishing on the bottom at distances from the coast of 42–192 km and depths between 55 and 118 m
Grouper	*Hyporthodus flavolimbatus* and *H. niveatus*	Bottom longline	Fishing on the bottom at distances from the coast of 52–213 km and depths between 54 and 245 m
Grouper and snapper nei	*Epinephelus* spp., *Lutjanus* spp., *Mycteroperca* spp., *Dermatolepis* spp.	Bottom longline	Fishing on the bottom at distances from the coast of 52–213 km and depths between 54 and 198 m
Southern stingray	*Hypanus americanus*	Bottom longline	Fishing on the bottom at distances from the coast of 4–129 km and depths between 3 and 68 m
Sharks	*Carcharhinus* spp., *Sphyrna lewini, S. mokarran, Galeocerdo cuvier*	Bottom longline and gillnet	Fishing on the bottom at distances from the coast of 6–204 km and depths between 7 and 126 m

(continued)

Table 17.1 (continued)

Fishery	Fishing target species	Fishing gear	Fishing operation zone
Little tunny	*Euthynnus alleteratus*	Gillnet	Fishing operation on the surface and mid-water to distances from the coast of 4–53 km and depths between 4 and 21 m
Jacks	*Caranx hippos*	Gillnet	Fishing operation on the surface water to distances from the coast of 3–46 km and depths between 7 and 14 m
Blue runner	*Caranx crysos*	Gillnet	Fishing operation on the surface water to distances from the coast of 4–21 km and depths between 4 and 10 m
Small sharks	*Rhizoprionodon terraenovae, Sphyrna tiburo*	Gillnet and bottom longline	Fishing operation on the bottom to distances from the coast of 3–148 km and depths between 5 and 99 m
Tripletail	*Lobotes surinamensis*	Gillnet	Fishing operation on the bottom and mid-water to distances from the coast of 4–12 km and depths between 3 and 10 m
Pompano	*Trachinotus carolinus*	Gillnet	Fishing operation on the surface and mid-water to distances from the coast of 2–25 km and depths between 3 and 25 m
Herring	*Cetengraulis edentulus, Harengula jaguana, Opistonema oglinum*	Gillnet	Fishing operation on the surface water to distances from the coast of 3–13 km and depths between 3 and 10 m
Tilefish	*Lopholatilus chamaeleonticeps*	Bottom longline	Fishing operation on the bottom to distances from the coast of 28–83 km and depths between 280 and 320 m
Weakfish	*Cynoscion nebulosus* and *C. arenarius*	Gillnet	Fishing operation on the bottom and mid-water to distances from the coast of 5 to 17 km and depths between 5 and 14 m
Atlantic seabob shrimp	*Xiphopenaeus kroyeri*	Trawlnet	Fishing operation on the bottom to distances from the coast of 4–14 km and depths between 3 and 8 m

Table 17.2 Hierarchical weighting of values based on layer analysis

Layer (*L*)	Weighting (*W*)
Oil ducts	0.25
Installations	0.3
Oil production zone	0.3
Anchorage vessels	0.1
Natural protected areas	0.05

The resulting interaction map was classified into five categories using the Natural Breaks method, which divides the data into classes in a way that maximizes differences between classes while minimizing differences within each class, thereby highlighting spatial patterns (Fariza et al., 2021). The interaction categories were defined as follows: very high, high, medium, low, and null. Data analysis was performed with the terra package (Hijmans et al., 2023) in the R programming language, version 4.3.2 (R Core Team, 2023).

17.3 Results

In general, the information from 1184 georeferenced fishing sets, covering 21 fisheries and 43 target species, showed that the potential interactions (PI) between small-scale fisheries and the oil marine infrastructure from sGoM indicated that 49.2% were mainly null PI, 30.1% were low PI, 15% were medium PI, 4.6% were high PI, and only 1.1% were very high PI.

The analysis by fishery indicated that two fisheries were found with null potential interactions (PI = 100%); these were: small sharks, herring, and weakfish fisheries. Other fisheries also had high percentages of null potential interactions, such as: small sharks (PI = 91.4%), grouper and snapper (PI = 89.7%), pompano (PI = 87.5%), blue runner (PI = 83.3%), southern stingray (PI = 75.4%), Spanish mackerel (PI = 69.8%), king mackerel (PI = 60%), sharks (PI = 53.9%), red snapper (PI = 53.3%), and Atlantic seabob shrimp (PI = 50%) fisheries. The main low potential interactions were found in the tilefish (IP = 100%), grouper (PI = 100%), tripletail (IP = 76%), jacks (IP = 66.7%), little tunny (IP = 42.9%), and gafftopsail sea catfish (IP = 38.1%) fisheries.

For the medium potential interaction category, the highest percentages were found in hardhead sea catfish (IP = 100%), vermilion snapper (IP = 35.3%), little

Fig. 17.2 Potential interactions between oil marine infrastructure and each small-scale fishery that operates in the southern Gulf of Mexico

tunny (IP = 28.6%), sea catfish (IP = 22.3%), and common snook (IP = 19.1%) fisheries. On the other hand, the high potential interactions were presented in the gafftopsail sea catfish (IP = 9.3%), red snapper (IP = 6.5%), and snook (IP = 4.6%) fisheries. Finally, very high potential interactions were observed in the red snapper (PI = 2.2%), southern stingray (PI = 1.5%), gafftopsail sea catfish (PI = 0.9%), and snook (PI = 0.4%) fisheries (Fig. 17.2).

The analysis of potential interactions between types of fishing gear and oil marine installations indicated that the fishing gears had very low potential interactions; only the bottom longline and gillnet had high interactions (PI = 6.7 and 3.6%, respectively) (Fig. 17.3).

The results for the three fisheries that most interact with the oil marine infrastructure in the sGoM indicate that the snook fishery had the most interactions with production area (PI = 28.7%) and anchorages (PI = 15.2%), for the red snapper fishery the main interactions were between the production area (PI = 18.5%) and pipelines (PI = 10.9%), and for the sea catfish fishery the interactions were with the pipelines (PI = 33.5%) (Fig. 17.4).

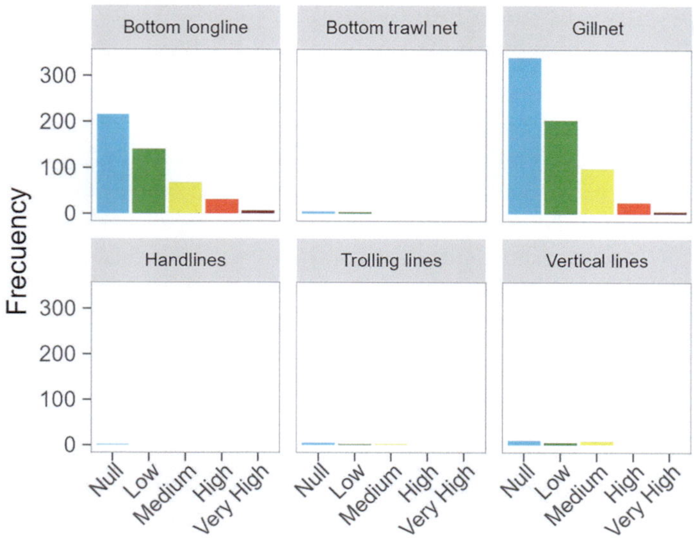

Fig. 17.3 Potential interactions between fishing gear type and oil marine infrastructure in the southern Gulf of Mexico

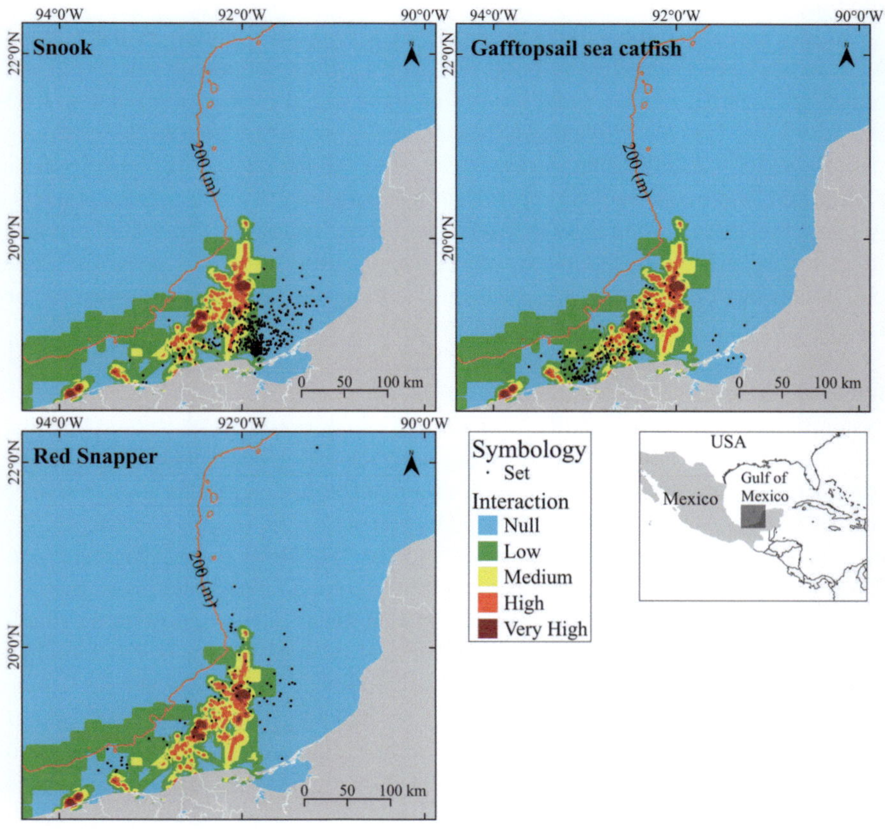

Fig. 17.4 Potential interactions between the three most important small-scale fisheries and the oil marine infrastructure in the southwestern Gulf of Mexico

17.4 Discussions and Conclusions

Coastal-marine zoning in the southern Gulf of Mexico (sGoM) is essential for managing interactions between economic activities and the conservation of coastal-marine ecosystems. According to Barragán and de Andrés (2015), an integrated zoning approach helps to reduce conflicts over use and promotes sustainable development, enabling proper management zones to protect biodiversity in areas of multiple uses. In regions of high oil activity, such as the sGoM, establishing exclusion zones around oil platforms can help protect marine species and fishing communities (Michel et al., 2005, 2013). Establishing a clear coastal-marine zoning framework is a priority measure in the region, as it allows territorial planning to minimize environmental impact, especially on sensitive ecosystems (Norse & Crowder, 2005; Lester et al., 2009).

The coexistence of the oil industry and small-scale fishing in the sGoM necessitates management policies that strike a balance between economic development and conservation. The oil industry significantly contributes to the economy but also impacts marine biodiversity and local fisheries (Fanning et al., 2011). Exclusion zones, though necessary for safety reasons, must be integrated into management frameworks that acknowledge the needs of fishing communities, avoiding the loss of productive areas and social exclusion.

In this context, Zalik (2009) examines how state intervention in the Gulf of Mexico has been channelled through the merging of community development and industrial security polices, particularly following the implementation of the 2003 Security Law. This law expanded exclusion zones around oil platforms, restricting fishing vessels' access to areas previously used for fishing. In response to these restrictions, some state agencies promoted financing schemes to encourage aquaculture as an economic alternative to traditional fishing zones. This interplay between industrial security and community development underscores the need for more comprehensive policies that acknowledge the socio-environmental impacts of the oil industry on coastal socio-ecological systems.

Along the same lines, Andrews et al. (2021) argue that oil development has profound and long-lasting environmental, social, and political implications. Through a systematic review, they identify four critical dimensions: (1) the environmental impacts of oil; (2) effects on small-scale fisheries and coastal livelihoods; (3) the spatial reconfiguration of coastal and marine areas; and (4) shifts in governance regime. In this context, where communities depend on the sea for subsistence, the economic benefits of hydrocarbons rarely offset the social and ecological costs. Thus, more equitable and context-sensitive intersectoral management is required.

Although the results do not reveal an intense direct interaction between small-scale fishing and oil infrastructure in operational or spatial terms, various authors have documented historical tensions and perceptions of exclusion among fishing communities, particularly in the context of recent energy reforms. Ramos-Muñoz et al. (2019) describe how, since the entry of the national oil industry into the shallow waters of the Gulf of Mexico, processes of spatial injustice emerged, in which fishers were not considered equal stakeholders in maritime spatial planning. While these tensions do not necessarily translate into everyday conflicts or direct confrontations, they do highlight a persistent need to enhance intersectoral coordination. In this regard, the existence of a new policy framework that recognizes the importance of social inclusion represents an opportunity to move toward more equitable and collaborative governance schemes between sectors.

Given this scenario, adopting co-management practices becomes crucial to prevent resource overexploitation and reduce intersectoral conflicts. Although small-scale fishing operates on a smaller scale, it is vital for coastal communities and contributes to food security, making its sustainability a regional management priority (Pauly et al., 2002; Charles, 2001). To balance these sectors, developing an integrated vision of conservation and zoning is essential, preventing oil exploitation from compromising biodiversity and the livelihoods of communities (Smith et al., 2011).

Research from Mexican institutions has been crucial in understanding and managing the interactions between small-scale fishing, the oil industry, and conservation in the Gulf of Mexico. Studies from The College of the Southern Border (ECOSUR) and the Gulf of Mexico Institute of Ecology, Fisheries, and Oceanography (EPOMEX) have highlighted the importance of coastal-marine planning to mitigate conflicts between small-scale fisheries and the energy sector, promoting conservation and community resilience along the coast of Campeche and Tabasco (Vega-Cendejas et al., 2023). Research conducted by the National Autonomous University of Mexico (UNAM) and the National Polytechnic Institute (IPN) has examined the environmental impact of oil activities on marine ecosystems, revealing that exclusion systems and environmental management plans can help minimize adverse effects on biodiversity and fisheries resources. Additionally, the Center for Research and Advanced Studies of the National Polytechnic Institute (CINESTAV) has studied the potential of decommissioned platforms as habitats for marine species, suggesting they could serve as fishery refuge zones. These studies underscore the need for inter-institutional collaboration to develop strategies that ensure a balanced use of marine and terrestrial resources in the region, integrating local community participation and promoting sustainable development.

Interactions between oil platforms and coastal fishing activities are limited due to exclusion zones surrounding the facilities, which restrict access for fishing vessels. Studies by Sumaila et al. (2020) confirm that access regulation minimizes physical interaction, reducing the risk of accidents and conflicts. Likewise, the restrictions help to protect both marine species and local fishers though these limitations can also negatively affect small-scale fishing by restricting traditional fishing areas (Fanning et al., 2011). Three exclusion zones, although necessary for safety, should be part of a management system that considers the needs of fishing communities, preventing the loss of productive fishing areas.

Oil platforms, considered national security installations, create tensions with coastal fishers, who face restrictions in their fishing areas., These restrictions affect the local economy, increasing conflicts between fishers and authorities. Additionally, studies on similar conflicts in other regions have shown that they can have lasting effects on local communities, especially when they are not included in management decisions. Establishing buffer zones and fostering dialogue with the relevant sectors are essential for mitigating these tensions.

The primary interactions of coastal fishing with the oil industry occur with support vessels rather than fixed platforms. These interactions include incidents such as fishing gear entanglements with propellers, generating risk and cost for both parties (Sumaila et al., 2020). These conflicts often intensify in areas where both activities share navigational spaces, especially when there are no clear regulations on shared water use. Management strategies that include buffer zones and shared navigation rules can effectively mitigate these conflicts and protect both fishing and oil operations (Fanning et al., 2011).

Decommissioned oil facilities are often an opportunity for marine conservation, acting as artificial reefs that foster biodiversity. Their structures can become alternative habitats that benefit various marine species. Reusing platforms to create fishery restoration zones is a strategy that has shown positive results in other regions (Claisse et al., 2014). Studies have also shown that artificial reefs help increase fish biomass and can generate revenue through ecotourism and recreational fishing activities (Lester et al., 2009; Smith et al., 2011). Implementing this strategy in the sGoM could strengthen both marine conservation and the local economy by establishing protected areas in strategic locations (Norse & Crowder, 2005).

Therefore, the challenges for coastal-marine planning in the sGoM must be understood and addressed from a multidisciplinary perspective, where the role and participation of fishers and primary users are fundamental for territorial solutions. Coronado et al. (2020) and Peña-Puch et al. (2021) have noted that the implementation of management strategies, it is important to consider the local context of fishers because the small-scale fisheries operate under the same fishery and maritime regulations, the fisheries in each localities are different in contexts and face different challenges to solve. For this reason, the information presented in this study is essential for understanding the complexity of the coupled social-ecological system involving small-scale fishing, the oil industry, and marine conservation.

Acknowledgments This study was part of the Shark and Ray Fisheries Research Project funded by the Mexican Institute for Sustainable Fisheries and Aquaculture Research (IMIPAS, by its Spanish acronym). Thanks to the small-scale fishers from the southern Gulf of Mexico for allowing us to obtain information about their fishing activities and to the anonymous reviewers.

References

Andrews, N., Bennett, N. J., Le-Billon, P., Green, S. J., Cisneros-Montemayor, A. M., Amongin, S., Gray, N. J., & Sumalia, U. R. (2021). Oil, fisheries and coastal communities: A review of impacts on the environment, livehoods, space and governance. *Energy Research & Social Science, 75*. https://doi.org/10.1016/j.erss.2021.102009

Arbo, P., & Thanh Thủy, P. T. (2016). Use conflicts in marine ecosystem-based management—The case of oil versus fisheries. *Ocean & Coastal Management, 122*, 77–86. https://doi.org/10.1016/j.ocecoaman.2016.01.008

Barragán, J. M., & de Andrés, M. (2015). Analysis and trends of the world's coastal cities and agglomerations. *Ocean & Coastal Management, 114*, 11–20. https://doi.org/10.1016/j.ocecoaman.2015.06.004

Cerón, F. (2024). *Mexico-Country commercial guide*. International Trade Administration of the U.S., Houston, Texas. https://www.trade.gov/country-commercial-guides/mexico-oil-and-gas. Accessed 2024.

Charles, A. T. (2001). *Sustainable fishery systems*. Blackwell Science.

Claisse, J. T., Pondella, D. J., Love, M., Zahn, L. A., Williams, C. M., Williams, J. P., & Bull, A. S. (2014). Oil platforms off California are among the most productive marine fish habitats globally. *Proceedings of the National Academy of Sciences of the United States of America, 111*(43), 15462–15467.

CNH. (2024). *Comisión Nacional de Hidrocarburos*. https://www.gob.mx/cnh. Accessed 2024.

CONAPESCA. (2021). *Programa Nacional de Pesca y Acuacultura 2020–2024*. Secretaría de Agricultura y Desarrollo Rural.

Coronado, E., Salas, S., Torres-Irineo, E., & Chuenpagdee, R. (2020). Disentangling the complexity of small-scale fisheries in coastal communities through a typology approach: The case study of the Yucatan Peninsula, Mexico. *Regional Studies of Marine Science, 36*, 101312. https://doi.org/10.1016/j.rsma.2020.101312

Coronado, E., Cheunpagdee, R., Salas, S., Torres-Irineo, E., Palomo-Cortés, L., & Espinoza-Tenorio, A. (2022). Exploración socioecológica de la pesca artesanal del sur del Golfo de México por medio del marco de análisis de gobernanza interactiva. *Ciencia Pesquera, 30*(1–2), 79–95.

Coronado, E., Zepeda-Domínguez, J. A., Espinoza-Tenorio, A., Cupido-Santamaria, D., Ramos-Muñoz, D., & Monzón-Alvarado, C. (2024). *Institutional mapping and implementation of the sustainable development goals across co-existing industries: The case of the fisheries-oil system in Mexico*. The Extractive Industries and Society. https://doi.org/10.1016/j.exis.2023.101390

DOF. (2003). *ACUERDO mediante el cual las secretarías de Marina, de Comunicaciones y Transportes, y de Agricultura, Ganadería, Desarrollo Rural, Pesca y Alimentación, establecen medidas de seguridad en la Sonda de Campeche*.

Fanning, L., Mahon, R., & McConney, P. (2011). *Towards marine ecosystem-based management in the Wider Caribbean*. Amsterdam University Press.

Fariza, A., Basofi, A., & Aryani, M. D. (2021). Spatial mapping of diphtheria vulnerability level in East Java, Indonesia, using analytical hierarchy process–natural break classification. *Journal of Physics: Conference Series, 1803*, 1. https://doi.org/10.1088/1742-6596/1803/1/012009

Hijmans, J. R., Bivand, R., Cordano, E., Dyba, K., Pebesma, E., Sumner, M. D. (2023). *terra: Spatial data analysis. R package version 1.7-55*. https://doi.org/10.32614/CRAN.package.terra.

Laaribi, A., Chevallier, J. J., & Martel, J. M. (1996). A spatial decision aid: A multicriterion evaluation approach. *Computers, Environment and Urban Systems, 20*(6), 351–366. https://doi.org/10.1016/S0198-9715(97)00002-1

Lara-Mendoza, R. E. (2020). *Puntos de referencia para la pesquería de tiburones y rayas del sureste del Golfo de México. Informe técnico (Documento interno)*. Instituto Nacional de Pesca y Acuacultura. Centro Regional de Investigación Acuícola y Pesquera de Ciudad del Carmen, Campeche.

Lester, S. E., Halpern, B. S., Grorud-Colvert, K., Lubchenco, J., Ruttenberg, B. I., Gaines, S. D., Airamé, S., & Warner, R. R. (2009). Biological effects within no-take marine reserves: A global synthesis. *Marine Ecology Progress Series, 384*, 33–46. https://doi.org/10.3354/meps08029

Liina-Maija, Q. (2019). Fishers' knowledge and scientific indeterminacy: Contested oil impacts in Mexico's sacrifice zone. *Maritime Studies, 18*, 65–76. https://doi.org/10.1007/s40152-018-0123-7

Mackenzie, W. (2017). *The rise and fall of black gold*. https://www.woodmac.com/news/feature/the-rise-and-fall-of-black-gold/. Accessed 2024.

Malczewski, J. (2006). GIS-based multicriteria decision analysis: A survey of the literature. *International Journal of Geographical Information Science, 20*(7), 703–726. https://doi.org/10.1080/13658810600661508

Michel, J., Gilbert, T., Etkin, D. S., Urban, R., Waldron, J., & Blocksidge, C. T. (2005). Potentially polluting wrecks in marine waters. *An Issue Paper Prepared for the 2005 International Oil Spill Conference*. Pp 1–40.

Michel, J., Owens, E. H., Zengel, S., Graham, A., Nixon, Z., Allard, T., Holton, W., Reimer, P. D., Lamarche, A., White, M., Rutherford, N., Childs, C., Mauseth, G., Challenger, G., & Taylor, E. (2013). Extent and degree of shoreline oiling: Deepwater horizon oil spill, Gulf of Mexico, USA. *PLoS One, 8*(6), e65087. https://doi.org/10.1371/journal.pone.0065087

Norse, E. A., & Crowder, L. B. (2005). *Marine conservation biology: The science of maintaining the sea's biodiversity* (p. 470). Island Press.

Pauly, D., Christensen, V., Guénette, S., Pitcher, T. K., Rashid Sumaila, U., Walters, C. J., Watson, R., & Zeller, D. (2002). Towards sustainability in world fisheries. *Nature, 418*, 689–695.

Peña-Puch, A. C., Pérez-Jiménez, J. C., Espinoza-Tenorio, A., Méndez-Loeza, I., & Williams-Jara, G. M. (2021). The sustainability of small-scale fisheries in oil-producing sections of the Gulf of Mexico. *The Extractive Industries and Society, 8*(4), 100975. https://doi.org/10.1016/j.exis.2021.100975

R Core Team. (2023). *R: A language and environment for statistical computing*. R Foundation for Statistical Computing. https://www.R-project.org/. Accessed 2024

Ramos-Muñoz, D. E., Ramos-Reyes, R., Zamora-Cornelio, L. F., Hernández-De la Cruz, A., & Espinoza-Tenorio, A. (2019). Exclusión en el Golfo de México: una visión desde los pescadores sobre la industria petrolera en Tabasco. *Cuadernos de Geografía: Revista Colombiana de Geografía, 28*(2), 357–372. https://doi.org/10.15446/rcdg.v28n2.73511

Smith, A. D. M., Brown, C. J., Bulman, C. M., Fulton, E. A., Johnson, P., Kaplan, I. C., Lozano-Montes, H., Mackinson, S., Marzloff, M., Shannon, L. J., Shin, Y. J., & Tam, J. (2011). Impacts of fishing low-trophic level species on marine ecosystems. *Science, 333*, 1147–1150.

Sumaila, U. R., Cheung, W. W. L., Lam, V. W. Y., Pauly, D., & Herrick, S. (2020). Climate change impacts on the biophysics and economics of world fisheries. *Nature Climate Change, 11*, 204–209. https://doi.org/10.1038/NCLIMATE1301

Uribe-Martínez, A., Aguirre-Gómez, R., Zavala-Hidalgo, J., Ressl, R., & Cuevas, E. (2019). Unidades oceanográficas del Golfo de México y áreas adyacentes: La integración mensual de características biofísicas superficiales. *Geofisica Internacional, 58*(4), 295–315. https://doi.org/10.22201/igeof.00167169p.2019.58.4.2059

Vega-Cendejas, M. E., Hernández de Santillana, M., Chi Espinola, A., & Aguilar Medrano, A. (2023). Fish community structure along a depth gradient at the Western Gulf of Mexico. *Regional Studies in Marine Science, 68*. https://doi.org/10.1016/j.rsma.2023.103237

Zalik, A. (2009). Zones of exclusion: Offshore extraction, the contestation of space and physical displacement in the Nigerian Delta and the Mexican Gulf. *Antipode, 41*(3), 557–582. https://doi.org/10.1111/j.1467-8330.2009.00687.x

Chapter 18
Diagnosis of the Energy Sector in the Campeche State

J. S. Narro-Ríos

18.1 National Energy Context

México is the seventeenth country in the world generating energy, with 1.05%. In 2022, the country produced primary energy in the amount of 7468.99 PJ. For 2023, production decreased by 1.39%, reaching 7365.30 PJ. Hydrocarbons accounted for 82.09% of the national production, primarily oil, with a share of 52.65%. Renewable energy production decreased from 2022 to 2023, falling from 1150.13 to 1014.39 PJ, representing 13.77% of the national production, except for eolic and solar energy, which all experienced a reduction in that year. Nuclear energy presented an increase, passing from 152.77 to 175.70 PJ, mainly due to an increase in production at the Laguna Verde nuclear plant, which represents 2.39% of the national production. In the case of coil, it reduces from 137.59 to 129.37 PJ, which represents 1.76% of the national production (SENER, 2025).

18.2 Campeche Summary

Campeche is one of the 31 states of the United Mexican States, located in eastern Mexico on the Yucatan Peninsula. It is bordered by Tabasco and Guatemala to the south and southwest, Yucatan to the north, Quintana Roo and Belize to the east and southeast, and the Gulf of Mexico to the west (Fig. 18.1). The state was widely known for its fishing industry, which became the country's main shrimp producer. It

J. S. Narro-Ríos (✉)
Secretaría de Medio Ambiente, Biodiversidad, Cambio Climatico y Energía (SEMABICCE), Campeche, México
e-mail: jorge.narro@campeche.gob.mx

A. Espinoza-Tenorio et al. (eds.), *The Fishing and Oil Industries in Coexistence*,
https://doi.org/10.1007/978-3-032-02621-7_18

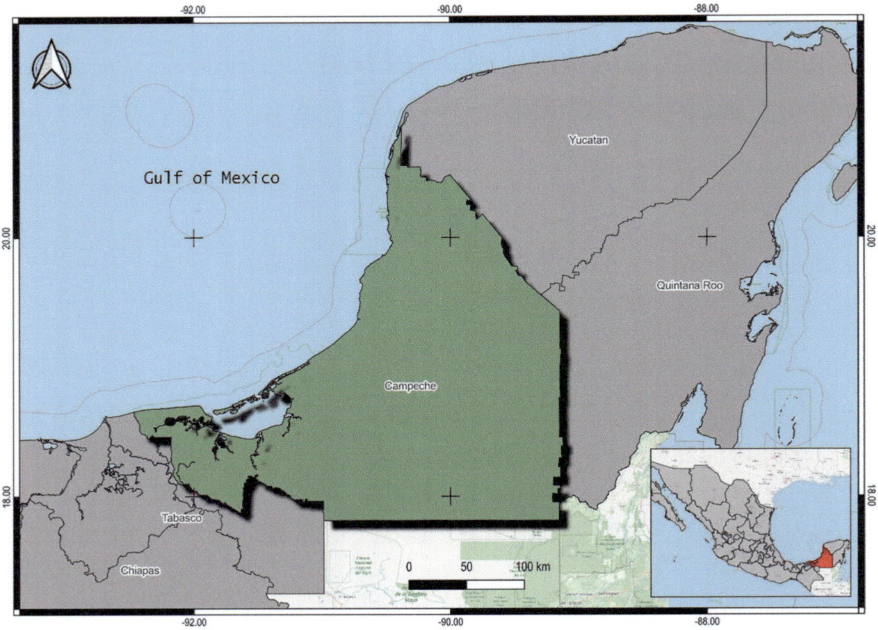

Fig. 18.1 Campeche's map

later became Mexico's leading oil producer and remains so today, although production from the Cantarell well, located in the waters of Campeche, has declined.

Campeche has a high potential for solar power generation, and it is an active generator, in addition to sustained wind speeds that meet the requirements for wind energy use. In the case of wind energy, many of these sites are near the sea, and others are directly over the sea, so there is great potential to take advantage of them. This potential opens the door to developments where green hydrogen or its derivatives can be produced, creating a value chain of global importance.

18.3 Renewable Energy Potential

Campeche has great potential for generating renewable energy, particularly in the case of solar photovoltaic and wind energy. In these cases, the National Renewable Energy Laboratory (NREL) has conducted a regional study for the Yucatan Peninsula to assess the potential for photovoltaic and wind energy generation. This study reveals a greater potential for solar energy generation, but also highlights a high potential for wind energy (Table 18.1).

In the case of solar energy, the state currently has one photovoltaic farm with an installed capacity of 300 MWp. However, several projects are under development in different municipalities, with some in the pre-feasibility, approval, or

Table 18.1 Yucatan Peninsula wind and photovoltaic energy potential

Resource	Maximum quantity	Comments
Photovoltaic potential	720 GW	The value is 10 times all the actual generation in México (70 GW) (Bracho et al., 2021)
Eolic potential	430 MW	The potential is lesser than photovoltaic but still big (Bracho et al., 2021)

redevelopment stage of project development. The sum of the peak installed capacity of these projects would already exceed one GWp. It is worth mentioning that the procedures for obtaining permits for installation, generation, and interconnection to the electricity transmission grid are federal and correspond to the Energy Regulatory Commission (CRE, by its Spanish acronym), the Ministry of Energy (SENER, by its Spanish acronym) and the Ministry of Environment and Natural Resources (SEMARNAT, by its Spanish acronym), among others. On the other hand, there is another scenario for small generators, known as distributed generation, which is connected directly to the distribution grid and does not require all these permits. Distributed generation has a maximum installed capacity limit of 0.499 MWp. By the second half of 2023, Campeche had installed 22.80 MWp (https://www.gob.mx/cre/documentos/pequena-y-mediana-escala) under this scheme, and this number is expected to increase soon due to several installation projects in public buildings and drinking water pumps throughout the state, led by the Government of Campeche. The technology of generating electricity through the photovoltaic effect is the most mature and advanced of the others, with the advantage of being suitable for small-scale use, making it very useful for all types of users.

In the case of wind energy, there is currently a project in Campeche with just over 100 MWp on standby; however, studies indicate that it has the capacity to generate this power. In addition, there is another project in the state that aims to create more than 300 MWp to be used for the production of green hydrogen, so the NREL study on the potential for wind power generation is probably conservative in its estimate, although it should be noted that there is also great potential for offshore generation, were the wind velocity is due the lack of obstacles in the trajectory. It is also advantageous that the seabed is relatively low due to the continental shelf of the Gulf of Campeche, so it would be more feasible to build this type of plant close to the coast. However, it should be noted that the construction costs are much higher than onshore.

The coastline of the state of Campeche is 523.3 km long, so it is feasible and even logical to try to generate electricity from the sea in some way, and for this reason, some universities and research centres have taken on the task of studying and reviewing this issue. The *Centro Mexicano de Innovación en Energía Océano* (CEMIE-Océano) has carried out studies on the generation of electricity by waves, currents, tides, and thermal gradients, in which they found that in the case of thermal gradients, there is no great potential due to the marine platform on the Campeche side, which avoids having a high temperature gradient. In the case of ocean currents, it has been found that the Gulf Stream can provide sufficient and profitable energy, but studies of the Campeche area are still needed (Santana et al. 2021). Tidal energy in the territorial waters of Campeche is low, so trying to exploit it would not be the best option. In general, it can be said that there is a potential for energy from currents, but further studies would have to be carried out to have a clear idea of the potential in the state.

Campeche state has 16 rivers, 10 lagoons, and one estuary (INEGI, n.d.). One way of generating electricity is through the movement of water using turbines specially designed for rivers, which can produce several kilowatts of power depending on the turbine's size. Therefore, studying the flow of rivers is essential to understanding the potential electricity generation that can be achieved and utilised locally. Another way to generate electricity is through the use of saline potential. The saline gradient takes advantage of the difference in salt concentration across a membrane, allowing for the generation of charge carriers that produce an electric current. This technology is still in the development phase. Some studies have been conducted in the Champotón River to assess the energy potential in the saline gradient near the river's mouth (Martínez, 2019). For the case of Champotón River, it was found that this type of energy is feasible, but the technology is not yet commercially available. There is also a study of the Champotón River basin, which reports an average annual volume of 591.90 million cubic metres registered at the height of *Canasayab*, where CONAGUA has a hydrometric station. The Champotón River also has a main channel length of 58 km and an average gradient of 3.48, so there may be potential to use portable river turbines that could contribute to power generation in the region during rainy periods.

Another energy vector that can be used is hydrogen obtained by renewable methods; therefore, it is known as green hydrogen and its derivatives. This vector can be obtained in different ways; as such, it is usually generated by electricity from renewable sources and produced through the hydrolysis process. This process involves splitting the water molecule into its constituent atoms by applying an electric current, resulting in two hydrogen atoms and one oxygen atom for each water molecule. Once these constituents are obtained, the hydrogen can be stored for later use, making it a viable method for storing renewable electrical energy. However, like any other thermodynamic cycle, there are losses when using a fuel cell and generating electricity by bringing the O and H molecules together to produce the water molecule again, as well as an electrical potential. An essential part of this process is the production of renewable energy, and the need for sufficient water to support it. At present, this vector has gained a lot of strength worldwide, which has made it an important factor in the generation of electrical energy, one of its most interesting applications is in public transport and freight because it does not need such large batteries that can weigh several tonnes, which is a better option than electric vehicles powered by batteries. This type of vehicle requires a pressurised storage tank, a fuel cell, and an electric motor.

One of the most important derivatives of green hydrogen is green ammonia. Ammonia is a very important molecule because it can be used as an agricultural nitrogen fertiliser in the field, and research has already shown that it can be used as a fuel for ship engines. Green ammonia is typically produced through a process known as the Haber-Bosch process, which involves a chemical reaction between nitrogen extracted from the air and hydrogen. There is currently a project in the state to produce green ammonia for export to international markets, in which the state government and the German government have collaborated to enable the company to conduct pre-feasibility studies for this project in Campeche. One of the results obtained is the Guide for the Development of Green Ammonia Projects in Mexico (https://h2lac.org/archivos/guia-para-el-desarrollo-de-proyectos-de-amoniaco-verde-en-mexico/), in which the interested companies collaborated, as well as the German government through *Deutsche Gesellschaft für Internationale Zusammenarbeit* (GIZ) and the government of the state of Campeche through *Secretaría de Medio Ambiente, Biodiversidad, Cambio Climático y Energía* (SEMABICCE).

As the state's main economic activities are fishing, livestock farming, and agriculture, there is significant potential to utilise the waste from these sectors to generate electricity in a sustainable manner; this type of energy is known as bioenergy. Worldwide, the United States is the largest producer of bioenergy, accounting for 44% of the global total, followed by Brazil, Indonesia, and Germany, with 28%, 37%, and 30%, respectively. In Mexico, the leading producer of bioenergy is Veracruz, which has the largest installed capacity, at 264.06 megawatts (MW), followed by Jalisco, Tabasco, and San Luis Potosí, with 83.32, 41.7, and 40.7 MW, respectively. In Campeche, according to the latest update of the CRE (https://www.gob.mx/cre/acciones-y-programas/micrositio-de-permisos-en-materia-de-generacion-de-energia-electrica), dated December 2023, there are two projects that use sugarcane bagasse and biomass as energy sources in part of their production chains, through internal combustion steam turbine technologies; between the two projects there is an installed capacity of around 33.9 MW and an estimated annual generation of 78.52 GWH/year. There is still the possibility of generating more electricity from biomass or biogas, and SEMABICCE is currently updating the greenhouse gas inventory, where the potential for electricity generation using these methods will be discussed.

The state of Campeche is unique in that it has the largest protected natural area in Mexico and the second largest in Latin America, known as *Gran Calakmul*, with almost 50% of its territory under some form of environmental protection. For this reason, the need to develop the potential of a green Campeche with renewable energy generation is more than evident.

18.4 Oil and Gas Extraction

Campeche has been and remains the leading oil and gas producer in the United Mexican States. The total production is located in the area known as the Gulf of Campeche, which is a vast inland gulf forming part of the Gulf of Mexico, which is located along the western edge of the Yucatan Peninsula and from the north to south includes a small part of the state of Yucatan, the entire coastline of the states of

Campeche and Tabasco, as well as part of the coastline of the state of Veracruz. The coasts of the Gulf of Campeche are characterised by low, sandy shores, due to the continental shelf known as the Campeche Bank (Fig. 18.2). This gulf does not allow the existence of seaports or deep draughts although it has an excellent potential for shrimp and other marine species due to its coastal lagoons.

There are more than 100 oil fields on the continental shelf of the Gulf, including the top 10 producing fields at the national level (Fig. 18.3). The following are the 15 most important fields in Campeche that are active and have had at least one well active for at least 1 year between 2022 and 2024; it should be noted that there are no onshore fields or wells in the state. Within the territorial waters of Campeche is the CHUC field, which has nine associated wells and which in 2023 had a total production of 8.35 million barrels per day (MBD) and associated gas production of 8.75 million cubic feet per day (MMPCD); the EK BALAM field has 49 associated wells and in 2023 will have total production of 96.50 MBD and associated gas of 24.22 MMPCD; the TARATUNICH field has three associated wells and in 2023 will have total production of 2.33 MBD and associated gas of 10. 25 MMPCD; the AYATSIL field has 39 associated wells and in 2023 had a total production of 91.75 MBD of oil and 9.76 MMPCD of associated gas; the SIHIL field has 24 associated wells and in 2023 had a total production of 17.37 MBD of oil and 6. 71 MMPCD; the POL field has four associated wells and in 2023 will have a total oil production of 1.79 MBD and associated gas production of 5.22 MMPCD; the LUM field has three associated wells and in 2023 will have a total hydrocarbon production of 0.92 MBD and associated gas production of 0.12 MMPCD; the CHAC field has seven associated

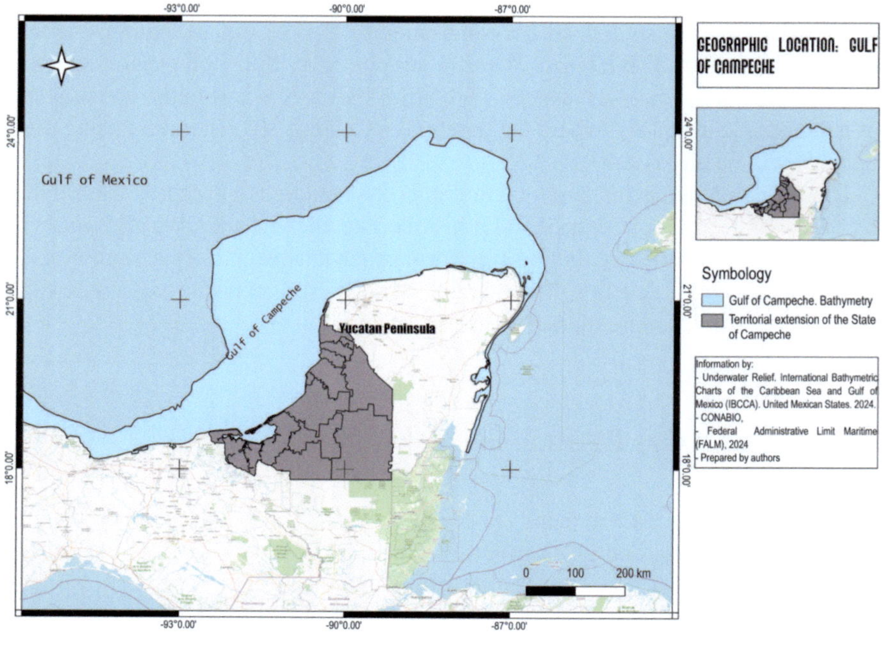

Fig. 18.2 Gulf of Campeche

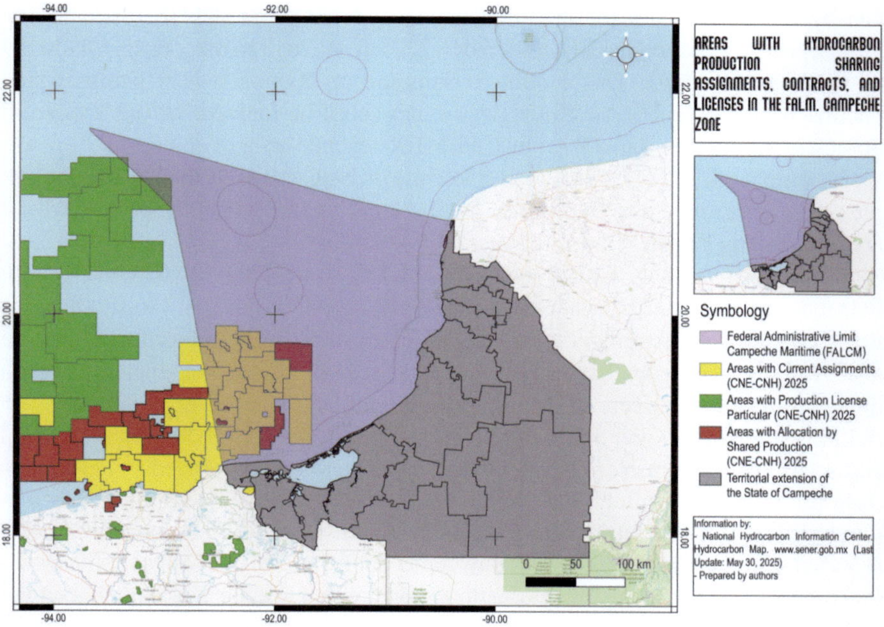

Fig. 18.3 Oil and gas platforms at Territorial Campeche's Sea

wells and in 2023 had a total hydrocarbon production of 4.54 MBD and associated gas of 1.94 MMPCD; the KU field consists of 45 associated wells and in 2023 had a total oil production of 20.18 MBD and associated gas of 201.57 MMPCD; the IXTOC field consists of nine associated wells and in 2023 had a total production of 4.13 MBD of oil and 13.7 MMPCD of associated gas; the CAAN field consists of six wells and in 2023 had a total production of 3.39 MBD of oil and 12.71 MMPCD of associated gas; the AKAL field consists of 76 wells and in 2023 had a total production of 21.55 MBD and associated gas of 821.31 MMPCD, making it the largest producer of associated gas in Campeche; the ABKATUN field consists of 13 wells and in 2023 had a total oil production of 9.45 MBD and associated gas of 13.63 MMPCD; the ZAAP field consists of 71 wells and in 2023 had a total oil production of 181.25 MBD and associated gas of 188.05 MMPCD; the MALOOB field is the last of the count and has 103 wells as constituents. In 2023, the field had a total production of 305.32 MBD of oil and associated gas of 356.52 MMPCD, making this field the largest oil producer in Campeche.

Some of the best-known complexes are the Cantarell, formed by the Nohoch, Chac, Akal, Kutz, Ixtoc, and Sihil fields, which was considered one of the 100 most important complexes in the world, contributing 45% of world production, but is currently in decline, producing only 15.4% of its peak in 2004 (Romo, 2015). The Ku-Maloob-Zaap complex is the largest in the country and, as its name suggests, is part of the Ku, Maloob, and Zaap fields. It was discovered by Pemex in 2002 and accounted for more than 40% of the state-owned company's oil production. Although production has been declining to date, by 2022, it had decreased to 31%, and by 2024, it is estimated that the complex will have lost 28.5% of its capacity. Pemex's oil production has also fallen by 12.1% from 2018 to date, from 1.81 MBD in 2018 to 1.59 MBD. This decline in domestic production is very similar to the decline in this complex, so Ku-Maloob-Zaap could be blamed for the decline in domestic production. On 14 August, PEMEX reached an agreement with oil services company CME Oil and Gas to reactivate drilling at two reservoirs in the complex in order to increase production. Both complexes are heavy oil producers and are operated by PEMEX Exploration and Production.

In Contract Area 4, Fieldwood Energy LLC is working in partnership with Petrobal S.A.P.I. de C.V., where the Ichalkil and Pokoch fields are located. These fields contain 2P (proved and probable) reserves of 68 million barrels of light oil and 92,000 million cubic feet of gas and have been in production since the second quarter of 2021. Between 2021 and 2041, the project is expected to produce 455 million barrels of oil and 567 billion cubic feet of gas. In April 2023, the National Hydrocarbons Commission (CNH, by its Spanish acronym) approved a plan to optimise production from fields that were in the second stage of declining production due to increased fractional water flow in the fields. In April 2024, production was reported at 13,800 barrels of oil per day, representing a 17% increase over the previous quarter. Recently, Lukoil International Upstream Holding B.V. announced the acquisition of Fieldwood. However, it will retain the commercial brand, Petrobal, similar to the case of Grupo Carso, which was recently acquired by another company. In Campeche, as far as is known, only these three companies are in the oil and gas production phase.

There is another group of about 190 wells in the exploration stage; some of these wells are Ek-Balam, Ket-1EXP, Alilkayol-1EXP, Moots-1EXP, Pit-100 1EXP, Yax-1EXP, Maat-1EXP, Hayabil-100 1EXP, Ochkan-1EXP, etc. The status of most of the total number of wells is abandoned, while others are in inactive status. A few are producers, and the least are temporarily abandoned. Almost all of the wells are located in shallow waters and within the territorial waters of Campeche. These wells typically utilise self-elevating, tensioned floating, and semi-submersible platforms that can be dismantled once the well's exploration has been completed. Exploration is a crucial phase of oil production, as it enables a certain level of production to be maintained when older wells are depleted and the explored wells need to be utilised for production.

Total liquid hydrocarbon production in the state in 2023 was 929.22 MBD, while for associated and non-associated gas, 2022.67 MMPCD were produced. Table 18.2 presents the percentages of oil and gas production in Campeche in relation to the rest of the country.

Table 18.2 Campeche state's contribution to national oil and gas production

Resource	Percentage	Comments
Oil	56.86% of the national production	Principal State producer of Mexico (https://sie.energia.gob.mx/inicio/)
Gas	38.43% of the national production	Within the first two producers in Mexico (https://sie.energia.gob.mx/inicio/)

Due to the possibility of accidents near oil and gas exploration and production platforms, there is a safety zone for navigation and overflight in the vicinity of oil facilities. According to the Federal Gazette of 11 October 2016, around wells, platforms, and other facilities for the exploration and production of hydrocarbons, a zone of 500 metres from the outer edge is to be maintained, in which only vessel or aircraft traffic necessary for the operation of such facilities is permitted. For offshore oil export facilities, such as monobuoys or floating production, storage, and offloading units, 5500 m must be maintained around them. In areas outside the safety zone where there are subsea installations such as headworks, cables or pipelines, anchoring is not permitted and only navigation is allowed; trawling, gillnets, or any other type of netting is prohibited and may be carried out at a distance of 1000 m from such installations.

As can be seen in the previous paragraph, there are restrictions on fishing areas. Considering that the areas surrounding the platforms have a diameter of 1 km and that there are hundreds of platforms, heads, cables, or pipelines, there is a significant reduction in the area or surface available for fishing. This can lead to the overfishing of certain regions of the sea, while others may remain free, which can lead to both favourable and unfavourable consequences. In general, it can be observed that fishing and oil and gas exploration and production activities must coexist in the Gulf of Campeche, which has led to disagreements among fishers. Many years ago, fishers had the entire territorial sea of Campeche for fishing, becoming the leading suppliers of shrimp in the country. Currently, there are large areas where fishing is prohibited; however, these areas can serve as sanctuaries for the reproduction of marine fauna.

18.5 Generation of Electric Energy at Campeche

The production and availability of energy are essential for the daily lives of people in Campeche, as well as for the overall development of a locality, community, or city. It is essential to continue working to expand coverage and alleviate energy poverty in the state. On the other hand, Campeche's geographical location means that temperatures in the summer can reach 45 degrees Celsius, not to mention the humidity, which creates a thermal sensation of more than 50 °C, leading to deaths from heat stroke. For this reason, Governor Layda Sansores and the federal government have managed to change the electricity tariffs throughout the state, with the support of the Autonomous University of Campeche, which carried out the study "The Effects of Humidity on Electricity Tariffs in the State of Campeche," which was used to support the change in electricity tariffs from 1C to 1F throughout the state. In addition, the Government of Campeche has a tariff support that provides half of the cost of electricity, before taxes and not exceeding 250 KWH per month, to Campeche households during the summer period from April to September, in order to help the population pay for the electricity needed to implement the use of cooling devices.

The generation of energy has always been a strategic area of vital importance for the development of any society. At the national level, energy consumption has been characterised by an almost total dependence on non-renewable energy sources, i.e. the extraction and processing of fossil fuels. However, in recent years, and due to the focus of new legislation on the use of renewable energy sources and ways of moving towards their sustainability, there has been an opening for the development of new types of energy generation projects, the reality is that, for the time being, the energy matrix at the national level will continue to depend on hydrocarbons. In this sense, Campeche should be considered not only the leading exporter of non-renewable resources but also a key player due to its potential for developing large-scale renewable energy projects.

There are different ways to generate electricity; the most common is the use of thermoelectric plants, which burn fossil fuels that heat water, which turns to a gaseous phase, and the steam is used to move turbines, which generate electricity. The state of Campeche is renowned for its oil and gas production, and recently, it has begun to explore the use of renewable energy sources, including solar and wind energy. The state's potential for producing electricity using these technologies has begun to be explored by various national and international companies, which have requested governmental information.

To gain an understanding of the electricity generation capacity currently available in the state of Campeche, we have referred to the statistics and figures provided by the CRE in its most recent update of December 2023, specifically the Installed Permitted Capacity. The CRE is the decentralised body of the federal government to which any activity that intends to produce energy (within a specific range) must report for its production authorisation process and registration before the aforementioned body; hence the term "authorised activity" is used to identify those that are included in the register and have the necessary authorisations for their operation.

Thus, according to the current federal regulatory framework (SEMARNAT-CRE), any energy project that intends to generate energy and has an installed capacity of 0.5 megawatts (MW) or more must be registered with the CRE. In this context, the state of Campeche has a licensed installed electricity generation capacity of approximately 1618.33 MW, according to the latest report from the CRE as of the end of December 2023. Figure 18.4 shows all the different methods to produce electric energy and their contribution to the net.

In this way, it can be seen that the most used technology in the entity is gas turbines/internal combustion with more than 50% of the total since these values are referred to by the oil activity in the platforms located in the region known as *Sonda de Campeche*, and to a lesser extent by the Federal Electricity Commission (CFE, by its Spanish acronym) for the operation of its facilities. As can be seen, the energy matrix also includes the production of energy by the solar plant called "La Pimienta," located in the municipality of El Carmen, which has a generating capacity of 300 MW on an area of around 700 hectares, contributing just over 20% of the total capacity in the state.

Although the data on installed licensed capacity is valid, it is still necessary to mention the activities with an installed capacity below the range required to be

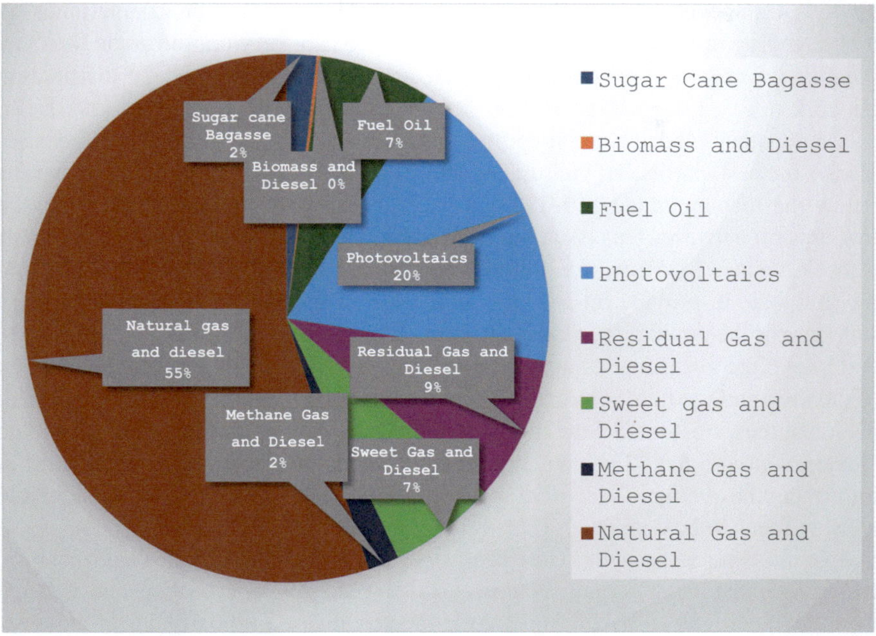

Fig. 18.4 Installed energy capacity in Campeche

included in the licensed activities, i.e. less than 0.5 MW. All power plants with a generating capacity of less than 0.5 MW using renewable energy sources are considered as distributed generation, which includes all activities intended to generate energy that do not exceed this capacity range; one of the characteristics of this type of generation is that it is exempt from the CRE authorisation process and therefore does not appear in its records; however, it is possible to obtain information on the status of installed capacity through the reports of interconnection requests to the CFE grid, which are made every 6 months. According to the latest report on interconnection requests to the CFE grid, issued for the first half of 2024, the installed capacity of distributed generation is 22.80 MW.

18.5.1 Electricity Distribution

The electricity distribution is a CFE responsibility; they own all the high, middle, and low-tension electricity networks in Mexico. At the end of the first quarter of 2024, the coverage reached 99.12%. Comparing this with the coverage at the beginning of 2022, which was 98.86%, the coverage increased by 0.26%, equivalent to providing electricity to 2356 new inhabitants. This increase in coverage has been a joint effort between the CFE, the state government of Campeche, and the municipalities. The aim is to continue to increase coverage to reach the national average of

99.46%. Campeche is the third most sparsely populated state in the country, making it challenging to achieve total electrification. There are very remote areas with few inhabitants, making it difficult to bring electricity to them through the distribution network.

On the other hand, the quality of the electricity provided in Campeche state is not always the best, particularly in rural or small city areas, where the voltage often falls below the recommended 110–127 V, which should be the average voltage in the low-tension network. This issue has been studied by various researchers (Nygren & Rabelo Avalos, 2025), who highlight some of the contradictions, including being the first state to provide oil and gas, while also noting the poor quality of energy services for the population. Constant efforts have been made by CFE, Campeche's, and municipal governments to improve electricity quality, but a significant portion of the low- and medium-tension grid is old and requires extensive maintenance, and the resources are insufficient to cover this topic. Recently, the Campeche State Government and the CFE have increased the funds allocated to projects aimed at improving the quality of electricity in various municipalities; however, considerable work remains to be done.

In this same sense, work has been done on non-conventional electrification, which involves using solar cells connected to batteries as a backup and isolated from the electricity distribution network, thereby making it possible to bring electricity to remote communities that are difficult to access. It is hoped that in the next 3 years the national average can be reached in a joint work between the three aforementioned actors, and in this way, the gap in inhabitants can be closed, which would be close to 4000 inhabitants who still suffer from energy poverty because they do not have access to it.

18.6 Conclusions

As could be seen in the different sections of this diagnosis, Campeche has a very close relationship with energy, being the leading producer of oil and gas in Mexico. However, this relationship has not always been in favour of the development of the state, and this can be seen in the loss of shrimp production in favour of oil, which led to the loss of many jobs for fishers that were not substituted by the oil activity, and therefore did not generate a more homogeneous economic distribution. Another important example is the case of the electricity distribution network, which is not even within the national average, which is why, through the different federal, state, and local actors, a more balanced relationship must be sought that guarantees access to clean, quality, and instantaneous energy for the population of Campeche.

On the other hand, there is a great potential in the state to generate renewable energy, led by photovoltaic and eolic, that allows this balance and even to be able to export this energy or to be used as a mean to obtain green hydrogen or green ammonia, without losing sight of the well-being of the local population, as well as the development of their activities that will allow them to escape from any kind of poverty and live in a state of well-being, but always with the well-being of nature in mind, which owns the second largest jungle reserve in Latin America, smaller only than Amazonia.

Acknowledgements The author wishes to thank E. R. Gómez-Couoh for his technical assistance generating the images.

References

Bracho, R., Flores-Espino, F., Morgenstein, J., Aznar, A., Castillo, R., & Settle, D. E. (2021). *The Yucatan peninsula energy assessment: Pathways for a clean and sustainable power system (no. NREL/TP-7A40–79680)*. National Renewable Energy Lab.(NREL).

Instituto Nacional de Estadística y Geografía (INEGI). (n.d.). *Conoce tu estado*. https://cuentame.inegi.org.mx/descubre/conoce_tu_estado/

Martínez, I. (2019). *Evaluación del potencial energético asociado a gradiente salino del río Champotón* [Tesis de maestría, Universidad Autónoma de Campeche]. Facultad de Ciencias Químico Biológicas, División de Estudios de Posgrado.

Nygren, A., & Rabelo Avalos, A. V. (2025). Oil extraction and indirect dispossession: Responsibility and resistance in southern sacrifice zones. *The Journal of Peasant Studies*, 1–32.

Romo, D. (2015). El campo petrolero Cantarell y la economía mexicana. *Problemas del desarrollo, 46*(183), 141–164. https://doi.org/10.1016/j.rpd.2015.10.007

Santana, J. S., Cabello Lugo, J., Mateos Farfán, E., & Pedroza González, E. (2021). *Dispositivos para la obtención de energía por corrientes: Situación actual* (CEMIE-Océano). Centro Mexicano de Innovación en Energías del Océano (CEMIE-Océano). https://doi.org/10.26359/EPOMEX.CEMIE112021

SENER. (2025). *Balance nacional de energía 2023*. Poder Ejecutivo de los Estados Unidos Mexicanos. https://www.gob.mx/cms/uploads/attachment/file/977268/Balance_Nacional_de_Energ_a_2023.FINAL06.02.2025.1.pdf

Chapter 19
Small-Scale Fishing and Oil Industry Interaction in the Southwestern Gulf of Mexico: A Proposed Assessment Method Based on SDG and Interdimensional Indicators

E. Núñez-Lara, A. Núñez-Laffón, and C. N. Martínez-Fernández

19.1 Interaction Between Small-Scale Fishing and Oil Industry Activities

The world's population did not exceed 1 billion inhabitants until approximately 1800 C.E. Even just 200 years ago, the extraction and marketing of natural resources did not have apparent irreversible environmental impacts. It did not affect the viability of human existence. As the Industrial Revolution began in earnest in the mid-1850s, conflicts between productive activities emerged. New technologies facilitated the large-scale exploitation of natural resources, consequently allowing the human population to grow exponentially, from one to seven billion inhabitants in just 150 years (United Nations, 2017).

At the turn of the twentieth century, the advent of petroleum as an energy source and the consequent establishment of an oil industry represented a watershed moment in the global economy. It brought significant benefits for societies, as well as

E. Núñez-Lara (✉) · C. N. Martínez-Fernández
Facultad de Ciencias Naturales, Universidad Autónoma del Carmen, Ciudad del Carmen, Campeche, México
e-mail: elara@pampano.unacar.mx; cmartinez@pampano.unacar.mx

A. Núñez-Laffón
Facultad de Ciencias Naturales, Universidad Autónoma del Carmen, Ciudad del Carmen, Campeche, México

Facultad de Ciencias de la Información, Universidad Autónoma del Carmen, Ciudad del Carmen, Campeche, México
e-mail: 191100@mail.unacar.mx

A. Espinoza-Tenorio et al. (eds.), *The Fishing and Oil Industries in Coexistence*,
https://doi.org/10.1007/978-3-032-02621-7_19

profound impacts on the environment and other productive activities occurring in the same space, such as small-scale fishing. Historically, competition for operational space and resources in fisheries had been between different fleets, including small-scale fleets, industrial fleets, and/or industrial and small-scale fleets. Territorial competition between different extractive activities, even between economic units of the same industry, is now a stronger driver of conflict than competition for target resources. Management tools based on sustainability principles can be used to mitigate these kinds of conflicts. Their application in some parts of the world, in conjunction with slowing human population growth and the ongoing transformation of the energy sector toward renewable sources, may provide a respite for humanity (Fig. 19.1).

Approximately 50 million fishers worldwide participate in the industry. Although use of modern fishing gear began in the 1880s, the greatest increase in fishing effort has occurred in the last 50 years (Anticamara et al., 2011). Primarily driven by unsustainable practices, this fishing effort has led to the overfishing or collapse of fish stocks (FAO, 2020; Graham et al., 2024). The catch from small-scale fishing fleets is almost entirely used for human consumption and thus contributes to food security at local to international scales. Small-scale fishing competes with other marine sector activities for resources, space, and government support, and, despite its importance, is often overlooked in marine planning strategies (Cohen et al., 2019). Indeed, small-scale fisheries are the largest employers in the marine sector (approximately 260 million people worldwide) and include a significant number of women (Harper et al., 2020). In some coastal regions, fisheries converge with offshore oil industry activities. The resulting competition for space is patently unbalanced, since, although small-scale fisheries directly benefit the largest number of people, the oil industry generates the highest overall revenues (34% of the total marine-based economy worldwide) (PRIO, 2019).

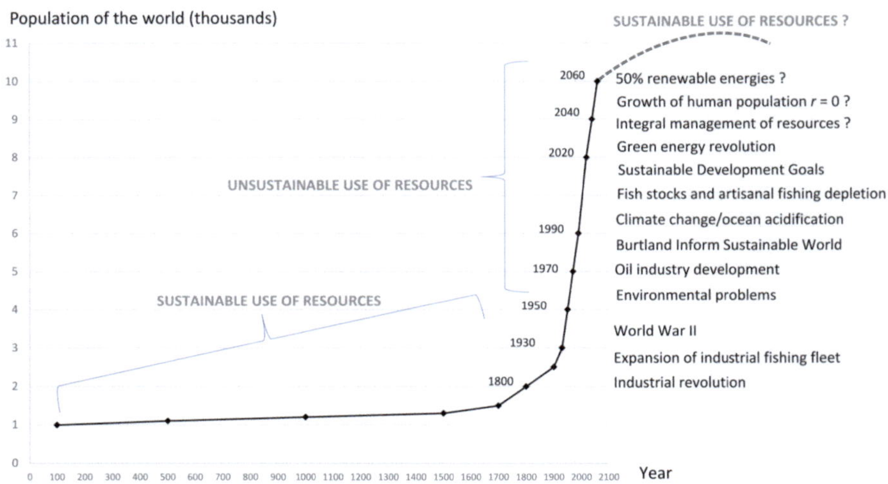

Fig. 19.1 World population, 100 to 2100 C.E. with selected relevant anthropogenic developmental milestones

The chains of activities in small-scale fisheries begin with catching fish (using different gears according to target species and regions), transport and landing of the catch and its products to the coast, and their marketing, mainly at the regional scale. Oil industry activities are multifold but include marine exploration, platform and submarine pipeline installation, and product transport and refinement, mainly in coastal facilities. Spatial overlap between these sectors begins in marine environments and extends to coastal zones. Exploration, rig, and platform construction often coincide with offshore fishing areas. Product transport to the coast occurs along common routes used by fishing boats, oil industry transports, and submarine pipelines. Finally, onshore oil terminals and refineries are near community residential areas. Across their entire sphere of activity, small-scale fisheries and the oil industry can have multiple interactions (Uhre & Leknes, 2017; Andrews et al., 2021).

The coast of the southwestern Gulf of Mexico has been fished for at least two millennia. Current fishing activities in the region include traditional extraction using artisanal gear, with the catch primarily intended for local consumption, and a middraft fleet focused on commercial target species, such as shrimp, which are distributed nationally and internationally (Martínez-Estrada et al., 2017; Ramos-Miranda et al., 2009). This portion of the Gulf, known as the *Sonda de Campeche*, covers approximately 8000 km^2. Along the coast (ca. 300 km), there are approximately 20 towns with populations of less than 20,000 people in which fishing is the main activity. In the same region, there are more than three hundred near-shore (< 200 km) oil platforms, some located in zones previously assigned to fishing (Fig. 19.2). In addition, there are about 2000 km of pipelines carrying oil to a pumping terminal located at *Cayo Arcas*, an area of minor sandy keys (< 2 km long and 1 km wide) surrounded by shallow coral reefs (1–25 m depth). These form part of the Banco de Campeche reef system, an essential ecological zone that interconnects the coral reefs of the Caribbean Sea with those of the Gulf of Mexico through larval flow (Núñez-Lara et al., 2017). More than 30 coral species and 100 fish species have been recorded there.

The oil industry has had myriad environmental impacts in the region, with pollution being the most evident. The most common type of pollution is the oil spill. Spills have a double environmental impact: the first one, a direct impact via organism mortality, and the second one, an indirect impact due to habitat damage. The effects of oil spills can vary depending on the ecosystem and species susceptibility, proximity to the spill source, spill size, and type of oil (Zang et al., 2019; Romo-Curiel et al., 2022; Uribe-Martínez et al., 2024). Spill impacts can be severe and long-lasting if oil installations are not in compliance with industry regulations. Under these circumstances, spills can persist for several years and have long-term impacts on biota, ranging from benthos and plankton to predators, including commercial fishery target species (Ruby & Ahilan, 2018; Zang et al., 2019). Some oil industry pollutants, such as heavy metals, can persist in the environment for decades (Uhre & Leknes, 2017). In the southwestern Gulf of Mexico, their proximity to oil industry infrastructure makes highly productive coastal ecosystems such as wetlands, mangroves, marshes, and coral reefs particularly vulnerable.

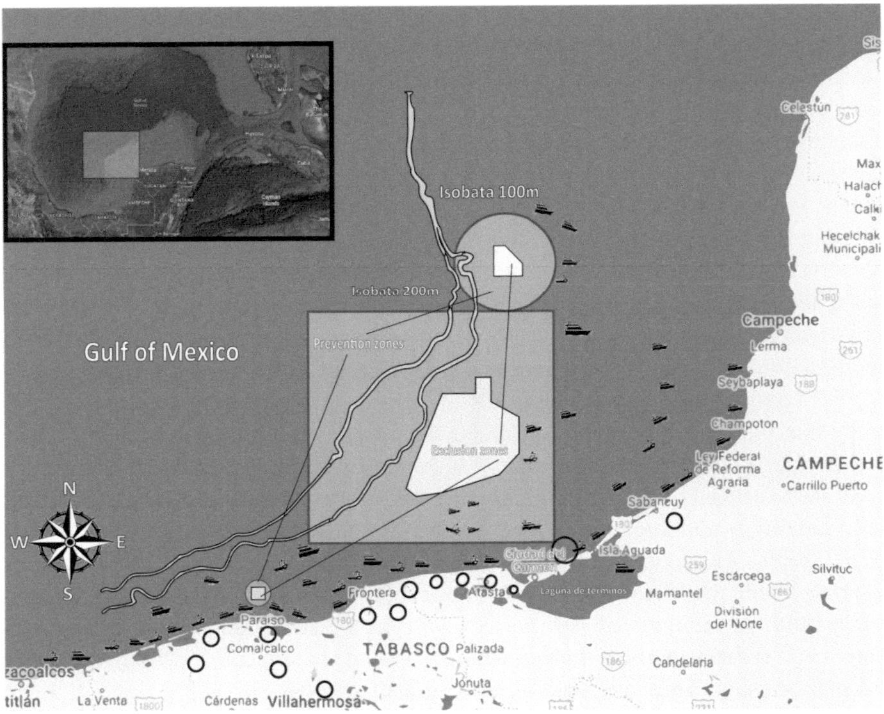

Fig. 19.2 Study area. The *Sonda de Campeche*, in the southwestern Gulf of Mexico. White shapes indicate exclusion zones, and gray shapes indicate prevention zones. Circles are towns: small circles ≤10,000 inhabitants, large circles ≥10,000 inhabitants. Boat icons indicate fishing areas, and size represents either small- or medium-scale fisheries

In 1979, drilling in the Ixtoc I exploratory oil well failed catastrophically, resulting in a massive spill that released approximately three million barrels (400 million liters) of oil into the *Sonda de Campeche* over nearly 10 months (NOAA, 1992). Multiple minor spills have occurred since then, highlighting the high frequency of these incidents and underscoring the need for a prompt and effective protocol for spill mitigation and subsequent monitoring (Uribe-Martínez et al., 2024). Following the Ixtoc I incident, the oil company *Petroleos Mexicanos* (PEMEX, a nominal public-private partnership) developed a spill response protocol. Among the measures adopted were security zones around oil installations for shipping and to enable the sustainable use of fishing resources. A total of 15,000 km^2 in the *Sonda de Campeche* has been left open for fishing activities. Restricted areas are defined as 5500 m around hydrocarbon exportation points, 500 m around each oil platform and well, and 2500 m around exploration and extraction installations for anchoring. These exclusion areas were established to avoid risks to fishers, but have done little to control the most conflictive area: the shallow marine zone (<200 m depth). In this zone, troublesome spatial overlap persists between fishery and oil industry activities, as it harbors the highest diversity of fishing species and yields the most abundant catches, while also containing the most extensive oil industry infrastructure (Fig. 19.2).

Regional fishers believe they have been excluded from their traditional fishing zones and are forced to fish in more distant, hazardous areas. They also perceive that they have been socially excluded, rejected for economic support or government compensation, and ignored in participatory planning processes (Ramos-Muñoz et al., 2019). Additionally, pollution control remains a persistent issue. Gas-flaring at oil wells affects coastal air quality across areas many kilometers from the emission source. This type of pollution contributes to acid rain, while the emissions of carbon dioxide, methane, nitrogen oxides, and other compounds contribute to climate change and ocean acidification, subsequently affecting shelled organisms and coral reef systems (Shavikin & Karnatov, 2018; Ologunorisa, 2001; Kingsley et al., 2024).

Five broader perceived social effects have been reported from the interaction between the oil industry and fishery activities in the *Sonda de Campeche*. First is the loss of traditional fishing occupations through replacement by oil industry jobs; income in the latter is higher, but jobs are not always permanent and do not necessarily lead to a better standard of living. Second, rural communities develop mixed economies in which some traditional activities persist, alongside modern ones. Commercial establishments tend to focus on those with higher incomes (i.e., oil industry workers), thereby raising the cost of living to the detriment of people experiencing poverty. Third, fisher households that had expected an improved quality of life from the extension of oil development have been disappointed by unplanned urban growth, classism, and delinquency. Fourth, air, water, and soil pollution are negatively impacting the health of fisher households and other residents. Common diseases in those living near oil installations include gastrointestinal problems, respiratory disorders such as asthma and other allergies, and cancer (Adekola et al., 2017). Finally, natives feel that their values, traditions, and cultural identities have been eroded by oil development, while immigrants are vulnerable to social segregation; both have been reported in other regions of the world where the oil industry has replaced traditional livelihoods (Ablo, 2015; Kamat et al., 2019).

The expansion of oil industry exploration and operation in the *Sonda de Campeche* over the last 50 years has displaced fisheries to remote sites and exacerbated territorial conflicts, despite a steady decline in oil production (Nyman, 2015). The Mexican federal government has consistently protected oil-related activities and infrastructure, prioritizing oil as an energy source over less contaminating sources (Quist & Rinne, 2017). Rather than working to resolve conflicts between the parties involved, politics has been instrumental in favoring the oil industry and shaping the regional asymmetries that prevent local populations from benefiting from oil industry development and controlling their fishing activities (Crona & Orjan, 2011; Peña-Puch et al., 2021). As in other emergent economies, fishers in Mexico who participate in marine spatial planning processes often feel that they are passive recipients of decisions made on their behalf without their needs being considered. However, their understanding of an industry's impact on marine life can be subjective and may be influenced by political motivations (Quist, 2019). There is a clear need for an explicit and transparent assessment process that integrates an interdisciplinary vision generated through collaboration between researchers, industry, and fishery stakeholders.

19.2 A New Proposed Approach to the Assessment of Fishing and Oil Industry Interaction

Significant progress has been made over the last 20 years by both government and non-governmental organizations (NGOs) in regulating productive activities and the use of natural resources in the Gulf of Mexico. In 2012, the Mexican federal government published the Gulf of Mexico and Caribbean Sea Regional and Marine Ecological Organization Program (*Programa de Ordenamiento Ecologico Marino y Regional del Golfo de Mexico y Mar Caribe*) (DOF, 2012). This program essentially absorbed the concepts of two other planning initiatives: the Strategic Action Program of the Gulf of Mexico Large Marine Ecosystem by the Global Environmental Facility of the United Nations Environmental Program—GEF 2016-2020-GoM-LME (GEF, 2018), and the National Sea and Coast Policy (Política Nacional de Mares y Costas) (CIMARES, 2011). These policies, along with other specific projects, aim to organize productive sector activities, thereby ensuring profits at all levels while preserving natural resources. However, old socioeconomic issues persist and new pressures on coastal ecosystems have emerged. Given this situation and the continued existence of unresolved conflicts, any contribution that facilitate integrated evaluations leading to sustainable solutions are valuable.

The United Nations Sustainable Development Goals (SDGs) constitute an extensively accepted and widely used framework intended to guide integrated assessments of progress toward these goals. The overall SDG framework builds upon the intentions of Agenda 21 (United Nations Division for Sustainable Development, 1992), the World Summit on Sustainable Development, and Transforming Our World: 2030 Agenda for Sustainable Development (United Nations, 2017), using them as fundamental guidelines. Innovative indicators and quantification tools are continually being developed to analyze progress toward the SDGs more effectively.

Our intention here is to propose a methodological approach that uses a reduced set of composite indicators derived from SDG targets to evaluate the socioeconomic, socioecological, and enviro-economic effects of the oil industry/small-scale fisheries interaction in the *Sonda de Campeche*. These would be interdimensional (ID) indicators, more integrated than individual indicators, but with less aggregation than indices (Table 19.1). The difference between ID indicators and indices (e.g., Vulnerable Population Index—VPI, Global Human Footprint Index, Human Development Index—HDI) is the number of variables they use. For example, the HDI comprises three main variables, each of which encompasses multiple subvariables. Interdimensional indicators, in contrast, are based on only two values, simplifying numerical quantification, analytical correlation, and interpretation.

Table 19.1 Examples of Sustainable Development Goals (SDG), targets, and indicators from UN General Assembly (A/RES/71/313, 2017), and proposed interdimensional (ID) indicators relevant to the small-scale fisheries/oil industry interaction in the *Sonda de Campeche*. Interdimensional indicators integrate environmental (a), social (b), and economic (c) variables

SDG	Target	Indicator	ID Indicator
1. End poverty in all its forms everywhere	1.3 Implement nationally appropriate social protection system and measures for all people	1.3.1 Proportion of population covered by social protection at a minimum level	1.3.1bc Proportion of population working in small-scale fishing sector covered by social protection
	1.5 Build the resilience of those in vulnerable situations, reducing their exposure and vulnerability to climate-related extreme events, environmental and socioeconomic disasters	1.5.1 Number of deaths, missing persons and directly affected persons attributed to disasters caused by natural or anthropogenic event per 100,000 population	1.5.1ca Capital spending by local governments to build resilience after natural and anthropogenic events
8. Promote inclusive and sustainable economic growth, full and productive employment and decent work for all	8.3 Promote development-oriented policies that support productive activities, decent job creation	8.3.1 Proportion of informal employment with respect to total employment	8.3.1cb Proportion of new formal jobs in fishing industry versus total local job creation
	8.6 Substantially reduce the proportion of youth Not in employment, education or training	8.6.1 Proportion of youth (15–24 years) not in education, employment, or training	8.4.1ba Proportion of scholarships and grants for youth addressing environmental protection and restoration
14. Conserve and sustainably use the oceans, seas, and marine resources for sustainable development	14.1 Prevent and significantly reduce marine pollution of all kinds, in particular from land-based activities	14.1.1 Index of coastal eutrophication and floating plastic debris density	14.1.1 ac Proportion of long-term contaminants in marine ecosystems from oil spills versus other pollution sources
	14.7b Provide access for small-scale fishers to marine resources and markets	14.7.1 Progress in degree of application of a legal framework which recognizes and protects access rights for small-scale fisheries	14.7.1bc Proportion of policy-institutional frameworks prioritizing small-scale fisheries rights compared to total regional policies directed at productive activities

These, in turn, differ from the individual indicators employed in the SDG context. The latter are intended for estimating specific sustainability themes. Although their methodological formulas can incorporate two or more datasets, they are designed to measure specific developmental conditions and facilitate comparisons between countries. ID indicators better reflect the rational interaction between environmental, social, and economic issues at the regional scale. This would make them more

Step 2

	TARGETS
1.3	Implement nationally appropriate social protection systems for all
1.5	Number of deaths, missing persons attributed to disasters for natural events
8.3	Achieve full and productive employment and decent work for all women
8.6	Substantially reduce the proportion of youth not in employment or training
14.1	Prevent and significantly reduce marine pollution of all kinds
14.7	Provide access for small-scale artisanal fishers to marine resources and markets

Step 3

	INDICATORS
1.3.1	Population on artisanal fishing sector covered by social protection
1.5.1	Capital spending of local governments to build resilience after natural events
8.3.1	Proportion of new formal jobs on fishing industry in relation to total employments
8.6.1	Proportion of scholarships and grants for youth concerning environmental jobs
14.1.1	Proportion of long-term contaminants on marine ecosystems from oil spills
14.7.1	Proportion of Policy-institutional frameworks prioritizing small-scale fisheries rights

Fig. 19.3 Assessment framework procedure based on SDGs (Credit: Azote for Stockholm Resilience Centre, Stockholm University CC BY-ND 3.0, n.d.) and Sustainability Targets and Indicators (United Nations, 2022), with a projected correlation circle generated by multiple factor analysis (MFA) (Borcard et al., 2018)

interesting to regional stakeholders and policymakers, while also making them easier to calculate, interpret, and explain (Fig. 19.3).

Our proposed approach begins by identifying the key SDG reflecting the main biophysical and socioeconomic aspects of the *Sonda de Campeche* region, as well as documenting the main concerns of regional residents related to the small-scale fisheries/oil industry interaction (Fig. 19.3—Step 1). To avoid duplicating effort, as recommended by the United Nations Division for Sustainable Development Goals (DSDG), it is particularly important to compile relevant official fisheries and oil industry data, as well as previous regional research (Coronado et al., 2024).

Following the wedding cake hierarchical ilustration, our proposed approach would incorporate those SDGs concerning natural resources and ecosystem services as a foundation. These could be prioritized in official decrees on conservation or restoration (CONANP, 2023a, b). Close attention needs to be paid to those ecosystem services that contribute to human well-being and provide substantial contributions to regional economies (Grima et al., 2023). Then, by focusing on the social context of regional communities, those SDGs should be selected that are related to local living conditions and linked to fishing and oil industry activities. This initial step is complemented by the selection of SDGs that correlate to the critical economic aspects of settlements within the *Sonda de Campeche* region, which depend on fishing and/or oil industry activities.

Next comes a selection of minimum target values in each of the chosen SDGs, which are related to the regional small-scale fisheries/oil industry interaction (Fig. 19.3—Step 2). This is an essential phase that will require the participation of stakeholders from the small-scale fishing and oil industries, in conjunction with an interdisciplinary group of experts. During this step, it is vital to consider the socioeconomic context that supports economic growth as broadly as possible, while applying a critical perspective on environmental and social concerns. This can be accomplished through numerous and broad-ranging workshops, employing a parsimonious approach and participatory principles guided by previously implemented methodological approaches (Said & Chuenpagdee, 2019; Coronado et al., 2024). This step will produce a matrix of targets linked to issues identified by participating actors, as well as those found in relevant research (e.g., Coronado et al., 2024; Peña-Puch et al., 2021; Uribe-Martínez et al., 2024; Peña-Puch & Rivera-Arriaga, 2025); these will be used to define a suitable set of indicators in the next step.

The target/issue matrix is then used to define a core indicator and an ID indicator for each SDG target (Fig. 19.3—Step 3). The core indicators could correspond to those already defined by UN departments and divisions (UN, 2017; United Nations, 2022) for a national scale. The ID indicators will be designed to reflect the regional situation of the *Sonda de Campeche* and will incorporate the explicit combination of two of the three sustainable properties (Table 19.1). Participation from various regional sectors (e.g., fishers, oil industry workers, residents from other sectors, scientists, local government, and NGOs, among others) will be crucial to creating a reasonable set of measurable aspects that can lead to sustainable commitments. Aspect selection should consider the minimum number of indicators and avoid including an exhaustive list of themes or subthemes. It should be explicitly focused on the most relevant aspects with the broadest possible consensus and have sufficient available data to produce valid quantifications.

Finally, we propose the use of multiple factor analysis (MFA) to analyze indicator data. This approach can correlatively incorporate more than two sets of variables, which can be measured in different units (quantitative, qualitative, or semi-quantitative). To date, MFA has been used in economics, sensory evaluation, and chemistry, but its potential in socioecological applications is promising (Abdi et al., 2013; Borcard et al., 2018). It is essentially principal component analysis (PCA), which is replaced by multivariate correspondence analysis (MCA) when data sets are compounded by non-quantitative variables. Each centered matrix generated by PCA or MCA is weighted so that all variables receive equal weights in the overall analysis, which accounts for different variances within groups of variables. In our proposal, the variables are represented by the indicator values, and the weighting is performed by dividing all the indicator values in each centered matrix by the first singular value obtained from the PCA or MCA (Borcard et al., 2018). The different variable subsets are then projected as an overall result, called a correlation circle (Fig. 19.3—Step 4). This graphic representation reveals common correlative structures, i.e., the relationships between indicators and groups of indicators. Packages and functions for MFA computation are incorporated in the R environment (R Core Team, 2023). The expected outcome is the formation of indicator groups via vector proximity, which concurrently represent aspects (socio-environmental, socioeconomic, and enviro-economic). In addition to illustrative graphics, this integrated analysis of the ID indicators could facilitate identification of developmental priorities or conflicting areas, allowing subsequent design of specific strategies by individual governing jurisdictions.

19.3 Final Reflections

Around the world, multiple effects have been reported in regions where fishing had been the predominant activity and was replaced by oil industry activities (e.g., Garcia-Cuellar et al., 2004; Ackah-Baidoo, 2013; Ramos-Muñoz et al., 2019; Andrews et al., 2021). When these two productive activities coincide spatially, they tend to conflict noticeably. This aligns with the information cited herein, as well as the widespread undocumented views of local populations in the *Sonda de Campeche*, which embrace numerous adverse effects from oil industry activities, particularly those related to the oil spills. Regrettably, there is limited research available detailing the real magnitude of past oil spill impacts on ecosystem health, the regional economy, and social activities. Besides, existing protocols implemented by PEMEX have been insufficient to prevent, mitigate, or monitor impacts, and the information the company shares with the public is neither transparent nor convincing (Quist & Rinne, 2017). Relevant Mexican law generally provides only a reference framework for conflict resolution, whereas protocols generated from academia are frequently ignored at the implementation phase. In this kind of situation, with clearly contrary interests, conflicts are more likely to be resolved through direct information exchange between small-scale fishing and oil industry stakeholders than through a proper application of rules and laws. The current imbalance in government attention between the fisheries and oil industries stems from a lack of regulatory guidelines and well-defined institutional responsibilities based on a common development agenda, such as the SDGs (Coronado et al., 2024). Analytical reviews of the dynamics in the *Sonda de Campeche* (Garcia-Cuellar et al., 2004; Peña-Puch & Rivera-Arriaga, 2025) have identified two main obstacles in the path to a sustainable living: one relates to the contexts of incompatibility in which fishing and oil industry activities have coexisted in the regional space, along with many other less prominent activities, without clear rules for coexistence. The second one is related to the difficulty of attaining an objective evaluation of the multiple interactions among social, economic, and environmental factors, as well as linking these factors to a regional development plan and monitoring. This condition has driven the creation of initiatives, strategies, policies, and methodologies (e.g., CIMARES, 2011; DOF, 2012; GEF, 2018; Coronado et al., 2024) aimed at organizing economic activities in the region to help communities progress toward sustainable development. The methodology proposed in this chapter differs from previous ones in the sense that it incorporates the use of indicators linked to the SDG framework, explicitly combining two of the three sustainability dimensions outlined in the framework. The conceptual intent behind our proposal is to offer an evaluation tool that integrates thematic, individual indicators, which are less complex than multivariate indices. We believe that implementing this assessment framework would yield outcomes that provide a clear picture for authorities to use when developing actions designed to mitigate regional conflicts between small-scale fishing and the oil industry.

References

Abdi, H., Williams, L. J., & Valentin, D. (2013). Multiple factor analysis: Principal component analysis for multi-table and multi-block data sets. *WIREs Computing Statistics, 5*, 149–179.

Ablo, D. (2015). Local content and participation in Ghana's oil and gas industry: Can enterprise development make a difference? *The Extractie Industries and Society, 2*(2), 320–327.

Ackah-Baidoo, A. (2013). Fishing in troubled waters: Oil production, seaweed, and community-level grievances in the Western region of Ghana. *Community Development Journal, 48*, 3.

Adekola, J., Fischbacher-Smith, M., Fischbacher-Smith, D., et al. (2017). Health risks from environmental degradation in The Niger Delta, Nigeria. *Environmental Planning Politics and Space, 35*(2), 334–354.

Andrews, N., Bennett, N. J., Le Billon, P., Green, S. J., Cisneros-Montemayor, A. M., Amoing, S., Gray, N. J., & Sumaila, R. U. (2021). Oil, fisheries and coastal communities: A review of impacts on the environment, livelihoods, space and governance. *Energy Research & Social Science, 75*, 1–15.

Anticamara, J. A., Watson, R., Gelchu, A., & Pauly, D. (2011). Global fishing effort (1950–2010): Trends, gaps, and implications. *Fisheries Research, 107*(1–3), 131–136.

Azote for Stockholm Resilience Centre, Stockholm University CC BY-ND 3.0. (n.d.). *The SDGs Wedding cake.*

Borcard, D., Gillet, F., & Legendre, P. (2018). *Numerical ecology with R*. Springer International Publishing AG, Springer Nature.

CIMARES *Comisión Intersecretarial para el Manejo Sustentable de Mares y Costas* (2011).

Cohen, P. J., Allison, E. H., Andrew, N. L., Cinner, J., Evans, L. S., Fabinyi, M., Garces, L. R., Hall, S. J., Hicks, C. C., Hughes, T. P., Jentoft, S., Mills, D. J., Masu, R., Mbaru, E. K., & Ratner, B. D. (2019). Securing a just space for small-scale fisheries in the blue economy. *Frontiers in Marine Science, 6*.

CONANP. (2023a). *Comisión Nacional de Áreas Naturales Protegidas: Programa para la Protección y Restauración de Ecosistemas y Especies Prioritarias (PROREST).* https://www.gob.mx/conanp/acciones-y-programas/programa-para-la-proteccion-y-restauracion-de-ecosistemas-y-especies-prioritarias-prorest.

CONANP. (2023b). *Comisión Nacional de Áreas Naturales Protegidas: Dirección Regional Planicie Costera y Golfo de México.* https://www.gob.mx/conanp/documentos/region-planicie-costera-y-golfo-de-mexico?state=published.

Coronado, E., Zepeda-Domínguez, J. A., Espinoza-Tenorio, A., Cupido-Santamaria, D., Ramos-Muñoz, D., & Monzón-Alvarado, C. (2024). Institutional mapping and implementation of the sustainable development goals across co-existing industries: The case of the fisheries-oil system in Mexico. *The Extractive Industries and Society, 17*.

Crona, B., & Orjan, B. (2011). Power asymmetries in small-scale fisheries a barrier to governance transformability? *Ecology and Society, 15*, 4.

DOF. (2012). *Acuerdo por el que se expide la parte marina del Programa de Ordenamiento Ecológico Marino y Regional del Golfo de Mexico y Mar Caribe*. https://dof.gob.mx/nota_detalle.php?codigo=5279084&fecha=24/11/2012#gsc.tab=0

Food and Agriculture Organization of the United Nations, FAO. (2020). *The State of the World Fisheries and Aquaculture*. Rome.

Garcia-Cuellar, J., Arreguin-Sanchez, F., Hernandez, S., & Lluch-Cota, D. B. (2004). Impacto ecológico de la industria petrolera en la sonda de campeche, México, tras tres décadas de actividad: una revisión. *Interciencia, 29*, 6.

GEF. (2018). *Large marine ecosystems strategic approach toolkit*. Paris, France: 85. pp. Online: https://iwlearn.net/resolveuid/7699de40-3ee5-41f1-80d8-ec64333758c7.

Graham, J. E., Bates, A. E., Krueck, N. C., Baker, S. C., Stuart-Smith, R. D., & Brown, C. J. (2024). Stock assessment models overstate sustainability of the world's fisheries. *Science, 385*, 860–865.

Grima, N., Jutras-Perreault, M. C., Gobakken, T., Ole Ørka, H., & Vacik, H. (2023). Systematic review for a set of indicators supporting the common international classification of ecosystem services. *Ecological Indicators, 147*, 109978.

Harper, S., Adshade, M., Lam, V. W., Pauly, D., & Sumalia, U. R. (2020). Valuing invisible catches: Estimating the global contribution by women to small-scale marine captures fisheries production. *PLoS One, 15*, 3.

Kamat, V. R., LeBillon, P., Mwaipopo, R., & Raycraft, J. (2019). Natural gas extraction and community development in Tanzania: Documenting the gaps between rhetoric and reality. *The Extractive Industries and Society, 6*(3), 968–976.

Kingsley, E. U., Uyiosa, O. A., Robert, B. O., Hussain, A., Benedict, O., Paul, A. A., Benita, E. S., & Ismail, H. (2024). Environmental influence of gas flaring: Perspective from The Niger Delta region of Nigeria. *Geofluids, 1*.

Martínez-Estrada, A., Melgoza-Rocha, A., Mascareñas-Osorio, I., & Cota-Nieto, J. J. (2017). *Overview of the fishing sector in Mexico: Part II. dataMares*. Interactive Resources.

National Oceanographic & Atmospheric Administration (NOAA). (1992). *Oil spills, case histories 1967–1991. Summaries of Significant US and International Spills*. Hazardous Material Response and Assessment Division. National Oceanic and Atmospheric Administration. Washington, DC, EEUU. 224 pp.

Núñez-Lara, E., González-Salas, C., & Pérez-España, H. (2017). The reef fish community of Obispos Reef, Campeche Bank Reefs, Gulf of Mexico: Exploring poorly-know sites. *Reef Encounter, 32*, 1.

Nyman, E. (2015). Offshore oil development and maritime conflict in the 20th century. A statistical analysis of international trends. *Energy Research & Social Science, 6*, 1–17.

Ologunorisa, T. (2001). A review of the effects of gas flaring on The Niger Delta environment. *The International Journal of Sustainable Development & World Ecology, 8*, 249.

Peña-Puch, A. C., & Rivera-Arriaga, E. (2025). Challenges for good governance in coastal socio-ecological systems of Campeche: Analysis from the expert users' perspective. *Marine Policy, 176*, 106634.

Peña-Puch, A. C., Pérez-Jiménez, J. C., Espinoza-Tenorio, A., Méndez-Loeza, I., & Williams-Jara, G. M. (2021). The sustainability of small-scale fisheries in oil producing sections of the Gulf of Mexico. *The Extractive Industries and Society, 8*, 4.

PRIO. (2019). *Geographical and resource dataset. Petroleum dataset*. Peace Research Institute Oslo.

Quist, L. M. (2019). Fishers' knowledge and scientific indeterminacy: Contested oil impacts in Mexico's sacrifice zone. *Maritime Studies, 18*, 65–76.

Quist, L. M., & Rinne, P. (2017). The politics of justification: Newspaper representation of environmental conflict between fishers and the oil industry in Mexico. *Environmental Values, 26*(4), 457–479.

R Core Team. (2023). *R: A language and environment for statistical computing*. R Foundation for Statistical Computing.

Ramos-Miranda, J., Flores-Hernández, D., & Do-Chi, T. (2009). *Assessment of the white shrimp fishery (Litopenaeus setiferus) in the Campeche Bank* (p. 38). Universidad Autónoma de Campeche.

Ramos-Muñoz, D. E., Ramos-Reyes, R., Zamora-Cornelio, L. F., & Espinoza-Tenorio, A. (2019). Exclusion en el Golfo de Mexico: Una vision desde los pescadores sobre la industria petrolera en Tabasco. *Cuadernos de Geografia: Revista Colombiana de Geografia, 28*(2), 357–372.

Romo-Curiel, A. E., Ramírez-Mendoza, Z., Fajardo-Yamamoto, A., Ramírez-León, M. R., García-Aguilar, M. C., Herzka, S. Z., Pérez-Brunius, P., Saldaña-Ruiz, L. E., Sheinbaum, J., Kotzakoulakis, K., Rodríguez-Outerelo, J., Medrano, F., & Sosa-Nishizaki, O. (2022). Assessing the exposure risk of large pelagic fish to oil spills scenarios in the deep waters of the Gulf of Mexico. *Marine Pollution Bulletin, 176*, 113434.

Ruby, P., & Ahilan, B. (2018). An overview of climate change impact in fisheries and aquaculture. *Climate Change, 4*(13), 87–94.
Said, A., & Chuenpagdee, R. (2019). Aligning the sustainable development goals to the small-scale fisheries guidelines: A case for EU fisheries governance. *Marine Policy, 107*, 103599.
Shavikin, A., & Karnatov, A. (2018). Main development problems of vulnerability mapping of sea-coastal zones to oil spills. *Journal of Marine Science and Engineering, 6*(4), 115.
Uhre, A. N., & Leknes, E. (2017). When the oil and fishing industries live side by side. *Marine Police, 75*, 108–115.
United Nations. (2017). *Transforming our world: the 2030 Agenda for Sustainable Development.* General Assembly, seventieth session. https://docs.un.org/en/A/RES/70/1.
United Nations. (2022). *Statistic Division of the Department of Economic Affairs: SDG Indicators Database. Update to the Global SDG Data Platform.* https://unstats.un.org/sdgs/indicators/indicators-list/
United Nations Division for Sustainable Development. (1992). *Agenda 21*. http://www.un.org/esa/sustdev/agenda21sp/index.htm
Uribe-Martínez, A., Espinoza-Tenorio, A., Cruz-Pech, J. B., et al. (2024). An affordable operational oil spill monitoring system in action: A diachronic multiplatform analysis of recent incidents in the southern Gulf of Mexico. *Environmental Monitoring Assessment, 196*, 1069.
Zang, B., Matchinski, E. J., Chen, B., Ye, X., Jing, L., & Lee, K. (2019). Marine oil spills-oil pollution sources and effects. In *World seas: An environmental evaluation* (pp. 391–406). Academic Press.

Glossary

Basic Geostatistical Area (AGEB, Spanish acronym for *Área Geoestadística Básica***)** The smallest territorial subdivision within Mexican municipalities, formally designated by the National Institute of Statistics and Geography (INEGI) for the systematic organization of census and geographic data. AGEBs serve as fundamental spatial units for socioeconomic analysis and urban planning.

Bienpesca (previously PROSPERA) A Mexican constitutional welfare program launched in 2019 and currently administered by CONAPESCA and the General Coordination of Welfare Programs, providing annual MXN 7,500 stipends to artisanal fishers and aquaculturists. Established to improve livelihoods in small-scale fisheries and aquaculture sectors, boost productivity among beneficiary producers, and enhance household food security.

Cantarell complex Discovered in the Sonda de Campeche in 1973, the Cantarell complex was the largest offshore oil field in Mexico and one of the largest globally, peaking at 2.1 million barrels per day in 2004. However, production declined significantly until the present day.

Community and Environmental Support Program (PACMA, Spanish acronym for *Programa de Apoyo a Comunidades y Medio Ambiente***)** A PEMEX initiative aimed at fostering socioeconomic development in oil-producing communities while mitigating the environmental impact of operations. PACMA implements Programs, Works, and Actions (PROA, *Programas*, *Obras y Acciones*) to support community and environmental efforts within PEMEX's contracts and projects.

Dos Bocas Port An industrial, commercial, and international oil port located in Paraíso, Tabasco, Mexico. Built by PEMEX in 1979, it began operations in 1982. Together with Ciudad del Carmen Port, Dos Bocas is one of Mexico's most critical oil ports, serving as a logistical base for oil offshore operations in the Sonda de Campeche.

A. Espinoza-Tenorio et al. (eds.), *The Fishing and Oil Industries in Coexistence*,
https://doi.org/10.1007/978-3-032-02621-7

Energy Reform (2013) A constitutional amendment under President Enrique Peña Nieto that opened Mexico's oil sector to private and foreign investment, ending PEMEX's monopoly. It introduced competitive bidding rounds for exploration/production contracts. Post-2018 policy shifts under the López Obrador and Claudia Sheinbaum administrations prioritized reversing privatization, strengthening state control through PEMEX and CFE, and promoting energy sovereignty through renewable projects.

Fishing Production Cooperative Societies Associations of fishers who collectively pool resources, labor, and expertise to advance shared economic interests in the fishing industry. In Mexico, these cooperatives operate under the legal structure of cooperative societies, granting members democratic control over joint fishing operations and resource management. Their organizational models vary significantly, encompassing: (1) small-scale artisanal fisheries, (2) industrial-scale fishing enterprises, and (3) integrated aquaculture production systems.

Garrulera Women in the fishing community of *Barra de San Pedro*, Tabasco, who specialize in processing (evisceration, cleaning, and preparation) and marketing fish and seafood.

Gulf of Mexico and Caribbean Sea Regional Marine Ecological Organization Program A policy instrument that regulates land uses and productive activities while promoting environmental protection, sustainable resource utilization, and ecosystem preservation through evidence-based management, analyzing both degradation trends and resource potential to balance ecological and economic objectives in marine-coastal zones.

Henequen (*Agave fourcroydes*) A species of the *Agave* genus that has been cultivated in the Mayan region for fiber since pre-Hispanic times. Its leaves are tipped with long spines, and its fibers are finer. Its cultivation has been of great importance in the *Yucatán Peninsula*, having more than 40 different uses.

Industrial and Semi-Industrial Fishing Fleet (ISIF) Vessels measuring 20–25 m in length that conduct offshore operations in waters deeper than 50 m, with fishing trips lasting 15–20 days. In the southern Gulf of Mexico, these vessels utilize diverse fishing methods, including longlines, traps, jimbas, and hookah systems to target multispecies catches such as red grouper, octopuses, and lobsters.

Ixtoc I spill A catastrophic offshore oil blowout that occurred in 1979, in the *Sonda de Campeche*, Gulf of Mexico. The exploratory well, operated by PEMEX, suffered a blowout and released approximately 3.5 million barrels (475,000 metric tons) of crude oil over 290 days before being capped in 1980. This event remains one of the largest accidental marine oil spills in history, significantly impacting marine ecosystems, coastal communities, and regional fisheries.

Marine Diesel Program A Mexican federal subsidy program administered by the National Commission of Aquaculture and Fisheries (CONAPESCA, by its Spanish acronym) that provides economic support to registered fishing and aquaculture producers through partial reimbursement of fuel costs.

Natural Oil Seepage (*Chapopotera*, from Nahuatl *chapopotl* **["pitch"])** The gradual, continuous release of crude oil and gas from submarine geological formations through seafloor fractures or sedimentary layers. It is a significant source of oil entering the marine environment. While the slow discharge allows for localized ecosystem adaptation through specialized microbial communities, chronic exposure impacts surrounding benthic and pelagic environments.

Oil boom in southeastern Mexico A transformative economic period beginning in the 1970s and marked by accelerated petroleum industry development across Mexico's Gulf states, particularly Tabasco, Campeche, and Chiapas. Initiated by major hydrocarbon discoveries, this boom fundamentally reshaped the region through industrial expansion (e.g., rapid scaling of oil extraction infrastructure and operations), economic impacts (e.g., became Mexico's primary oil-producing zone and created employment shifts from agriculture to oil sector), and socio-environmental consequences (e.g., coastal urbanization patterns and ecosystem disruption in sensitive marine/terrestrial habitats, and cultural changes in traditional communities).

Oil Exclusion Zones Safety perimeters established around offshore oil installations (both surface platforms and subsea infrastructure) that extend radially from the facility's edge. These restricted areas serve to prevent unauthorized vessel access or operations, mitigate collision risks and operational accidents, and reduce potential pollution incidents through controlled activity buffers.

Olmeca Refinery A PEMEX-operated crude oil processing facility located in *Dos Bocas*, Tabasco, forming one of seven refineries in Mexico's National Refining System. Originally named *Dos Bocas* Refinery, it was rebranded to honor the Olmec civilization's legacy in the region. Construction commenced in 2019, with operational inauguration in 2022. The refinery represents Mexico's first major downstream infrastructure project in decades, designed to process 340,000 barrels per day (bpd) of Maya heavy crude upon reaching full capacity.

PEMEX (*Petróleos Mexicanos*) Mexico's state-owned petroleum company, established through the historic 1938 oil expropriation, represents a cornerstone of Mexican energy sovereignty. Functioning as a vertically integrated monopoly until 2013, it maintained exclusive control over all segments of Mexico's oil and gas value chain, from upstream exploration to downstream retail fuel distribution. PEMEX continues to dominate Mexico's energy sector, operating crude oil refineries, petrochemical production complexes, and active offshore drilling platforms.

Riparian Pact (*Pacto Ribereño*) A grassroots peasant movement originating in 1976 within Tabasco's riverine communities to resist the socio-environmental consequences of petroleum industrialization, specifically targeting PEMEX operations. Organized as a collective response to uncompensated ecological and economic damages, its demands included restitution for oil-related destruction of aquatic ecosystems and agricultural lands, as well as accountability for oil spills affecting riparian ecosystems.

Small-Scale Fisheries (SSF) Fishing operations characterized by minimal infrastructure, low technological inputs, basic safety provisions, and limited regulatory oversight. These fisheries typically employ small vessels with reduced tonnage capacity, utilizing traditional fishing techniques and operating in coastal or near-shore areas. In the Gulf of Mexico, SSF predominantly target demersal species (e.g., groupers, snappers), as well as pelagic species (e.g., tuna, sharks), shrimp, octopus, and other species.

Sonda de Campeche A semi-enclosed basin of the Gulf of Mexico located primarily off Mexico's Campeche coast. This region constitutes one of the nation's most vital oil production zones, containing the highly productive Cantarell and *Ku-Maloob-Zaap* complexes. Its shallow carbonate platform enables extensive offshore drilling while hosting sensitive coral and mangrove ecosystems as well as important fishing areas.

Southwestern Gulf of Mexico (SWGoM) This book refers to the marine region encompassing the *Sonda de Campeche* and the coastal waters of Tabasco state. However, a broader definition can also involve the Mexican states of Tamaulipas, Veracruz, and Yucatán.

The Mexican Oil Industry Mexico is a significant oil-producing nation with a crucial role in global energy markets. Ranking among the world's top petroleum producers, the country derives substantial output from offshore operations in the Sonda de Campeche, where two distinctive crude types are extracted: Maya (heavy) and Isthmus (light) oils. These varieties exhibit fundamentally different properties that directly influence their environmental impacts, including differential toxicity levels and degradation rates. The sector faces pressing challenges, including production declines at mature fields, aging infrastructure, and urgent requirements for expanded investment in exploration and next-generation field development.

Trust Fund for Fishing Reconversion in the Gulf of Mexico (FIFOPESCA) Established in 2004, this financial mechanism supports the transformation of sustainable fisheries and local economic development in the region. Its core objectives focus on helping affected fishers retrain, create value chains, and develop productive projects that mitigate the impact of exclusion zones on the region's coasts.

UN Global Compact Launched by the United Nations in 2000, this voluntary corporate sustainability initiative urges businesses worldwide to align their strategies and operations with ten universal principles spanning four key areas: human rights, labor standards, environmental protection, and anti-corruption. Additionally, it mobilizes private sector action toward achieving the UN Sustainable Development Goals (SDGs) through responsible business practices and collaborative partnerships.

Index

A. Espinoza-Tenorio et al. (eds.), *The Fishing and Oil Industries in Coexistence*, https://doi.org/10.1007/978-3-032-02621-7

MIX
Papier aus verantwortungsvollen Quellen
Paper from responsible sources
FSC® C105338
FSC
www.fsc.org

If you have any concerns about our products,
you can contact us on
ProductSafety@springernature.com

In case Publisher is established outside the EU,
the EU authorized representative is:
Springer Nature Customer Service Center GmbH
Europaplatz 3, 69115 Heidelberg, Germany

Printed by Libri Plureos GmbH
in Hamburg, Germany